BIO-BASED COMPOSITES FOR HIGH-PERFORMANCE MATERIALS

From Strategy to Industrial Application

Edited by **Wirasak Smitthipong**
Rungsima Chollakup • Michel Nardin

CRC Press
Taylor & Francis Group
Boca Raton London New York

CRC Press is an imprint of the
Taylor & Francis Group, an **informa** business

CRC Press
Taylor & Francis Group
6000 Broken Sound Parkway NW, Suite 300
Boca Raton, FL 33487-2742

First issued in paperback 2021

First issued in hardback 2019

© 2015 by Taylor & Francis Group, LLC
CRC Press is an imprint of Taylor & Francis Group, an Informa business

No claim to original U.S. Government works

ISBN 13: 978-1-4822-1448-2 (hbk)
ISBN 13: 978-1-03-224092-3 (pbk)

Visit the Taylor & Francis Web site at
http://www.taylorandfrancis.com

and the CRC Press Web site at
http://www.crcpress.com

Publisher's Note
The publisher has gone to great lengths to ensure the quality of this reprint but points out that some imperfections in the original copies may be apparent.

Contents

Preface

Most synthetic plastics are made from petroleum and its by-products. Plastics derived from fossil resources are mostly non-biodegradable. The increased use of plastics over the years has resulted in increases in waste that have become significant concerns because of their negative impacts on the environment. The need to develop non-petroleum-based and sustainable feedstocks is urgent, and the need has shifted the attention of many academic and industrial researchers toward bio-based materials.

This book provides an overview of the state of the art and emerging trends in the area of bio-based composites. In recent years, interest in the development of natural fibers as reinforcements in composites has increased greatly. Many reports in the field of bio-based composites have been published but they usually focus on a single aspect of biocomposites. The objective of this book is to contribute to the overall knowledge of bio-based composites. The recent developments in technology make the understanding of bio-based composites vital for ensuring an eco-friendly environment. This book discusses what is known about bio-based composites and, just as important, what is new. We have attempted to maintain a good balance of reporting developments in academic, industrial, and governmental laboratories and reflecting international views.

A bio-based composite is basically defined by its biocompatibility properties based on its origin. To meet this definition, a bio-based composite must be biodegradable and foster an eco-friendly environment in addition to exhibiting other practical properties. A variety of solutions exist in practice, and their applications require interdisciplinary understanding, particularly in the fields of materials science and engineering.

An understanding of bio-based composites involves analysis of filler–matrix interactions, marketing and political strategies, raw materials and their characteristics, basic design principles, properties and applications, life cycle assessments, and future trends. All these issues are discussed in the chapters of this book. We hope this book will be beneficial to scholars, academics, regulatory agencies, research and development communities, and industries worldwide. It fills important gaps in our knowledge of the bio-based materials that constitute environmentally friendly technologies for the world's population by presenting the latest relevant academic knowledge and industry expertise.

Wirasak Smitthipong
Rungsima Chollakup
Michel Nardin

Editors

Wirasak Smitthipong earned a PhD in chemistry in 2005 with a scholarship from Institut de Chimie des Surfaces et Interfaces, Conseil Générale du Haut Rhin and Fondation de l'Ecole Nationale Supérieure de Chimie de Mulhouse, University of Haute Alsace, France. He was awarded at a highlight lecture at the 2005 EUROMAT conference in Prague.

Dr. Smitthipong was awarded a National Science Foundation post-doctoral fellowship at the Materials Research Laboratory and College of Engineering at the University of California at Santa Barbara in the U.S.A. from 2006 to 2008. He spent 6 years as a chemist at Thai Caprolactam Public Co., Ltd., and 5 years working for Michelin as a lead rubber formulation designer in France and in Thailand.

He now performs research at the Agricultural and Agro-Industrial Product Improvement Institute of Kasetsart University in Bangkok, Thailand, and is an invited lecturer at the International School of Engineering at Chulalongkorn University also in Bangkok.

Dr. Smitthipong's areas of interest are surface and interface phenomena, rubber, adhesion, composite materials, biomaterials, polymer science and engineering, supramolecular and nanoscale materials, and materials engineering and development. He is the author or co-author of more than 60 scientific articles and papers, 3 invited book chapters, and 45 communications for national and international conferences and edited 2 books.

Rungsima Chollakup joined Chulalongkorn University as a food chemist in 1995. Since then she has performed research at the Cassava and Starch Technology Research Unit in Thailand, attended a 6-month training course on "Polymer Blends for Biodegradable Plastics" at the National Institute of Bioscience and Human Technology in Japan, and was awarded a scholarship to pursue a PhD program in textile science at the University of Haute Alsace in Mulhouse, France. Dr. Chollakup has been a senior researcher at the Agricultural and Agro-Industrial Product Improvement Institute of Kasetsart University in Bangkok since 2004.

Dr. Chollakup received the Excellent Paper Award from the *Journal of Textiles and Apparel* in 2005. She then participated in a post-doctoral fellowship program at the College of Engineering at the University of California at Santa Barbara and served as a visiting researcher at Laboratoire de Photochimie Moléculaire et Macromoléculaire, Université Blaise Pascal in France from 2009 to 2012.

Her professional interests include natural fibers, polymer characterizations, composite materials, and biomaterials. She has written more than 50 scientific articles and papers, 3 invited book chapters, and 52 communications presented at national and international conferences.

Michel Nardin started his career as a chemical engineer of the National High School of Chemistry of Paris in 1975. He earned a PhD in physical chemistry in 1980 from the Universities of Strasbourg and Mulhouse, France, and pursued post-doctoral studies in the Polymer Group of the Department of Physics at University of Leeds in the United Kingdom in 1985 and 1986.

Dr. Nardin joined the French National Center for Scientific Research (CNRS) as a researcher in 1982 and was named director of research in 1994. He is interested in the fundamental and practical aspects of adhesion, physical chemistry of surfaces and interfaces, wettability, capillary impregnation of powders and fabrics, thermodynamics and micromechanics of fiber–matrix interfaces in composite materials, tack and adhesive properties of elastomers, interactions of surfaces and living matter. Dr. Nardin has written more than 170 scientific articles and papers, and 11 book chapters, and presented 65 invited lectures and 180 communications at national and international conferences.

Contributors

Alireza Amini
Department of Civil and
 Environmental Engineering
University of Massachusetts
Amherst, Massachusetts

Sanjay R. Arwade
Department of Civil and
 Environmental Engineering
University of Massachusetts
Amherst, Massachusetts

Haroutioun Askanian
Institut de Chimie
 de Clermont-Ferrand
Clermont Université
Clermont-Ferrand, France

Ansou Malang Badji
Section Physique Appliquée
Université Gaston Berger
 de Saint Louis
Dakar, Senegal

Caroline Baillie
School of Environmental
 Systems Engineering
University of Western Australia
Perth, Australia

Christophe Baley
LIMATB
Université Bretagne Sud
Lorient, France

Alain Bourmaud
LIMATB
Université Bretagne Sud
Lorient, France

Roman Cermak
Faculty of Technology
Tomas Bata University
Zlín, Czech Republic

Jiye Chen
Faculty of Technology
University of Portsmouth
Portsmouth, UK

Rungsima Chollakup
Kasetsart Agricultural and
 Agro-Industrial Product
 Improvement Institute
Kasetsart University
Bangkok, Thailand

Arkadiusz Chworos
Centre of Molecular and
 Macromolecular Studies
Polish Academy of Sciences
Lodz, Poland

Peggi L. Clouston
Department of Environmental
 Conservation
University of Massachusetts
Amherst, Massachusetts

Sophie Commereuc
Institut de Chimie
 de Clermont-Ferrand
Clermont Université
Clermont-Ferrand, France

Anicet Dasylva
Départment de Physique
Université Cheikh Anta Diop
Dakar, Senegal

Peter Davies
IFREMER
Marine Structures Group
Plouzané, France

Florence Delor-Jestin
Institut de Chimie
    de Clermont-Ferrand
Clermont Université
Clermont-Ferrand, France

Brent Erickson
Biotechnology Industry Organization
Washington, DC

Ali Faghihnejad
Department of Chemical and
    Materials Engineering
University of Alberta
Edmonton, Alberta, Canada

Ya Feng
Institut de Chimie
    de Clermont-Ferrand
Clermont Université
Clermont-Ferrand, France

Solène Gaudin
Institut de Chimie
    de Clermont-Ferrand
Clermont Université
Clermont-Ferrand, France

Alexandre Govin
Ecole Nationale Supérieure
    des Mines
Saint Etienne, France

Mamadou Gueye
Départment de Physique
Université Cheikh Anta Diop
Dakar, Senegal

René Guyonnet
Ecole Nationale Supérieure
    des Mines
Saint Etienne, France

Hyun-Joong Kim
Laboratory of Adhesion
    and Biocomposites
Seoul National University
Seoul, Korea

Antoine Le Duigou
LIMATB
Université Bretagne Sud
Lorient, France

Valérie Massardier
Ingénierie des Matériaux
    Polymères
INSA
Lyon, France

Michael May
Fraunhofer Institute for
    High-Speed Dynamics
Ernst Mach Institute
Freiburg, Germany

Deborah Mohrmann
Fraunhofer Institute for
    High-Speed Dynamics
Ernst Mach Institute
Freiburg, Germany

Kristyna Montagova
Department of Polymer
    Engineering
Tomas Bata University
Zlín, Czech Republic

Claudine Morvan
Laboratoire PBS
Université de Rouen
Mont Saint Aignan, France

Michel Nardin
Institut de Science des Materiaux
    de Mulhouse
Mulhouse, France

Diène Ndiaye
Section Physique Appliquée
Université Gaston Berger
   de Saint Louis
Dakar, Senegal

Kuakoon Piyachomkwan
National Center for
   Genetic Engineering
   and Biotechnology
National Science and Technology
   Development Agency
Pathum Thani, Thailand

V.P. Sharma
Indian Institute of
   Toxicological Research
Council of Scientific and
   Industrial Research
Lucknow, India

Rina Singh
Biotechnology Industry
   Organization
Washington, DC

Wirasak Smitthipong
Kasetsart Agricultural and
   Agro-Industrial Product
   Improvement Institute
Kasetsart University
Bangkok, Thailand

Klanarong Sriroth
Faculty of Agro-Industry
Kasetsart University
Bangkok, Thailand

Jackapon Sunthonrvarabhas
Cassava and Starch Technology
   Research Unit
National Center for
   Genetic Engineering
   and Biotechnology
Bangkok, Thailand

Potjanart Suwanruji
Department of Chemistry
Kasetsart University
Bangkok, Thailand

Rattana Tantaherdtam
Kasetsart Agricultural and
   Agro-Industrial Product
   Improvement Institute
Kasetsart University
Bangkok, Thailand

Thimothy Thamae
Department of Chemistry and
   Chemical Technology
National University of Lesotho
Roma, Lesotho

Coumba Thiandoume
Départment de Physique
Université Cheikh Anta Diop
Dakar, Senegal

Adams Tidjani
Départment de Physique
Université Cheikh Anta Diop
Dakar, Senegal

Vincent Verney
Institut de Chimie de
   Clermont-Ferrand
Clermont Université
Clermont-Ferrand, France

Sittochoke Wanlapatit
National Center for
   Genetic Engineering
   and Biotechnology
Kasetsart University
Pathum Thani, Thailand

Paul Winters
Biotechnology Industry Organization
Washington, DC

Hongbo Zeng
Department of Chemical and
    Materials Engineering
University of Alberta
Edmonton, Alberta, Canada

1

Bio-Based Composites: An Introduction

Rungsima Chollakup, Wirasak Smitthipong, and Michel Nardin

CONTENTS

1.1 Introduction

Many composites used today are at the leading edge of materials technology, with performance and costs appropriate for high-level applications such as spacecraft. However, nature has used heterogeneous materials combining the best aspects of dissimilar constituents for millions of years. Nowadays, bio-based composites are preferred materials that present a good compromise between final performance and environmentally friendly properties.[1-3] The objective of this chapter is to define clearly the scope and overall organization of this book.

Composites have emerged as a valuable class of engineering materials because they offer many attributes not attainable with other materials. Their light weights coupled with high stiffness characteristics and selectable properties have fostered their use for many years in satellites, high-performance aircraft, world-class sailboats, and even submarines.[4,5] Now these materials demonstrate their worth in the mundane but equally demanding consumer, infrastructure, and sporting goods arenas.

Conventional structural composites are blends of two or more components, one of which consists of stiff long fibers, and the other acts as a binder or matrix that holds the fibers in place. The fibers are strong and stiff relative to the matrix and are generally anisotropic (having different properties in different directions). The fiber strength and stiffness are usually much greater, often several times stronger and stiff than the matrix material.[6]

The matrix material can be polymeric (e.g., polyester resins, epoxies), metallic, ceramic, or carbon. When the fiber and matrix are joined to form a composite, they retain their individual identities and both directly influence the final properties of the composite whose construction involves layers of fibers and matrix stacked to achieve the desired properties in one or more directions.[2,7]

A composite can be tailored so that the directional dependence of strength and stiffness matches that of the loading environment. To do that, layers of unidirectional material are oriented to satisfy the loading requirements.[8] These layers (or plies) contain fibers and matrix. Because of the use of directional layers, the tensile, flexural, and shear properties of a structure can be disassociated from one another to some extent, for example, shear can be changed without changing the flexural or tensile stiffness.[9]

Fibers can be classified according to structure, diameter (or cross-sectional width), and length. In general, a material is classified as a fiber if its diameter or cross-sectional width is less than 0.0254 m and its aspect (length:diameter) ratio is greater than 10. Most commercial fibers meet these requirements easily. A fiber is called a whisker if its microstructure is predominantly a single crystal.[10]

Fibers or whiskers can be classified as continuous or discontinuous. Continuous fibers are capable of being manufactured to indefinite lengths.[2,11] Discontinuous fibers can be chopped from continuous fibers or short fibers and are called staples in this case.[2,11] Discontinuous whiskers are manufactured to a definite length in batch-type processing, while continuous whiskers are produced by melt processes such as the laser-heated floating zone technique. To summarize, the types of fibers are continuous (indefinite length), continuous whisker (single crystal, indefinite length), discontinuous (chopped continuous or staple), and discontinuous whisker (single crystal, definite length).[3]

If parallel and continuous fibers are combined with a suitable matrix and cured properly, unidirectional composite properties should be obtained.[12,13] The functions and requirements of the matrix are important for an effective composite (see Figure 1.1). A desirable matrix:

- Keeps the fibers in the structure in place
- Helps distribute or transfer loads
- Protects the filaments in the structure before and during fabrication
- Controls the electrical and chemical properties of the composite
- Minimizes moisture absorption
- Exhibits low shrinkage
- Wets and bonds to fiber
- Has low coefficient of thermal expansion
- Must flow to penetrate into fiber bundles
- Completely eliminates voids during compacting and curing
- Has reasonable strength, modulus, and elongation (elongation should be greater than fiber length)
- Must be elastic for transferring load to fibers

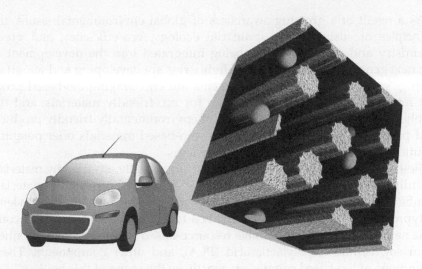

FIGURE 1.1
Cartoon model of natural fibers dispersed in polymer matrix that could be used for automotive applications.

- Has strength at elevated temperatures (depending on application)
- Has low temperature capability (depending on application)
- Has excellent chemical resistance (depending on application)
- May be processed into final composite shape easily
- Has dimensional stability (maintains shape)

1.2 Bio-Based Composites

Composite materials have existed for centuries, but only within recent years has the utilization of natural plant fiber–reinforced composites attracted the attention of the scientific community.[4,5] Interest in composites reinforced with natural fibers is growing because of their low cost, light weight, high specific strength, renewability, and biodegradability.

Composites are finding extensive uses in packaging, furniture, construction, and other industries. Natural fibers are emerging as viable alternatives to glass fibers either alone or combined in composite materials for various applications in automotive parts, building structures, and rigid packaging materials.[14–16] The advantages of natural fibers over synthetic or man-made fibers such as glass are low cost, low density, competitive specific mechanical properties, carbon dioxide sequestration, sustainability, recyclability, and biodegradability.

As a result of a growing awareness of global environmental issues, the principles of sustainability, industrial ecology, eco-efficiency, and green chemistry and engineering are being integrated into the development of the next generation of materials.[15,17] Industries are developing and manufacturing "greener" materials; governments are encouraging bio-based product research; academics are searching for eco-friendly materials; and the public is coming to value the benefits of environmentally friendly products and processes, but at affordable prices. Bio-based materials offer potential solutions to complex environmental problems.

Bio-based composites can be produced from partially eco-friendly materials or fully eco-friendly (green) materials.[15,17,18] Partially eco-friendly materials combine natural fibers with petroleum-based plastics such as polyethylene, polypropylene, polyester, and others. Eco-friendly or green materials combine natural fibers with renewable resource-based natural plastics obtained from soy, cellulose, polylactic acid (PLA), and other components. These definitions of bio-based composites constitute the scope of this book.

The combination of natural fibers and natural plastics (green composites) produces the necessary performance entirely or in combination with petroleum-based polymers and offers a path to achieve eco-friendly materials in the 21st century. However, the need to produce 100% green materials as substitutes for petroleum-based materials is not immediate. Biocomposites that contain significant amounts of bio-based materials can achieve this at an affordable cost-to-performance ratio to compete with petroleum-based materials and still maintain a positive balance among ecology, economy, and technology.

1.3 Scope and Organization

This book is intended to provide a comprehensive reference source for the latest advances in the area of bio-based composites that can substitute for and compete with traditional petroleum-based materials while reducing environmental harm and maintaining economic viability.

We have assembled chapters on topics ranging from the strategy of bio-based resources (food versus non-food markets and material versus energy applications), natural fiber materials (agricultural materials such as grass, straw, and traditional wood), raw materials for fabricating bio-based composites, filler–matrix interactions, and basic design principles, properties, and applications. In addition, we included two comprehensive chapters on life cycle analysis and future trends of bio-based composites that may emerge as the frameworks upon which sustainability of materials and processes will be established.

References

1. N. Rjiba, M. Nardin, J.Y. Dréan, and R. Frydrych. A Study of the Surface Properties of Cotton Fibers by Inverse Gas Chromatography. *Journal of Colloid and Interface Science,* 314 (2007): 373–380.
2. M.J. John and S. Thomas. Biofibres and Biocomposites. *Carbohydrate Polymers,* 71 (2008): 343–364.
3. R. Chollakup, W. Smitthipong, and P. Suwanruji. Environmentally Friendly Coupling Agents for Natural Fibre Composites. In *Natural Polymers, Volume I: Composites,* M.J. John and S. Thomas, Eds. Cambridge: Royal Society of Chemistry, 2012, p. 349.
4. A.K. Mohanty, M. Misra, and L.T. Drzal. *Natural Fibers, Biopolymers, and Biocomposites.* Boca Raton,FL: CRC Press, 2005.
5. D.N. Saheb and J.P. Jog. Natural Fiber Polymer Composites: A Review. *Advances in Polymer Technology,* 18 (1999): 351–363.
6. R. Chollakup, W. Smitthipong, W. Kongtud et al. Polyethylene Green Composites Reinforced with Cellulose Fibers (Coir and Palm Fibers): Effect of Fiber Surface Treatment and Fiber Content. *Journal of Adhesion Science and Technology,* 27 (2013): 1290–1300.
7. J. Summerscales, N. Dissanayake, A. Virk et al. A Review of Bast Fibres and Their Composites. Part 2: Composites. *Composites Part A: Applied Science and Manufacturing,* 41 (2010): 1336–1344.
8. J. George, S.S. Bhagawan, and S. Thomas. Effects of Environment on the Properties of Low-Density Polyethylene Composites Reinforced with Pineapple-Leaf Fibre. *Composites Science and Technology,* 58 (1998): 1471–1485.
9. B.F. Blumentritt, B.T. Vu, and S.L. Cooper. The Mechanical Properties of Oriented Discontinuous Fiber-Reinforced Thermoplastics. I. Unidirectional Fiber Orientation. *Polymer Engineering & Science,* 14 (1974): 633–640.
10. T. Elena, J. Long, and P. W. Michael. Strategies for Preparation of Oriented Cellulose Nanowhiskers Composites. In *Functional Materials from Renewable Sources.* ACS Symposium Series. Washington, DC: American Chemical Society, 2012, pp. 17–36.
11. K. Van De Velde and P. Kiekens. Effect of Material and Process Parameters on the Mechanical Properties of Unidirectional and Multidirectional Flax–Polypropylene Composites. *Composite Structures,* 62 (2003): 443–448.
12. M.Z. Rong, M.Q. Zhang, Y. Liu et al. The Effect of Fiber Treatment on the Mechanical Properties of Unidirectional Sisal-Reinforced Epoxy Composites. *Composites Science and Technology,* 61 (2001): 1437–1447.
13. G. Romhány, J. Karger-Kocsis, and T. Czigány. Tensile Fracture and Failure Behavior of Thermoplastic Starch with Unidirectional and Cross-Ply Flax Fiber Reinforcements. *Macromolecular Materials and Engineering,* 288 (2003): 699–707.
14. E. Bodros, I. Pillin, N. Montrelay et al. Could Biopolymers Reinforced by Randomly Scattered Flax Fibre Be Used in Structural Applications? *Composites Science and Technology,* 67 (2007): 462–470.

15. N. Graupner, A.S. Herrmann, and J. Müssig. Natural and Man-Made Cellulose
    Fibre-Reinforced Poly(Lactic Acid) (PLA) Composites: An Overview about
    Mechanical Characteristics and Application Areas. *Composites Part A: Applied
    Science and Manufacturing*, 40 (2009): 810–821.
16. B.C. Suddell and W.J. Evans. *In Biopolymers and Biocomposites: Natural Fiber
    Composites in Automotive Applications*, A.K. Mohanty, M. Misra, and LT. Drzal,
    Eds. Boca Raton, FL: CRC Press, 2005 p. <no.>.
17. M. Avella, A. Buzarovska, M. Errico et al. Eco-Challenges of Bio-Based Polymer
    Composites. *Materials*, 2 (2009): 911–925.
18. J.T. Kim and A.N. Netravali. Non-Food Application of Camelina Meal:
    Development of Sustainable and Green Biodegradable Paper–Camelina
    Composite Sheets and Fibers. *Polymer Composites*, 33 (2012): 1969–1976.

2

Bio-Based Strategy: Food and Non-Food Markets

Rina Singh, Brent Erickson, and Paul Winters

CONTENTS

2.1 Introduction

Two concurrent trends in the chemical industry are creating opportunities for manufacturers to adopt industrial biotechnology applications. First, abundant, low-cost natural gas resources are creating a feedstock revolution for the chemical industry, rapidly displacing the use of petroleum naphtha. Growth in natural gas production has generated a surplus of low-cost ethylene and a concurrent shortage of propylene, butadiene, and benzene.[1]

The shortages and accompanying high prices offer market opportunities for biomass-based chemicals. Second, major consumer product companies—responding to long-term trends in consumer demand for products that are healthier for the environment—have established sustainability programs aimed to reduce the use of harmful petrochemicals, increase the use of recycled and renewable materials, and increase energy efficiency.

Following a decade of intensified research and development, industrial biotechnology applications for renewable chemicals are reaching commercial readiness at an opportune moment. Many of the technologies are feedstock-flexible—able to use sugar from any source of biomass or solid waste. Some can even convert methane to target chemicals biologically.

The availability of capital has been one of the major roadblocks for the commercialization of advanced biofuels made from non-food resources. The renewable chemical industry, especially producers of low-volume, high-value renewable specialty chemicals, can pursue commercialization strategies that are less capital intensive.

2.2 Consumer Demand

Growing consumer demand for products that are healthier for the home environment is well documented. Consumer product manufacturers and retailers recognize the need to meet this consumer demand in order to protect their brands. Retail giant Walmart and consumer product titans such as Coca-Cola, Heinz, Ford, Nike, and Procter & Gamble have established annual sustainability reports and policies to obtain more renewable contents for their products and packaging. Fully 95% of the top 250 companies worldwide publish annual sustainability reports.[2]

Consumer brand owners also understand that product performance is a higher priority for their customers than sustainability.[3] Half of consumers polled in a 2012 survey were unaware that personal care, household cleaning, laundry, and baby care products contained petroleum-based ingredients. Once informed, though, more than 60% of consumer respondents were "concerned" or "very concerned" about the potential long-term health risks and negative impacts on the environment.[4]

Two additional studies of consumers in the United States, Canada, and China found that the number of respondents who would actively choose a bio-based product over a petroleum one increased as they gained information about the products. When respondents were given definitions of bio-based products as "green products" composed of renewable resources that are biodegradable, 75 to 80% indicated they would select them over petroleum products if they were available at a similar cost and performance.[6]

Second, anecdotal evidence suggests that brand owners are wary of renewable chemicals made from corn-derived sugars. In the debate over the sustainability of biofuels, many environmental non-governmental organizations (NGOs) have claimed that corn production is not sustainable.[7] Consumer product companies apparently wish to avoid any controversy or negative perception about the use of corn.

When product performance is equal, sustainability can help differentiate a product for consumers. Renewable chemicals can replace petroleum-derived chemicals without sacrificing cost or performance. Producers face the twin challenge of educating consumers about renewable chemicals and the sustainability of the raw materials or feedstocks used to produce them, whether food or non-food crops. At the same time, most product manufacturers are not readily familiar with the industrial biotechnology companies that are beginning to make renewable chemicals for household and personal care products.[8]

2.3 Market Potential

Consumer demand for affordable and sustainable products represents a considerable market opportunity. Sales of green household cleaning products in the United States, for example, more than doubled from $303 million in 2007 to $640 million in 2011. Projected growth will take this market segment to $825 million annually by 2016.[9]

Worldwide, the green cleaning product market could reach $9.72 billion by 2017. While the United States and Europe will remain the largest markets, Asia Pacific countries are expected to contribute to a 25% compound annual growth rate.[10] The global bath and shower products market reached $10.8 billion in 2010, shipping more than 3.8 billion items. It is projected to grow 18.6% to reach $12.85 billion by 2015. Europe accounts for more than half the demand in this market segment.[11]

Table 2.1 provides current and future estimates for sales of consumer products with renewable content, including renewable chemicals. With this growing demand for green products, the overall renewable chemical ingredient market is expected to reach $83.4 billion to $84.8 billion by 2018, with an annual growth rate of 7.7%.[12,13]

Renewable chemicals made with industrial biotechnology are already well positioned to meet this demand. The commercialization of new biotechnologies and capital investments in new manufacturing facilities over the past several years is making good on the efforts of federal programs to support the industry's emergence.[14]

2.4 Sustainability of Renewable Chemicals

Chemicals and bio-based products from renewable biomass can achieve significant environmental benefits. In 2007, the U.S. Environmental Protection Agency recognized that industrial biotechnology processes to manufacture monomers and polymers would produce fewer environmental impacts than

TABLE 2.1

U.S. and Global Market Estimates for Renewable Chemicals

Market Sector	Scope	Market Value (Billions/Year)	Forecast Value (Billions/Year)	Sources
Renewable chemicals	Global	$57 to $59.1/2013–2014	$83.4 to $84.8/2018	12,13
Food flavorings	Global	$19.8/2010		27
Green household cleaning products	Global		$49.72/2017	10
	U.S.	$0.640/2011	$0.825/2016	9
Lubricants	Global	$44/2011		28
Nutraceuticals	Global	$142.1/2011	$204.8/2017	29
	U.S.	$50.4/2010		30
	Europe	$35/2010		30
Organic personal (skin, hair, oral care) products and cosmetics	Global	$7.6/2012	$13.2/2018	31
Personal care (bath and shower) products	Global	$10.8/2010	$12.85/2015	11
Textiles and dyes	Global	$19.6/2013	$23.4/2018	32

petrochemicals because bio-based processes "typically operate on renewable resources, at low temperatures, in aqueous environments, and produce few byproducts."[15] Biological processes can also reduce the number of process steps in the production of a target chemical, creating economic efficiency.[16]

Displacing petrochemicals with renewable chemicals will further reduce U.S. reliance on petroleum and lessen industrial pollution. Commercialization of new chemicals would leverage biotechnology developed within the U.S. and create new markets for domestic agriculture, spreading the benefits of economic growth to rural areas of the country. Some renewable chemicals are already cost-competitive with petrochemicals.[15] Many companies are feedstock agnostic in commercializing industrial biotechnology applications for renewable chemicals.

In the United States, using corn-derived sugars for renewable chemicals is an important economic opportunity that can add significant value to the sugar fraction of the grain while preserving proteins, starches, and oils for food, feed, fuel, and other uses. By creating a new market for corn sugar, renewable chemical production can support profitability for grain farmers and add value to productive farmland. Further benefits would include preservation of farm lands and promoting increased productivity and fractionation of corn supplies. The creation of additional demands for agricultural biotechnology solutions will increase productivity for corn and other grains and also increase demands for new sources of sugars from biomass that will support the emergence of cellulosic sugars.

2.5 Economic Opportunities from Renewable Chemicals

The Obama administration officially released the National Bioeconomy Blueprint on April 26, 2012, after 6 months of gathering and analyzing public comments.[17] While the original request for comment issued by the White House Office of Science and Technology (OSTP) sought input on the grand challenges to be addressed by building a bioeconomy, the nation's blueprint as released addresses five federal bioeconomy strategic objectives to realize the full potential of the U.S. bioeconomy and highlights early government-supported achievements aimed at meeting those objectives.[18]

The National Bioeconomy Blueprint panel discussion held by academic and industry representatives at the White House on ways in which investments and innovations contribute to the U.S. bioeconomy was summarized recently.[19] The blueprint presents a vision of "a previously unimaginable future" enabled by the bioeconomy, including liquid biofuels derived directly from CO_2, biodegradable bio-based plastics, tailored food products, personalized medicine, and environmental monitoring. It also indicates that the bioeconomy has emerged as a priority for the Obama administration as a path to economic growth, prosperity, and societal benefits.

The blueprint further calls for "unlocking the promise of synthetic biology," which is supportive of the legislative program of the Biotechnology Industry Organization (BIO) calling for a Department of Energy (DOE) grant program for the use of synthetic biology technology to develop sustainable biofuels and renewable chemicals.[16,20]

Synthetic biology is the key green technology in industrial biotechnology, and today synthetic biology is enabling a new revolution in chemical production for advanced biofuels and bioplastics. Industrial biotechnology companies have developed automated methods for microbial strain engineering that enable the rapid development of microorganisms that convert inexpensive renewable raw materials into useful renewable chemicals by fermentation. This new technology provides opportunities to extract greater value from underutilized agricultural materials while using less petroleum for chemical synthesis.

The economic opportunities from renewable chemicals are outlined in a roadmap report issued by DOE's Energy Efficiency and Renewable Energy (EERE) program, showing the synthesis for top building block chemicals from renewable biomass.[21] Biological transformations account for most routes from plant feedstock to building blocks, but chemical transformations predominate in the conversion of building blocks to molecular derivatives and intermediates. The biological primary conversions need to occur first to develop the building blocks, which is where the research and development focus is currently for industrial biotechnology companies, before secondary transformations are developed.

A recent study led by the Oak Ridge National Laboratory (ORNL) projected that the United States would have 1.1 to 1.6 billion tons of available sustainable biomass for industrial bioprocessing by 2030. The finding was a highlight of the *2011 U.S. Billion-Ton Update: Biomass Supply for a Bioenergy and Bioproducts Industry* updating a 2005 study undertaken by DOE and ORNL. The report examines the nation's capacity to produce a billion dry tons of biomass resources annually for energy uses without impacting other vital farm and forest products such as food, feed, and fiber crops.

The study provides industry, policy makers, and the agricultural community with county-level data and includes analysis of current U.S. feedstock capacity and the potential for growth in crops and agricultural products for the production of renewable chemicals for advanced biofuels, bioplastics, and other everyday consumer bio-based products. According to DOE, "with continued developments in biorefinery capacity and technology, the feedstock resources identified could produce about 85 billion gallons of biofuels—which is enough to replace approximately 30% of the nation's current petroleum consumption."[22]

2.5.1 Renewable Chemicals in Composites

Unsaturated polyester resins (UPRs) are produced by the polycondensation of saturated and unsaturated dicarboxylic acids with glycols. Unsaturated polyester resins form highly durable structures and coatings when they are cross-linked with vinylic reactive monomers, most commonly styrenes. The properties of unsaturated polyester resins depend on the types of acids and glycols used and their relative proportions. On their own, cross-linked unsaturated polyester and vinyl ester resins have limited structural integrity, but when combined with such chemicals as fiberglass or mineral fillers, their mechanical strength is enhanced. When combined with fiberglass, the polymers transform into fiberglass-reinforced plastic (FRP) consumed primarily in the construction, marine, and land transportation industries.

Non-reinforced cross-linked unsaturated polyester resins are used to make cultured marble and solid surface countertops, gel coats, automotive repair putties and fillers, bowling balls, buttons, and other products. UPRs are the most widely used resin types for composites, constituting more than 70% of all thermoset resins. They are used in the production of fiber-reinforced plastics and non-reinforced filled products. Due to the easy processing of UPRs, they can be applied in a large variety of manufacturing processes. UPRs are well utilized in the marine (pleasure boat) and automotive industries and also in the production of wind turbine blades.

The largest global UPR producers include DSM, Ashland Inc., AOC Resins, and Reichhold and Cook Composites and Polymers. Maleic anhydride is the primary entry chemical in the production of UPRs. Because itaconic acid has a structure similar to maleic anhydride, it can also be used for the production of UPRs. Itaconic acid in this case cannot be used as a drop-in replacement but it can serve as a good substitute.

Among the companies most active in UPR production, DSM currently focuses on the task of using itaconic acid in the most effective way. The company is planning to start with the commercialization of an itaconic-based UPR and is developing a route for 100% bio-based polyester composites. In May 2012, it published its patent for the production of bio-based polyester composites from itaconic acid.

The UPR market is expected to grow in the future by 5.5%. The main growth is expected to be in the automotive industry where UPR components will replace metal parts. The qualities of UPRs include low carbon footprints, long durability, and low weight. The value added by low weight is important in the automotive industry. Cars constructed from lightweight components will weigh less, and the lower weight will have a strong positive impact on fuel consumption since fuel prices constantly increase.

A number of substitutes can replace maleic anhydride in the production of UPRs, for example, bio-based fumaric acid. Myriant is working currently on developing such a substitute. The market entry for bio-based fumaric acid probably will be much easier and faster as the compound may be used as a drop-in replacement for maleic anhydride. The estimated potential of replacing maleic anhydride in UPRs with itaconic acid is projected up to 5% of the maleic anhydride used in UPRs by 2020.

Genencor, now part of DuPont, has a long history in the design and operation of cell factories. Over the years, the company produced industrial enzymes, commodity chemicals, and renewable chemicals including 1,3-propanediol, ascorbic acid, and tryptophan. The production of BioIsoprene™ by fermentation is similar to previous metabolic engineering efforts made to develop and produce the enzyme, but exhibits some unique differences since the monomer is a highly flammable gas.

The key components for isoprene production are the isoprene synthase and isoprenoid enzymes. Natural plant sources for isoprene are known. Isoprene synthase can be prepared, but protein engineering is required to achieve an improved enzyme suitable for a viable commercial process.[23] A unique process to produce BioIsoprene™ from renewable biomass is in the development stage at DuPont and at the Goodyear Tire & Rubber Company.

Numerous products may be made with the C_5 hydrocarbon, one of which is a synthetic *cis*-polyisoprene (synthetic rubber). It has been recognized as a suitable replacement for natural rubber in many applications for decades, and there are advantages in creating biopolyisoprenes with improved physical properties. Other companies jointly developing synthetic *cis*-polyisoprene include Bridgestone Corporation, which recently polymerized Ajinomoto's bioisoprene. Currently, 2 billion pounds of isoprene are produced per year and used in tires and other products.

2.5.2 Commercialization of Renewable Chemicals

A recent sustainability survey shows that sustainability and the use of renewable chemicals are now components of mainstream thinking and action for chemical producers and users.[8] Companies in the chemical industry want sustainability as part of their message; they want to be seen as leaders and use sustainability to differentiate themselves. Many large, mature, and established chemical and biotechnology companies and emerging venture-backed companies are involved actively in the development and commercialization of renewable chemicals from a variety of renewable biomass sources.

In the United States, a growing number of emerging companies are pursuing the conversion of renewable biomass into sustainable sugar, whether from feedstock such as algae or terrestrial woody biomass. Some of these emerging companies are trying to follow a biorefinery model, similar to the practices of petrochemical refineries that co-produce large-volume fuels and high-value chemicals.

The oldest biorefineries operating in the United States have use corn wet milling since the mid-1800s. In this process, starch is recovered and sold or further converted by several processes into other value-added products. The starch is also enzymatically hydrolyzed and used as feedstock to produce high fructose corn syrup, bioethanol (for biofuels), lactic acid, lysine, citric acid, and a variety of other fermentation products.

One major difference between a biorefinery and petroleum refinery is that one of the main products of a biorefinery—at least from first-generation operations based on corn and sugarcane—is food for human and animal consumption, as shown in Figure 2.1. This practice has caused discussions and controversy and created a serious debate with policy makers about the

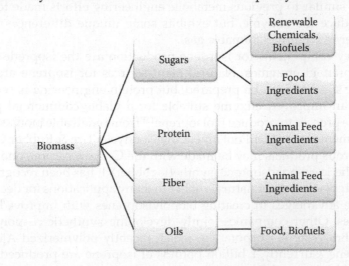

FIGURE 2.1
Biorefinery.

sustainability of first-generation biofuels in particular, and to a lesser extent renewable specialty chemicals produced as co-products. To address this issue, other sources of renewable biomass for future biorefineries are being developed, for example, non-food energy crops such as miscanthus, switchgrass, sorghum, cassava, camelina, jatropha, and willow.

Wastes are also being used as feedstocks, including oil palm waste, wood residues, rice husks, corn stover, sugarcane bagasse, coconut husks, maize cobs, and industrial and municipal solid wastes. The general consensus is that the biorefineries still in the nascent stage will initially focus on large-volume fuels followed by high-value chemicals similar to the evolution of petrochemical refineries. A wide range of renewable building blocks can be made from biomass, and the planned introduction of over 2 million tons of renewable chemicals in 2013 is shown in Figure 2.2.

A number of conversion technology processes for utilizing various renewable biomasses available to value-added products are being explored. Two long-used primary technology platforms for converting renewable biomass to renewable chemicals and fuels are thermochemical and biochemical processes. *Thermochemical platforms* like gasification and pyrolysis use heat and uncatalyzed reactions with steam and/or oxygen to break down biomass structures. *Biochemical platforms* use natural and synthetic biological and/or chemical means of converting starch, cellulose, and hemicelluloses in biomass to sugars to feed fermentation operations.

Now other types of platforms are emerging. Examples are heterogeneously catalyzed pyrolysis (Anellotech Inc.), heterogeneous catalysis converting

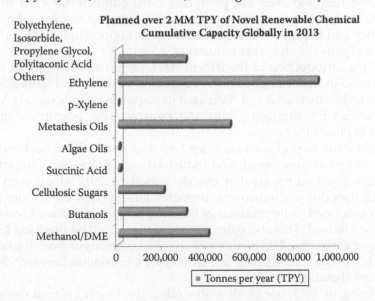

FIGURE 2.2
Manufacturing novel renewable chemicals.

sugar solutions to hydrocarbons (Virent Inc.), hybrid fermentations of syngas from biomass gasification or other stack sources to make renewable alcohols and other chemicals and fuels (Lanza Tech, Coskata Inc.), and chemical (strong acid or base) digestion of biomass to make intermediates that are then converted to final chemical or fuel products by hydrogenation (ZeaChem Inc.).

2.5.2.1 Shifting Patterns of Corn Grain Demand and Supply

Much of the criticism of the biofuel industry's use of corn starch stems from the support it receives from the federal Renewable Fuel Standard (RFS), a policy to reduce U.S. reliance on imported petroleum and greenhouse gas emissions by opening the domestic fuel market to renewable fuels. Because the RFS regulates market activity, the EPA must assess the indirect market and environmental impacts of the law.[24] At present, renewable chemical production is not supported by equivalent federal policy.

Renewable fuel made from corn meets environmental goals and does not adversely impact U.S. food production according to available data. The EPA conducts thorough life cycle analyses of each proposed feedstock, energy source, and technology pathway for biofuel production and has determined that corn ethanol meets the policy standard by reducing greenhouse gas emissions in comparison to petroleum gasoline.[24]

Moreover, agricultural productivity data demonstrates that the market has additional room to use corn sugars to produce renewable chemicals. Renewable fuel and chemical producers must compete for corn supplies with other users of corn; their competitive advantage stems from the higher value they add to grain supplies and the market opportunities they present to regional growers and other industries.

Since the introduction of the RFS in 2005, patterns of corn grain supply and demand in the United States have changed, as shown in Figure 2.3. Corn production has increased and decreased in response to price signals. Market opportunities for alternate grains and crop rotation practices prompted farmers to plant other crops.

U.S. domestic uses of corn grain for feed and residual use declined from 2005 peaks while food, seed, and industrial uses increased, primarily for renewable fuels (although that use decreased in 2012 along with other domestic uses due to a nationwide drought). However, dried distiller grains (DDGs) generated as by-products of biofuel production replaced some corn grain used in feed. This changing pattern of using mill by-products for feed in place of grain parallels previous changes in patterns of overall grain use according to U.S. Department of Agriculture's Economic Research Service (ERS) data shown in Figure 2.4.

Beginning in 1985, use of all grains other than corn for feed decreased, while the use of oil seed meal and other by-product feeds increased. As the use of corn grain started declining in 2005, the use of corn gluten, meal, and

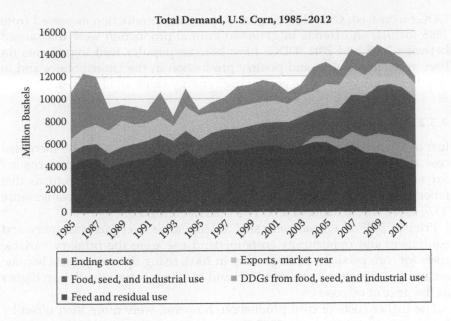

FIGURE 2.3
Total demand for U.S. corn, 1985–2012.

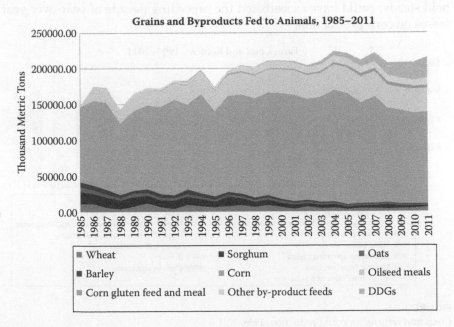

FIGURE 2.4
Patterns of grains and by-products used in livestock feed, 1985–2011.

DDGs increased. Overall, U.S. grain-fed animal production increased from 1985 to 2011, and trends in grain-fed animal production were unchanged between 2005 and 2011. DDGs have become popular feed ingredients for beef and dairy, swine, and poultry production in the United States and in 50 other countries because of their high energy value and low cost.[25]

2.5.2.2 Profitability for Farmers

Just as changing patterns of demand for corn are driven by the increased cost and the availability of lower-cost substitutes, production patterns are driven by costs and potential returns. After a decade of low returns that failed to recover costs, the return on corn production has been positive since 2007, as shown by national level ERS data in Figure 2.5.

Prior to 2005, overhead costs such as capital recovery of machinery and equipment and opportunity costs for land use were the primary variable costs for corn production. Beginning in 2005, rising operating costs became an additional primary variable. Fuel and fertilizer costs were driven higher as the price of oil rose.

The higher costs of corn production, however, were more than offset by higher returns beginning in 2007. Corn prices were up to 30% higher on average between 2006 and 2010 than they would have been if ethanol production remained at 2005 levels according to one study.[2] Without the biofuel market, the increase in overhead costs for fertilizer and fuel (even if all other factors held steady) could have exacerbated the preceding decade of year-over-year losses on corn production.

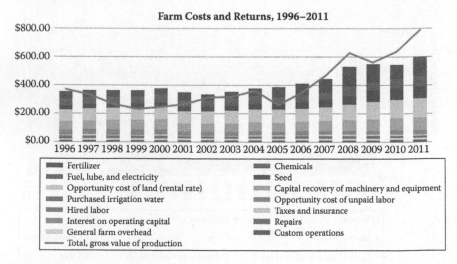

FIGURE 2.5
Costs and returns for corn production, 1996–2011.

2.6 Biorefinery: Part of the Solution

With rising costs, corn producers need additional markets. While biofuels represent one new market and by-products of biofuel production are filling a former market for grain, other uses of corn grain such as high fructose corn syrup (HFCS), cereal, glucose, dextrose, and starch have remained at relatively static levels for nearly a quarter century as shown in the ERS data in Figure 2.6. The number of corn biorefineries capable of producing these product streams has also remained fairly static. The market for HFCS is expected to decline in future years as other sweeteners take market shares and demands for sweeteners decline.[18]

A biorefinery could manufacture 22.4 pounds of biopolymer in addition to the oil, feed, and meal normally produced with ethanol or sweeteners from each bushel of corn. If current and projected growth of biopolymers were met entirely with corn sugars, the demand for corn would grow from 42 million bushels in 2011 to 112 million bushels in 2016. This demand represents 0.5 to 1% of corn production in 2012, which was reduced by drought. It is approximately half the amount of corn used for glucose and dextrose or starch, and about a quarter of the amount used for HFCS in 2012. It would also represent approximately 4% of consumption of corn for ethanol production.

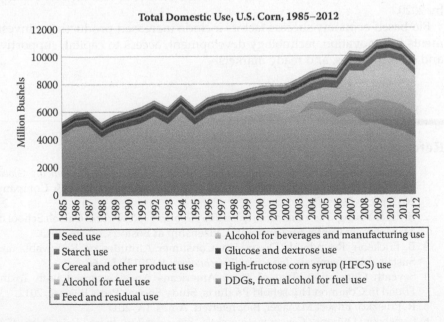

FIGURE 2.6
Feed and industrial uses of corn, 1985–2012.

2.7 Conclusion

Industrial biotechnology's role in enabling the growth of the sustainable chemistry sector is obvious, since it is the core technology that will enable growth of this sector. Wide interest in sustainable chemistry continues based on opportunities for developing cost-effective businesses, often combining biochemical and petrochemical capabilities. Newer conversion technologies will provide more process technology choices.[27-32] Companies are increasingly branding their products green to promote sustainable images.

Growth of the United States sustainable chemistry industry is driven by consumer demand and government policy and regulation. Retail market pull is driving demand. Walmart announced an environmental labeling program for all products carried in its stores and wants to be a "good steward for the environment" by using only renewable energy sources and producing zero waste. Customers will see a rating system for each product that shows its full environmental cost.

The U.S. Navy will produce half of its fuel from renewable sources by 2020. DuPont has set a goal of generating $1 billion by 2015 from renewable materials and fuel technologies. Procter & Gamble said it eventually will use only recycled or renewable materials to make and package its products. Virgin America CEO David Cush wants his airline to run on 10% biofuels by 2020.

Bio-based economy success factors include increased productivity, investments in innovation, technology development, access to capital, supportive and stable policies, and ready markets.

References

1. A. Horncastle, A. Sastry, S. Singh et al. *Future of Chemicals: Rebalancing Global Feedstock Disruptions with "on-Purpose" Technologies.* Dubai: Booz & Company, 2012, pp. 1–20.
2. Ernst & Young. *The Value of Sustainability Reporting.* New York: Carroll School of Management Center for Corporate Citizenship at Boston College, 2013.
3. B. Erickson. Roundtable Discussion: Consumer Attitudes on Renewable and Sustainable Chemicals. *Industrial Biotechnology*, 9 (2013): 55–60.
4. Seventh Generation. Two of Three Americans Are Concerned with Toxins Found in Common Household Products, Study Shows. Press Release, 2012.
5. R. Pruszko. Project Manager, Biopreferred. Ames, IA, 2011.
6. Genencor/Dupont. *Genencor Household Sustainability Index.* Palo Alto, CA: Environics Research, 2011.
7. J. Earley, and A. Mckeown. *Worldwatch Report #180. Red, White and Green: Transforming U.S. Biofuels.* Washington, DC: Worldwatch Institute, 2009.

8. K. Burr. Sustainability Research. Sutton, Surrey, UK, 2012.
9. Packaged Facts. *Green Cleaning Products in the U.S.* Rockville, MD, 2012.
10. Global Industry Analysts. *Household Green Cleaning Products.* San Jose, CA, 2011.
11. Marketline. *Bath and Shower Products: Global Industry Guide.* Chicago, 2011.
12. Marketsandmarkets.com. *Renewable Chemicals Market: Alcohols (Ethanol, Methanol), Biopolymers (Starch, PLA, PHA, Polyethylene, and Others), Platform Chemicals, and Others. Global Trends and Forecasts to 2018.* Rockville, MD, 2013.
13. Research and Markets. *Renewable Chemicals Market: Global Trends and Forecasts to 2018.* Dublin, 2013.
14. B. Erickson, J.E. Nelson, and P. Winters. Perspective on Opportunities in Industrial Biotechnology in Renewable Chemicals. *Biotechnology Journal*, 7 (2012): 176–185.
15. D. Ahmann and J. Dorgan. *Bioengineering for Pollution Prevention through Development of Biobased Energy and Materials.* Washington, DC: National Center for Environmental Research, U.S. Environmental Protection Agency, 2007, pp. 117–129.
16. R. Singh. Facts, Growth, and Opportunities in Industrial Biotechnology. *Organic Process Research & Development*, 15 (2011): 175–179.
17. White House. *National Bioeconomy Blueprint.* Washington, 2012.
18. MECAS. *Alternative Sweeteners in a High Sugar Price Environment.* London, 2012.
19. R. Singh. The National Bioeconomy Blueprint: Meeting Grand Challenges. *Industrial Biotechnology Journal*, 8 (2012): 94–96.
20. B. Erickson, R. Singh, and P. Winters. Synthetic Biology: Regulating Industry Uses of New Biotechnologies. *Science*, 333 (2011): 1254–1256.
21. W. Todd and G. Petersen, Eds. *Top Value Added Chemicals from Biomass. Vol. I: Results of Screening for Potential Candidates from Sugars and Synthesis Gas.* Washington, DC: Pacific Northwest National Laboratory and National Renewable Energy Laboratory, Department of Energy, Office of Biomass Program, 2004.
22. R.D. Perlack and B.J. Stokes. *U.S. Billion-Ton Update: Biomass Supply for a Bioenergy and Bioproducts Industry.* Oak Ridge, TN: U.S. Department of Energy, 2011.
23. G.M. Whited, F.J. Feher, D.A. Benko et al. Development of a Gas-Phase Bioprocess for Isoprene Monomer Production Using Metabolic Pathway Engineering. *Industrial Biotechnology*, 6 (2010): 152–163.
24. U.S. Environmental Protection Agency. *Renewable Fuel Standard Program Regulatory Impact Analysis* (EPA-420-R-10-006). Washington, DC, 2010, pp. 299–300.
25. G.T. Shurson. Impact of United States Biofuels Co-Products on the Feed Industry. In *Biofuel Co-Products as Livestock Feed: Opportunities and Challenges*, H. P. Makkar, Ed. Rome: United Nations Food and Agricultural Organization, 2012.
26. C.R. Carter. *The Effect of the U.S. Ethanol Mandate on Corn Prices.* Davis, CA: Department of Agriculture and Resource Economics, University of California, 2012, p. 40.
27. Ubic Consulting. *The World Biotech Flavor Market.* Newport Beach. CA, 2013. http://www.ubic-consulting.com/template/fs/Biotech%20Flavor.pdf
28. Transparency Market Research. Technology Trends in Lubricants Market for Turbine Oil, Compressor Oil, Gear Oil, Hydraulic Oil, Bearing Oil, and Heat Transfer Fluid Lubricant Applications. Albany, NY, 2013a.
29. Bourne Partners. *Sector Report: Nutraceuticals Industry.* Charlotte, NC, USA, 2013.
30. Frost & Sullivan. Global Nutraceutical Industry: Investing in Healthy Living. Paper presented at Third International Conference on Nutraceuticals, Dietary Supplements, and Functional Foods, Mumbai, India, 2011.

31. Transparency Market Research. Organic Personal Care Products Market for Skin Care, Hair Care, Oral Care and Cosmetics: Global Industry Analysis, Size, Share Growth, Trends, and Forecast, 2012–2018. Albany, NY, 2013b.

32. Transparency Market Research. Textile Chemicals Market: Global Industry Analysis, Market Size, Share, Growth, and Forecast, 2009–2018. Albany, NY, 2012.

3

Strategy of Bio-Based Resources: Material versus Energy

Kuakoon Piyachomkwan, Sittichoke Wanlapatit,
Wirasak Smitthipong, and Klanarong Sriroth

CONTENTS

3.1 Introduction

Bio-based resources are renewable, widely distributed, available locally, mold-able, recyclable, easily available in many forms, biodegradable, combustible, and reactive. Bio-based fibers have high aspect ratios, high strength-to-weight ratios, are relatively low in energy conversion, and have good sound and thermal insulation properties.[1-5] The fiber structure is hollow and laminated with molecular layers and an integrated matrix.

Some may consider some of these properties (e.g., biodegradability and combustibility) as problems, but they provide a means of predictable and programmable disposal not easily achieved with other resources. A lot of discussion surrounds the pros and cons of the use of bio-based resources and the material-versus-energy debate continues. This chapter describes strategic and political issues (plants producing energy versus material) that may hit a few nerves in the current ecological climate.

A good bio-based resource could have the potential (with good material properties) to be used in composite applications. However, it could also have potential to be developed for energy applications. Should we follow the material or energy applications for bio-based resources? The answer depends on several parameters: application demand, technology cost, resource production and management, political policy, and other issues.

Materials are deemed important or induce high impacts based on the specific properties that make them well suited for certain applications. For bio-based composites, an important factor might be an eco-friendly environment or biodegradation.[6] Bio-based resources are good choices to be used for material applications. On the other hand, energy is a major challenge of the 21st century. Meeting the world's growing demand for new energy sources in a sustainable way will require the deployment of more resources than are in place currently. The world's energy economy is built upon a base of fossil oils. The sustainable energy economy is more critically dependent on a wider array of materials.[7,8] It is imperative that the materials necessary to implement these potentially game-changing technologies are readily available to the diverse industries engaged in the markets.

It may be possible to alleviate some supply issues simply by re-evaluating the supply chain. Sourcing strategies include diversification, hedging, strategic inventory reserves (stockpiling), and buying in volume. However, for most materials, sourcing will require a long-term global vision in view of exponentially growing technology. Corporations with major investments in energy technologies may choose to engage in some stockpiling of materials to ensure stable supplies as part of a broader strategy. In general, however, government stockpiling of these materials is a short-term stopgap measure that will likely be insufficient to address materials issues over the long term.

Governments and industries are engaged in worldwide explorations to identify new possibilities for diversifying materials sources. Merely increasing production is not sufficient or sustainable. The supplies of bio-based resources, especially starches and fibers, are not infinite. Instability exists in the marketplace because some starches are produced primarily as food products and fibers are produced for textile applications. Nevertheless, from the composite application view, starches can be used for the matrix and fibers can be used as reinforcement materials for matrices.[9] From the energy application view, starches can be transformed to bio-ethanol and fibers can be burned to generate gas for producing electricity.[10]

Generally, an increase in demand will provide a strong enough economic incentive to increase the production of a bio-based resource. Boosting demand for bio-based resources would make them more cost effective; but meeting the demand also depends on the political policy of each country or zone (the European Union (EU), for example).

Another supply-side improvement strategy is to analyze the production and manufacturing processes, breaking them down into component steps to identify troublesome bottlenecks and improve efficiencies, thereby increasing yields. The greatest difficulty in improving process yield is the variability arising from facilities issues, weather conditions, and plant diseases. This variability demonstrates the lack of a scientific workforce that must be improved. Even so, many young scientists are reluctant to focus on biomaterials and energy disciplines because they are not considered "hot topics" by the most prestigious research journals.

Environmental strategy should include reduction of waste, improving manufacturing efficiency, and recycling of end-of-life products. Each of these operations can exert a significant impact on increasing the supply of bio-based resources.[11] Finally, a bio-based economy should include a low waste production chain starting from the use of land and fresh water through the transformation and production of bio-based products adapted to the requirements of end users. More precisely, a bio-based economy integrates the full range of natural and renewable biological (land and water) resources, biodiversity, and biological (plant, animal, and microbial) materials from processing to consumption.

A bio-economy encompasses the agriculture, forestry, fisheries, food production, and biotechnology sectors along with a wide range of industrial sectors spanning the production of energy and chemicals and the construction and transport industries. It comprises a broad range of generic and specific technological solutions (already available or to be developed) capable of being applied across these sectors to enable growth and sustainable development. Two examples are food security and meeting requirements for industrial materials for future generations.[12,13]

3.2 Bio-Based Resources for Composite Application

We have used bio-based resources for so long that we tend to accept their performance limitations such as swelling, shrinking, rotting, burning, and ultraviolet radiation degradation. By learning to live with these limitations, we have also limited our expectations of performance and as a result limit our ability to accept new concepts for improved performance and expanded markets. Bio-based composites are very familiar materials that have been used by consumers for centuries, particularly in low-cost, medium- to low-performance markets. We may have limited our expectations of bio-based composites based on past discoveries and not kept abreast of recent advances in chemistry and materials science research.

One of the best ways to deal with the durability of bio-based composites is to design for improved durability. In housing, this is done by installing large roof overhangs to minimize or prevent moisture, biological, and ultraviolet degradation. In construction, bio-based composites can be placed on durable foundations so that biocomponents do not come into direct contact with the ground. In some uses, it is not possible to solve durability problems by design so additional measures must be taken. To expand the use of agrofiber-based composites in adverse environments, it is necessary to interfere with nature's recycling chemistry. Industries that treat bio-based composites to perform better in adverse environments continue to develop.

For protection from biological degradation, current research focuses on several areas. One involves investigation of the mechanisms involved in degradation pathways and finding ways to stop key reactions from occurring. Knowing the reaction products that develop during degradation makes it possible to block these reactions from taking place.

A second area concentrates on stopping the oxidation reactions that are responsible for severe strength losses in bio-based composites. This involves the use of antioxidants to stop initial fungal attacks. An initial fungal attack is characterized by an organism's need to colonize. If colonization cannot take place, the organism cannot survive. We know that biological degradation occurs when an organism reduces the pH of a bio-based resource. If the buffering capacity of the resource is increased, the lowering of pH is prevented.

Finally, chemical modification of the cell wall polymers is being studied as a means of protection against biological attack. This is based on the theory that organisms require a certain level of moisture to hydrolyze glycosidic bonds and chemical modification reduces the equilibrium moisture content (or fiber saturation point) below that required for attack. Research in fire retardants concentrates mainly on insolubilizing or bonding effective fire retardants into bio-based resources. Fire retardant can be included in matrix structures.

Water repellency research centers on developing effective surface coatings to exclude moisture while treatments for dimensional stability mainly involve chemical reactions to bulk cell walls to green volume so water cannot swell the composites. High temperature steaming of bio-based resources for dimensional stability is also under investigation. Researching methods for stabilizing bio-based composites from degradation by ultraviolet (UV) radiation focuses on stabilizing lignin polymers since they are the cell wall components most susceptible to UV energy.

Degradation can also be reduced by adding polymers to cell matrices to help hold degraded fiber structures together to prevent water leaching of the undegraded carbohydrate polymers. Polymer coatings that reflect UV radiation are also being developed using cold plasma modification. Biotechnology that can modify bio-based resources using enzymes and genetic engineering that can alter plant chemistry are exciting technologies now under study to improve durability and other performance properties of bio-based composites.

A lot of new composite materials show potential for future development. Markets for existing composites will expand, but whole new markets are possible. A partial list of new possibilities based mainly on fibers includes filters, sorbents, structural composites, non-structural composites, molded products, packaging, and combinations with other resources.

Biomass consists of the most abundant carbohydrate compounds in the world. It should be noted that 75% of the total organic matter or more than 90% of the carbohydrates found in nature are in the form of polysaccharides. Polysaccharide polymers can be classified into several groups based

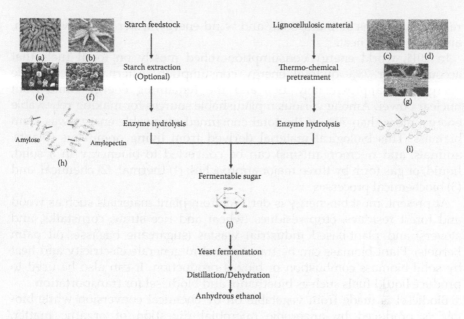

FIGURE 3.1
Bioethanol production from starch crops and plant fibers starting from starch feedstocks such as corn grains (a), cassava roots (b), or lignocellulosic materials such as cereal straw (c) and sawdust or wood residues (d). The starch feedstocks can be processed optionally to extract starch granules: corn (e) and cassava (f) or directly used for the next step without starch extraction. Lignocellulosic materials must be pretreated by thermochemical processes to liberate cellulose fiber (g). The helical starch (h) and ribbon-like cellulose (i) can then be hydrolyzed enzymatically to glucose (j), which is fermented by yeast to ethanol. The fermented mash is distillated and dehydrated to eventually yield anhydrous ethanol. Hexagonal symbols in (h) and (i) represents anhydroglucose units linked by α- and β-glycosidic linkages in starch and cellulose, respectively. Arrows indicate positions of enzyme hydrolysis.

on use criteria. Starch, a polysaccharide polymer obtained from plants in the form of minute granules (Figure 3.1), has characteristics that make it a potential bio-based composite candidate. Chapter 5 provides details about natural fibers.

3.3 Strategic and Political Issues Surrounding Energy Applications of Bio-Based Resources

The production of bio-based energy has emerged in many regions as a means to reinforce energy security, conserve the environment, diminish global warming, and strengthen socio-economic self-reliance of communities, thus promoting economic and ecological sustainability. Bio-based energy or bioenergy is renewable, i.e., it comes from continually replenished resources such as biomass, solar and wind energy, water and wave action, and geothermal heat.

In 2011, world energy consumption relied mostly on fossil fuels that accounted for 78% of total energy consumption. Alternative renewable energy was approximately 19% and the remaining energy came from nuclear power. Among various replenishable sources for making renewable energy, more than 70% of the total consumed renewable energy was from biomass. This biological material derived from living organisms (plants, animals, and microorganisms) can be converted to bioenergy in a solid, liquid, or gas form by three major approaches: (1) thermal, (2) chemical, and (3) biochemical processes.

At present, most bioenergy is derived from plant materials such as wood and forest residues, crop residues (wheat and rice straw, cornstalks, and stovers) and plant-based industrial wastes (sugarcane bagasse, oil palm kernels). Plant biomass can be transformed to generate electricity and heat by solid biomass combustion or biogas production. It can also be used to produce liquid fuels such as bioethanol and biodiesel for transportation.

Biodiesel is made from vegetable oil by chemical conversion while biogas is produced by anaerobic microbial digestion of organic matter. Bioethanol is derived from a biochemical conversion of sugar and starch feedstock or lignocellulosic biomass (Figure 3.1). The commercialization of bioethanol to date involves only sugar and starch crops that are recognized as first-generation feedstock. The industrial conversion of lignocellulosic biomass to bioethanol is now under development and is considered a second-generation biofuel. At present, lignocellulosic or fibrous biomass is converted to bioenergy by direct combustion or under a limited amount of oxygen to produce synthesis gas (syngas).

3.3.1 Starch for Energy

The two types of biomass primarily used for industrial production of bioethanol are sugar and starch crops. Both crops can be converted to bioethanol via a biochemical process known as yeast fermentation. The main sugar and starch crops used for industrial bioethanol production are sugar cane and corn, respectively. Other starch crops such as wheat, rice, potato, cassava, sweet potato, and rye can be converted to bioethanol directly without starch extraction. The key producer of bioethanol in the world is the United States; corn accounts for 61% of total global production, followed by using sugarcane from Brazil (26% of total global production, Figure 3.2). The remaining bioethanol production comes from the European Union, China, Canada, and Thailand, as summarized in Table 3.1.

Bioethanol is an alternative liquid fuel for the automotive industry. It can be used in the form of a pure hydrous (95%) product called E100 or as anhydrous ethanol (99.5%). The pure hydrous material can be used directly, but only by modified engine or flexible fuel vehicles (FFVs). With the exception of Brazil,

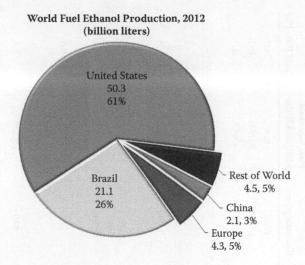

World Fuel Ethanol Production, 2012
(billion liters)

FIGURE 3.2
World bioethanol production (billion liters) in 2012.[16]

most countries primarily use anhydrous ethanol in a gasoline blend named gasohol.

A number of common blends of anhydrous ethanol and gasoline are used around the world (Table 3.2). They are classified by an international nomenclature consisting of the letter *E* followed by a figure to indicate the percentage (by volume) of bioethanol content. Blends having 20% or more bioethanol must be used with modified engines or FFVs that limit its consumption. In most countries, mandatory or optional policies to promote bioethanol as an alternative fuel are based on consumption of blends with low proportions of anhydrous ethanol (Table 3.2).

More importantly, the availability of bioethanol supplies must be considered by a country before it announces an alternative energy policy. Shortages of bioethanol and price increases can dissatisfy consumers. Most often in a country where bioethanol is produced to serve local demand according to a national policy, the government must develop a feedstock plan concurrently.

As noted previously, the main starch crop for bioethanol production is corn; others are wheat and cassava. These crops are exploited both for biofuel production and also as human staple foods. Many controversial issues surround food versus fuel security; sometimes feed security concerns are also raised. That is why lignocellulosic or algae-based bioenergy is considered to be a long-lasting sustainable bioenergy source in many countries.

Figure 3.3 illustrates the world production of major starchy crops and bioethanol from 2000 through 2012. It is interesting that increases in crop production are far smaller than increases in bioethanol production. Between 2001 and 2010, crop production did not increase greatly. Increases only of 38.06, 32.29, 16.87, 10.52, and 7.23% were noted for corn, cassava, rice, wheat,

TABLE 3.1

Present Bioethanol Production from Various First-Generation Feedstocks in Leading Countries

Country	Feedstock			Fuel Ethanol (Million Liters)				Fuel Gasoline (Million Liters)	Blend Rate (%)
	Type	Amount[17] (× Ktons)	No. of Factories	Production	Import	Export	Consumption		
United States	Corn	130,000	194	50,346	1,853	2,807	49,011	n.a.	n.a.
Brazil	Sugarcane	297,900	408	20,739	553	2,500	18,590	39,698	46.8
EU-27	Sugar beet	9,206	69	4,620	827	86	5,633	115,420	4.9
	Corn	4,215							
	Wheat	4,195							
	Rye	453							
	Barley	387							
China	Corn	5,000	5	2,509	3	7	2,509	99,356	2.3
	Wheat	1,050							
	Cassava	336							
India	Cane molasses	8,975	115	2,154	34	22	305	22,132	1.4
Canada	Corn	3,307	14	1,725	1,005	56	2,396	42,799	5.6
	Wheat	940							
Thailand	Cane molasses	2,218	19	655	0	0	389	7,705	5.0
	Sugarcane	654							
	Cassava	468							
Australia	Wheat	545	3	440	0	0	440	18,228	2.4
	Sorghum	200							
Colombia	Sugarcane	4,450	6	362	0	0	366	n.a.	n.a.
	Cassava	8							

n.a. = Not available.

TABLE 3.2

Production Plans and Policies of Bioethanol Consumption in Some Regions

Country	Policy	Ethanol Production and Consumption (Million Liters)	
		2012 Production	Target Consumption
North and Central America			
United States	E10 (10 States) and E15 by EPA approval	50,346	136,000 in 2022 with 60,000 from cellulosic
Canada	E5 (2010), Saskatchewan E7.5 and Manitoba E8.5	1,725	42,628 of gasohol
South America			
Brazil	E20, E25, Anhydrous and hydrous ethanol blend	20,739	53,204 of gasohol
Colombia	E10	362	n.a.
Argentina	E5	253	9,800 of gasohol
Peru	E7.8	235	2,570 of gasohol
Europe			
EU-27	E5, E10	4,620	n.a.
Asia–Pacific			
China	E10 (6 provinces)	2,509	223,215 of gasohol
India	E5	2,154	47,000 of gasohol
Thailand	E10, E20	655	9 million liters/day in 2021
Australia	E10 (optional)	440	19,053 of gasohol

n.a. = Not available.
EPA = U.S. Environmental Protection Agency.

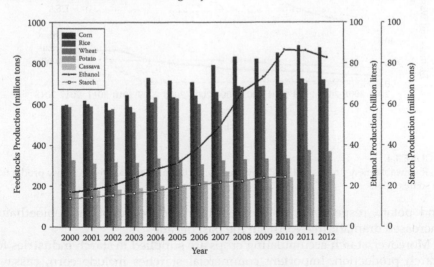

FIGURE 3.3
World production of major starchy crops and bioethanol production from 2000 to 2012.

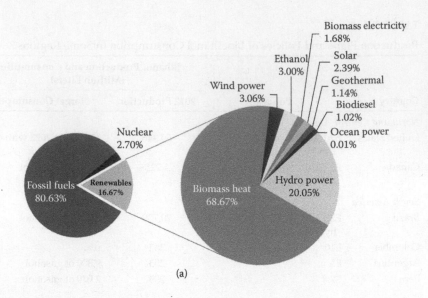

(a)

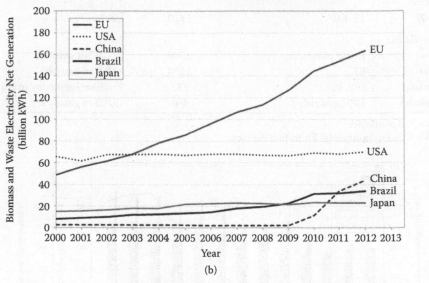

(b)

FIGURE 3.4
(a) Renewable energy consumption (%) by sources. (b) Biomass and waste electricity production in some regions.

and potato, respectively. In contrast, the world production of bioethanol increased dramatically—370% in the past decade.

Moreover, starch-accumulating crops are supplied to starch industries for starch production. Important commercial starches include corn, cassava, wheat, and potato, with production around 13, 8, 1.5, and 1 million tons per annum, respectively. The global market size for starch has enlarged

TABLE 3.3

Conversion Ratios of Raw Materials for Starch and Bioethanol Production

Raw Material Type (% Wet Basis)		Conversion Ratios (Kg of Raw Material)	
	Starch Content	1 kg of Starch	1 liter of Bioethanol[18]
Corn grain	65 to 76	1.7[19]	1.9[20] to 2.58
Cassava root	25 to 35	4.4[21]	5.5 to 6.0
Potato tuber	17 to 19	5.7[22]	8.50
Wheat grain	66 to 82	2.4[23]	1.8[20] to 2.60

continuously as illustrated in Figure 3.4. The reason is that starch has remarkable functional attributes and can fill many applications in food and non-food industries after modification by physical and chemical approaches and hydrolysis treatment by enzymes. The largest starch consumer is sweetener production by enzyme methods. Other applications for starch are in bakery, dairy, and meat products, desserts, confectionary, soups, sauces, dressings, beverages, papers, glues, textiles, cosmetics, pharmaceuticals, ceramics, construction, coloring, printing, and biodegradable materials. Between 2001 and 2010, world starch production increased by 70%—far less than the growth of ethanol production. Table 3.3 presents the estimated ratios for conversions of starch raw materials into 1 kg of extracted starch and 1 liter of ethanol. Remarkably, consumption of starch crops has increased far faster than crop production. That can inevitably lead to supply shortages and price increases.

Accordingly, some countries have issued national policies on crop production improvement. The sustainable strategy is to improve crop productivity, i.e., the amount of product produced per unit of planting area. By raising crop productivity, the yield of bioethanol produced per unit of land will also increase, as demonstrated in the cases of corn and cassava (Table 3.4).

TABLE 3.4

Estimated Bioethanol Yields from Various Feedstocks

					Cassava		
	Corn					Thailand	
	World Average Yield[24]	U.S.	Wheat[15]	Potato[15]	World Average Yield[15]	Current Yield	Targeted[24] Yield
Yield (tons/ha)	5.0	9.0	2.8	17	12	22.5	31.25
Starch content (% wet basis)	62	62	60	19	25	25	25
Starch yield (tons/ha)	3.1	5.6	1.7	3.2	3.0	5.5	7.8
Ethanol yield (liters/ha)*	2,050	3,700	1,120	2,140	1,990	3,640	5,170

*Conversion ratio of ethanol = 0.567 kg ethanol per 1 kg starch; ethanol density = 0.789 g/ml.

3.3.2 Plant Fibers for Energy

Plant fibers are lignocellulosic materials consisting primarily of cellulose (a glucose polymer), hemicellulose (a mixture of polysaccharides), and lignin. The fibers can be derived from various sources such as wood, forest residues, plant agricultural wastes, and agro-industrial wastes. The characteristics of plant fibers from different sources vary greatly in composition, structure, and mechanical properties that which make them highly useful in production of textiles, paper and pulp, and composite materials.

Plant fibers can be employed also as energy sources. They can be converted to generate heat, power, and biofuels by different conversion processes. The conversion of biomass to heat and power can be achieved by thermochemicals. The production of biofuels from biomass is a biochemical process. Thermochemical conversion of biomass involves combustion, gasification, and pyrolysis.

The most common practice for generating biomass energy is the direct combustion of solid biomass (having moisture content <50%) to generate heat that can be used directly or used to produce power by creating steam to drive turbines for electricity production. Combined heat and power (CHP) facilities, also called cogeneration plants, are used in modern utility plants. Alternatively, biomass can be transformed to synthesis gas by thermal gasification or to liquid biofuel (bioethanol) by yeast fermentation.

Lignocellulosic materials are used presently to generate heat and electricity rather than in biofuel production. Biomass heat accounts for the most bioenergy produced from renewables (Figure 3.4a). The European Commission (EU 27) generates the most biomass heat, followed by the U.S., Canada, Russia, and China. Total world wood pellet consumption was 22.4 million tons in 2012. In addition, the most produced biopower region is EU 27, led by Germany, followed by Sweden, United Kingdom, Finland, and Italy. The second-largest producer of biopower is the United States, followed by Brazil, China, and Japan.

Types of biomass used for bioenergy production vary greatly by geographic condition, but they can be classified simply as forest, agricultural, and agricultural processed residues. In some regions, energy crops have been proposed for commercial farming. The crops should be short rotation coppice arrangements that provide high yield and good quality, require low energy input and low cost, and grow readily. Examples are woody crops such as poplar, willow, and eucalyptus and grasses and herbaceous plants such as sweet sorghum, sugarcane, miscanthus, switch grass, and cord grasses, all of which yield high levels of dry matter per unit of land.

The selection of biomass used in bioenergy production depends on several factors: supply availability and affordability, biomass characteristics, type and efficiency of conversion technology, and the form of energy required for end applications. Unlike material application that relies greatly on physical attributes, the efficiency of the biomass conversion process depends considerably on chemical properties such as moisture content, proportions of fixed carbon and volatiles, ash and residue contents, alkali metal levels, cellulose:lignin ratios, and calorific values.

Some physical properties of biomass that must be considered for thermo-chemical conversion include density, particle size, and friability. The heat value of biomass can be expressed as the calorific value (CV) that indicates the amount of energy released per unit of mass or volume when burnt in air. Two types of CV are used to describe the heat value of biomass: (1) high heating value (HHV) or gross calorific value and (2) lower heating value (LHV) or net calorific value. HHV is defined as the amount of heat released when biomass is burnt in air, including the latent heat contained in water vaporization. LHV does not include the latent heat content of water vaporization. Therefore, LHV is the preferred term for expressing the actual available energy of a biomass.

Tables 3.5 and 3.6 report types of biomass used for bioenergy production in some regions and the calorific values of some biomass materials, respectively. Nevertheless, the actual energy recovered can vary based on conversion technology and type of recovered energy.

The conversion of biomass to bioheat and power is projected to increase continuously, especially in EU regions. In some countries like the UK and Denmark, targets of 100% renewable electricity are planned (Table 3.7). The major sources of biomass are still forest residues, agricultural residues, and organic and manure wastes. The amount of biomass energy available depends on land availability and biomass productivity and both are subjected to influence by several factors.

Currently, the total land devoted to forestry and agriculture is approximately 69% of the total surface land and expansion is almost nonexistent (Table 3.8). Energy yield then can increase only by improved biomass productivity. Biomass crops are affected by genetic and environmental factors such as soil fertility, climate, water availability, and management. Improvement in managing genetic and environmental factors represents an efficient means of providing sustainable biomass supplies.

3.4 Pros and Cons for Using Bio-Based Resources: Material versus Energy

Using environmentally sound technologies to make bio-based composites that are cost effective is the direction we need to pursue. At present, many of composites that are cost effective may not be the best from an environmental perspective and vice versa. Many companies talk about moving toward green technologies, but this concept has yet to be defined effectively. In fact, many concepts in the environmental arena are not defined clearly or understood.

TABLE 3.5

Common Biomasses Used for Heat and Power Production by Major Producers

Country or Region	Type of Biomass	Biomass and Waste Production[25] (1,000 kTOE)	Biomass and Waste Electricity Net Generation[26] (GWh)	Renewable or Biomass Energy Policy[25]
China	Crop residue (51%): stalks, grain-processing residues, and bagasse Animal dung (23%) Firewood (14%) Industrial waste (10%) MSW (2%)	215	43.6	Renewable capacity 30 GWh biomass in 2020
India	Fuel wood: wood logs, chipped wood, pellets, saw dust Crop residue: bagasse, rice, and wheat husks, straw, groundnut shells, cotton, maize, and soybean stalks Dry dung	184	4.1	12th Five-Year Plan: Add 30,000 MW of renewable energy capacity 2012 to 2017
Europe	Forestry residue: black liquor, sawdust, bark, roundwood, and others Energy crop: salix, poplar, miscanthus, hemp Agricultural waste: straw, stover	112	163.3	Multiple renewable energy sources
U.S.	Wood and wood-derived: bark, sawdust, wood chips, wood scraps, and paper mill residues MSW: biogenic sources, landfill gas, sludge waste, agricultural by-products, and others	65	70.3	State renewable portfolio standards; each state chooses to fulfill its mandate using a combination of renewable energy sources or other renewable sources
Brazil	Agricultural residue: sugar cane bagasse, straw, bark, and wastes Fire wood: wood residue, sawdust MSW	65	34.0	Biomass: from 5.4 GW in 2010 to 8.5 GW by 2019

Country	Biomass sources			Policy
Indonesia	*Agricultural residue:* EFB from palm oil, bagasse, rice husk *Fuel wood:* forest residue, wood industry residues, coconut shells, saw timber, small rubber tree logs	52	0.1	Multiple renewable energy sources
Thailand	*Wood and wood-derived:* fuel wood, wood chips and bark, sawdust, charcoal, fast growing trees *Crop residue:* bagasse, rice husks, oil palm residues, coconut shells MSW	21	3.1	Renewable Energy Development Plan, 2008–2022 *Electricity sector:* Biomass 3,700 MW and MSW 160 MW in 2020 *Heating sector:* Biomass 6,760 kTOE and MSW 35 kTOE in 2020
Canada	*Woody biomass:* forest products, pulp and paper, fire wood, bark, wood chips and pellets, sawdust *Agricultural residue:* cereal straw, corn and soybean stovers, crop residues from oilseed MSW	15	0.06	Clean Energy Fund aims to reduce greenhouse gas emissions by demonstrations of green technologies to support increased integration of renewable and clean power and heat technologies
Vietnam	*Agricultural residue:* rice husks and straw, sugar cane bagasse, other straws, barks, and wastes *Wood and wood-derived:* forest wood, rubber wood, logging and saw mill residues, coconut shells MSW	11	6.4	National Power Development Plan biomass generation targets 500 MW in 2020 and 2,000 MW in 2030
World		1,311	355	

MSW = municipal solid waste. EFB = empty fruit bunches.

TABLE 3.6

Calorific Values of Some Potential Biomass Sources[14]

Biomass	LHV (MJ/kg)	HHV (MJ/kg)	% Moisture Content
Wood	18.6	n.a.	20
Wheat straw	17.3	n.a.	16
Barley straw	16.1	n.a.	30
Fir (dry basis)	n.a.	21	6.5
Danish pine (dry basis)	n.a.	21.2	8.0
Willow (dry basis)	n.a.	20.0	60
Poplar (dry basis)	n.a.	18.5	45
Cereal straw (dry basis)	n.a.	17.3	6
Miscanthus (dry basis)	n.a.	18.5	11.5
Switchgrass (dry basis)	n.a.	17.4	13 to 15
Farmed trees (dry basis)	19.55	20.59	n.a.
Herbaceous biomass (dry basis)	17.21	18.12	n.a.
Corn stover (dry basis)	16.37	17.41	n.a.
Forest residues	15.40	16.47	n.a.
Cane bagasse (dry basis)	15.06	16.35	45 to 50

n.a. = Not available. LHV = Low heating value. HHV = High heating value.

TABLE 3.7

Current and Target Shares of Electricity Production from Renewable Resources of Some Countries in EU Region

Country	Biopower Generation (Terawatt Hours/Year)	% Share	
		Current	Target
Germany	37	21	35 by 2020
			50 by 2030
			65 by 2040
			80 by 2050
United Kingdom	12	10.3	50 by 2015
			100 by 2020
Italy	10	14	26 by 2020
Denmark	5.2	40	50 by 2020
			100 by 2050
Belgium	4.6	10.8	20.9 by 2020
France	4.5	12	27 by 2020
Portugal	3.0	47	59 by 2020

Many products made today from recycled resources are costly and have reduced performance properties as compared to the original products made using virgin resources. An interesting aspect of using biodegradability as a promotional element in advertising is that we do not want all products to biodegrade, for example, airplanes or critical components of high-rise

TABLE 3.8

Total Forestry and Agriculture Land Usage (x kHa), 2011 and 2012[15]

Region	Total Land	Forestry	Agriculture				
			Total	Cereals	Pulses	Tubers	Sugars
World	13,003	4,027	4,912	703.1	77.6	55.3	30.6
Africa	2,965	671	1,170	105.5	23.0	26.0	1.7
North and Central America	2,133	705	608	87.7	6.5	1.4	2.9
South America	1,756	861	608	38.1	3.8	3.5	10.6
Asia	3,094	594	1,634	337.0	38.3	18.2	11.7
Europe	2,207	1,006	470	115.3	3.7	6.0	3.4
Oceania	849	190	423	19.6	2.4	0.3	0.4

buildings. Biodegradability will have a place in future products, but we must apply the concept where it can be used to our advantage.

From an environmental view, one important issue in bio-based composites is the nature of the adhesive used. Most industrial adhesives used today are petroleum-based, i.e., they contain phenol, formaldehyde, urea, isocyanates, or other petroleum products. Concerns about using petroleum-based adhesives focus on releases of volatiles during production and use of composites, exposure to resin toxicity during production, use, recycling, and disposal of composites, and costs.

Research is underway to develop new adhesive systems that are not petroleum-based, are less toxic, involve the use of renewable resources, and are based on a better understanding of the mechanisms of adhesion. These systems also require a much better understanding of interface and interphase relationships between bio-based elements and matrices. Some of the latest research centers on enzymes, surface activation, biotechnology, chemical modification, and cold plasma technology.

A lot of pro and con discussion centers on the use of bio-based resources for energy. We must realize that the ultimate end point of recycling bio-based resources is composting, burning, or land filling. Unless we devise new technologies to control the degradation of resources caused by recycling, we must carefully consider the full life cycles of bio-based composites. It is interesting that a group of people may be very much in favor of composting and strongly opposed to burning. What they do not understand is that the two processes produce approximately the same amounts of carbon dioxide.

Burning bio-based resources is not a major factor in global concerns about carbon dioxide production. Burning bio-based resources can be considered a cyclic phenomenon, that is, the carbon dioxide produced by burning or composting bio-based resources is equal to the carbon dioxide consumed in the production of the resources. The real carbon dioxide problem arises from the burning of fossil fuels. The production of bio-based resources cannot consume the vast amounts of carbon dioxide generated by the burning of sequestered carbon millions of years old.

3.5 Conclusions

The future of bio-based composites will be very exciting and dynamic. It will be driven by traditions, trends, costs, performance, availability of resources, and legislation. Of these, the most critical issue is cost. Logical, creative, and futuristic ideas will have few chances of success if the economics are not positive. In the area of bio-based resource utilization, several competing ideologies today are driving public opinion. On the one hand, we have a growing need to develop new bio-based composites to support the requirements of material engineering. On the other hand, we have concerns about energy consumption, expanding world populations, maintaining wilderness areas, cleaning up the environment, and excessive consumption of natural resources.

There is no question that renewable, recyclable, and sustainable resources will play a major role in future world developments. Bio-based resources and composites will be important parts of this dynamic future. We must not allow ourselves to be locked into a mental framework tied to past technologies or close our eyes to exciting new possibilities. By considering all bio-based resources as raw materials for bio-based composites or energy, we are not limited to a single type of resource. The world is full of renewable bio-based resources of many varieties At present, only a few are used for composites.

References

1. European Commission. *Bio-Based Economy in Europe: State of Play and Future Potential, Part 2. 27.* Luxembourg: European Union, 2011.
2. Swedish Government Commissioned Formas. *Swedish Research and Innovation Strategy for a Bio-Based Economy.* Stockholm, 2012, p. 36.
3. S.M. Ogbomo, K. Chapman, C. Webber et al. Benefits of Low Kenaf Loading in Bio-Based Composites of Poly(L-Lactide) and Kenaf Fiber. *Journal of Applied Polymer Science,* 112 (2009): 1294–1301.
4. S.H. Lee and S. Wang. Biodegradable Polymers/Bamboo Fiber Biocomposite with Bio-Based Coupling Agent. *Composites Part A: Applied Science and Manufacturing,* 37 (2006): 80–91.
5. R. Chollakup and W. Smitthipong. Evolution of Mulberry and Non-Mulberry Silk Fibers for Textile Applications. In *Silk: Properties, Production and Uses,* P. Aramwit, Ed. New York: Nova Science, 2012, pp. 41–86.

6. R. Chollakup, W. Smitthipong, and P. Suwanruji. Environmentally Friendly Coupling Agents for Natural Fibre Composites. In *Natural Polymers. Volume I: Natural Polymer Composites*, M.J. John and S. Thomas, Eds., Cambridge: Royal Society of Chemistry, 2012, pp. 161–168.
7. Confederation of European Paper Industries. *Biobased for Growth, a Public–Private Partnership on Biobased Industries*. Brussels, 2013.
8. U.S. Department of Agriculture. *Bioenergy and Biobased Products: Strategic Direction 2009–2014*. Washington, 2010, p. 12.
9. R. Chollakup, W. Smitthipong, W. Kongtud et al. Polyethylene Green Composites Reinforced with Cellulose Fibers (Coir and Palm Fibers): Effect of Fiber Surface Treatment and Fiber Content. *Journal of Adhesion Science and Technology*, 27 (2012): 1290–1300.
10. R. Pecenka, C. Furll, C. Idler et al. Fibre Boards and Composites from Wet Preserved Hemp. *International Journal of Materials and Product Technology*, 36 (2009): 208–20.
11. R. Chollakup, W. Smitthipong, and K. Sriroth. *Green Biomaterials*. Nontaburi, Thailand: Manusfilm Partnership, 2009.
12. European Commission. *European Strategy and Action Plan toward a Sustainable Bio-Based Economy by 2020*. http://ec.europa.eu/research/consultations/bioeconomy/introductory_paper.pdf
13. B. Eickhout. A Strategy for a Bio-Based Economy. In *Green New Deal*. Brussels: Green European Foundation, 2012.
14. P. Mckendry. Energy Production from Biomass (Part 1): Overview of Biomass. *Bioresource Technology*, 83 (2002): 37–46.
15. Food and Agricultural Organization. *Food and Agricultural Commodities Production*. Rome. http://faostat.fao.org/site/339/default.aspx
16. Center for Sustainable Systems. *Biofuel Factsheet*. Ann Arbor: University of Michigan, 2013.
17. Office of Energy Efficiency & Renewable Energy. http://energy.gov/eere
18. F.O. Licht. *World Ethanol and Biofuels Report*. 2012, p. 323.
19. Wikimedia Foundation. Corn Syrup. http://en.wikipedia.org/wiki/Corn_syrup
20. C. Drapcho, J. Nghiem, and T. Walker. *Biofuels Engineering Process Technology*. New York: McGraw Hill, 2008, Chap. 4.
21. North Eastern Tapioca Trade Association. 2012. http://nettathai.org
22. Wikimedia Foundation. Starch Production. http://en.wikipedia.org/wiki/Starch_production
23. AGRANA. Agrana Opens Wheat Starch Plant at Site in Pischelsdorf, Lower Austria. Press Release, http://www.agrana.com/en/agrana-group/news/?newsID=1197&cHash=edd7bb6f99a7d08f426b5fdabd598dde
24. National Science and Technology Development Agency. *R&D Strategy for Thai Cassava Industry, 2012–2016*. Bangkok: Ministry of Science and Technology, 2022, p. 62.
25. U.S. Energy Information Administration. EIA Standard. http://www.eia.org
26. U.S. Energy Information Administration. Monthly Generator Capacity Factor Data Now Available by Fuel and Technology. http://www.eia.gov

4

Bio-Inspired Materials

Arkadiusz Chworos and Wirasak Smitthipong

CONTENTS

4.1 Introduction

Natural selection is the process of biological evolution that has kept plant and animal species in balance for more than one million years. Scientists have embraced these evolution refinements to create new technologies based on the studies of materials science and engineering that originated in material mimicking nature (biomimetic materials).[1]

The formation of inorganic materials naturally depends on the temperature and pressure of the environment. In living organisms, these types of materials are formed by extremely precise and complex self-assembly processes that lead to desirable structures with advanced functionalities. However, due to the necessary equilibrium (live processes) between formation and degradation of the constituent materials provided in nature, materials are often brittle and weak. Thus, the strength of hard tissues such as shells and bones is based on structure rather than material selection. As material selection for new technologies reaches its limitations, engineers look to nature's examples of structural optimization to develop the next generation of such technologies.[2–6]

Generating bio-inspired materials and structures follows two major approaches. First, materials are selected, extracted, or derived from a natural source, then refined, remolded, or altered in other ways to create useful tools. Examples are unlimited, starting from stone shaping, wood carving, and leather tanning to naturally extracted medications (hormones and antibiotics).

In the second, technologically more advanced approach, the idea but not the product is derived from nature, and we then reproduce, mimic, or apply the formation and organization of materials in a natural environment. This approach utilizes synthetic materials to produce performances that mimic those of biological precursors. The current frontier in bio-inspired materials goes further and uses bio-inspired processes at the molecular level to generate new materials and structures. Examples are tissue engineering, biocompatible implants, bone-fused materials, and other molecular-based approaches.

Bio-inspired materials can be viewed as materials mimicked or derived from natural biological systems and used for specific applications. These types of materials can have natural origins as noted earlier or be made fully from synthetic components. Observation and understanding of natural processes such as self-assembly and self-organization, molecular recognition, hydrogen bonding, electrostatic introduction, hydrophobic interaction, complementarity, and specific recognition are fundamental for creating bio-inspired materials via a so-called bottom-up approach[7-11] Multi-component interactions can be exploited in the designs of new functional and bio-based composites.

4.2 Hierarchical Structures in Nature

We could argue for or against the premise that all biological materials in nature are hierarchically structured. A natural environment represents a balance among different species, but at the molecular level nature maintains a balance of various types of molecules. The designs of the structures and materials are intimately connected in biological systems. In synthetic materials science, a disciplinary separation based largely on the traditional approach of viewing materials and structures as distinct is apparent. As an example, we illustrate the structure of bone at different scales (Figure 4.1).

Bone is an organ with a high density and low water content; it is relatively hard, yet lightweight, and consists of marrow, blood vessels, nerves, and other structures. Bone tissue is a distinct organ; it strengthens the body and protects its vital organ. It consists of a high-density outer layer (around 80 wt% of all bone tissue) known as compact bone. The inner layer presents a porous layer with consecutive meshes (around 20 wt% of bone tissue) and is called cancellous bone.[12,13]

Bone tissue can be viewed as a composite material based on two main components: hydroxyapatite (HAP) and collagen. HAP is an inorganic

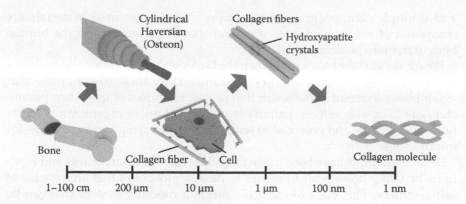

FIGURE 4.1
Structure of bone at various size scales. A bone consists of osteon that looks like a cylinder. It contains bone cells (osteoblasts) that help build bone tissue and interact with collagen fibers. The fibers are bonded together with hydroxyapatite (HAP) that looks like a tube with a diameter about 70 to 100 nm and a thickness about 1 nm by volume ratio between collagen and HAP at 6:4. The structure of bone is an example of a complex bio-based composite from nature. During bone tissue formation, osteoblast cells are created first. They accumulate minerals based on calcium phosphates for use in bone tissue preparation.

calcium phosphate mineral responsible for hardness. Its general chemical formula is $Ca_{10}(PO_4)_6(OH)_2$. Collagen is a fibrous protein that provides bone with flexibility. HAP is the main component that provides strength, but it also stimulates bone cell formation. Hence, in this example, we can use HAP in artificial bone or as a coating material in prosthetic organ engineering.[14–16]

4.3 Self-Assembly

A design strategy for a bio-inspired material that imitates nature and is amenable to specific interactions is the use of self-assembling molecules whose structures contain biologically active portions and application fractions. The self-assembly systems mimic biological architectures by using a modular set of small molecules capable of associating into a supramolecular structure, for example, molecules like peptides and nucleic acids. These types of molecules can be assembled into controlled structures such as micelles, vesicles, tubules, and other more complex architectures. These assemblies have potential as functional materials or therapeutic drug delivery carriers.

Significant challenges surround the applications of such technologies. One can expand the complexity of new structures made from biomolecules such as peptides or nucleic acids. Their potential is limitless due to the sequence variability. However, beyond the fundamental knowledge of folding processes and multiple component assembly, many such bio-inspired materials are composed of new sets of molecules and thus their impacts on the human body remain to be determined.

We must value the balance between the biological applications and toxicities of such materials. Several unknown parameters for these nascent molecular assemblies can result in the restriction of use. Examples of unknown parameters are toxic side effects, immunological reactions of organisms, stability in cellular media, and potential to scale-up complex drug carrier systems for multiple dose delivery.

Several attempts have been made to imitate biological structures and functions by using bottom-up strategies to design molecules that are capable of self-assembly. This class of molecule offers a modular system that can be tailored to interact with cellular elements.[17-19] Two types of biocompatible supramolecular assemblies described here can be developed for encapsulation or complex formation that would facilitate delivery of drugs: (1) peptide amphiphiles and (2) nucleic acids. Additionally, therapeutic sequences can be embedded within the supramolecular scaffolds.

4.3.1 Peptide Amphiphiles

Proteins are biomolecules that exist in all living organisms. Several types of scaffold proteins are responsible for various biological functions, for example, collagen in animal skin, tendons and connective tissues, keratin in nails, fibroin in silk, and soft functional proteins. All proteins are composed of mixtures of different amino acids synthetized by the translation machinery inside cells. Each amino acid building block is an organic molecule with two functional groups: (1) a negative potential amino group ($-NH_2$) and a positive carboxylic group ($-COOH$).[20]

In the translation process, amino acids are connected according to a specific RNA sequence blueprint by a peptide bond ($-CONH$). Amino acids fewer than 50 units long are called polypeptides. Polypeptides are usually straight structured, without branching points. Polypeptides retain functional groups at both termini, and in solution the amino group and carboxylic group are designated as the N-terminal and C-terminal of the molecule, respectively. Proteins are considered polypeptides; the number of amino acids exceeds 50 units. Proteins are usually responsible for various structural and enzymatic functions in living organisms.

Recent advancements in biomolecular engineering and cell signaling analysis show how different cells signal their functions within whole tissues. For instance, neuronal or bone cells that have different locations and functions may differentiate according to the needs of neighbor cells and whole organs. The human body contains a plethora of different types of cells, but how all of them interact over distance is still not fully understood.

We know that cell–cell interactions between those in close proximity is expressed and exerted through cells, ligands, and cell surfaces, for example,

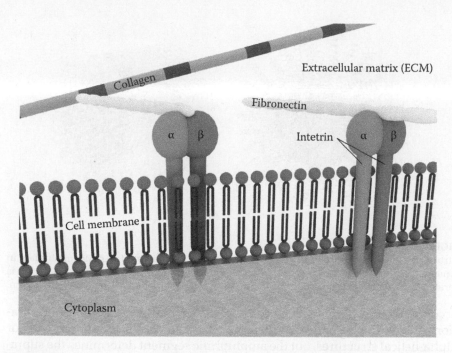

FIGURE 4.2
Cell surface receptors or integrins consists of alpha and beta types that pull out of the cell membrane and interact with collagen and fibronectin at the extracellular matrices.

the fibrous proteins (collagen and fibronectin) outside cells or extracellular matrices (Figure 4.2). Closer examination of such fibrous proteins at the molecular level shows that they are polypeptides. Protein sequence determines the specificity of ligands to cell surface receptors and defines cell positions and functions.[21–23]

Peptides can be derived from natural sources by controlled proteolysis. However, in the case of short sequences, chemical synthesis is more efficient because it allows precise control of the sequences of amino acids and the lengths of the final products. Using a solid-phase methodology, pioneered by Robert B. Merrifield with a temporary 9-fluorenylmethyloxycarbonyl (Fmoc) or tert-butyloxycarbonyl (t-Boc) protecting group, peptides can be synthesized on polystyrene resins with relatively high efficiency.[24]

To achieve different functionalities, short peptides started to be modified with other pendent groups. One example is adding a lipid or alkyl molecule at the N-terminal of a peptide using a piperidine linker. Lipid-functionalized peptides are called peptide amphiphiles. They present the dual properties of being polar (hydrophilic of peptide) and non-polar (hydrophobic of alkyl) in a single molecule.[25]

Peptide amphiphile properties depend on the polypeptide sequence that may be polar, non-polar, or neutral. They also depend on the type of lipid (single or double molecular chain). Peptide amphiphile molecules may

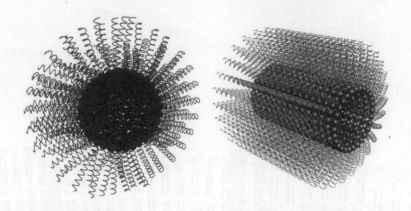

FIGURE 4.3
Structure of spherical micelle (left) and cylindrical micelle (right) of peptide amphiphile in aqueous solution. The N-terminal of a polypeptide with an alkyl hydrophobic molecule is located at the center of the micelle; the C-terminal of polypeptide contacts the aqueous solution.

accommodate different structures depending on the sequence, concentration, and buffer composition. Polypeptides can fold into random coils or alpha-helical structures, but the amphiphilic segment determines the supramolecular arrangement.

With the increased concentration of molecules in solution typical of lipids, we can reach critical micelle concentration (CMC) in which peptides amphiphiles self-organize into spherical micelles.[26] The alkyl (lipid) part of the peptide amphiphile also affects the CMC; a single molecular chain will normally lower the CMC more than a double one. Moreover, if the temperature of such spherical micelles is lowered below 10°C but higher than the freezing point, the micelle structures can transform from spherical to cylindrical (Figure 4.3).[27]

From an application view, peptide amphiphiles appear to be very versatile. We can imagine using the inside of a micelle as a drug carrier for non-polar molecules that otherwise would be toxic or unstable before reaching their therapeutic destination. However, such a molecule should not alter the structure of the peptide amphiphile nanoparticle.

The peptidic shells of micelles can be functionalized with hydrophilic groups such as DNA or RNA fragments. This type of functionalization can be used for cell surface recognition (with aptameric structures) or as a regulatory sequence in a gene or antisense strategy.[28] Peptide amphiphiles can be used also to change surface properties for cell adhesion using cell integrin compatibility with the peptide surface.[29] There is an apparent effect and relationship between the molecular weights of such hybrids and their compatibilities with living cells. However, this is a relatively new technology and requires proper implementation to reduce side effects and immune responses.

4.3.2 Nucleic Acids

Nucleic acids are molecules; like proteins, they are fundamental components of all living organisms. They play an essential role in passing genetic information from one generation to the next. All vital information that determines structure and function is embedded in the genetic code whose sequence is based on bases denoted by letters: adenine (A), cytosine (C), guanine (G), thymine (T), and uracil (U). Whether DNA or RNA was the first molecule is arguable, but evolutionary DNA has been chosen to be responsible for storage and inheritance of genetic information in all developed organisms. Only some retroviruses and viroids (deficient of protein envelopes) retain the information in the form of single- or double-stranded RNA.

DNA and RNA are very similar chemically; one difference is the presence of 2'-OH on the sugar ring and methyl group present in T and not in U. However, the most pronounced difference is in the structure and function. DNA usually folds into a monotonous double stranded helix according to base-pair complementarity (A:T and G:C).[30] RNA can form higher ordered structures with additional non-canonical interactions and can thus perform multiple functions. RNA can act similar to a protein in that it is able to catalyze enzymatic reactions (formation and/or cleavage of chemical bonds). More recently, it has been discovered that short fragments of RNA are also responsible for regulatory and signaling processes.

From a bioengineering view, DNA appears more suitable for implementation in biological materials; the materials are more stable and structurally simple and chemical synthesis is more efficient.[31] However, we can also imagine using RNA for particular tasks in which nucleic acid components can be used as structural scaffolds and regulatory sequences in molecules. Control of the nucleobase sequence allows precise programmability of the structure and potential for assembly into a desirable three-dimensional scaffold.[4,8,32]

To reinforce the strength and durability of a material, nucleic acids can be combined with other biocompatible polymers (Figure 4.4). Because DNA and RNA are generally negatively charged due to a phosphate group at each nucleotide, they can be combined easily with positively charged counter-partners. Researchers have applied nucleic acids to create biocompatible materials using electrostatic interactions with cationic surfactants.[32]

Nucleic acids were combined with cationic lipids to take advantage of regulatory sequences (gene delivery, antisense therapy, and siRNA technology) and biofilm structure potential. It has been demonstrated that the ratio of a mixture of DNA or RNA and a lipid surfactant dictates the type of the structure and the properties for such biofilms. For instance, long DNA (200 to 2000 base pairs) mixed with didecyldimethylammonium bromide (DDAB) tends to form a regular film with a lamella type of structure that has suitable mechanical properties. This is probably due to the potential movement (sliding) and stretching of the long DNA polymers among different layers.[33]

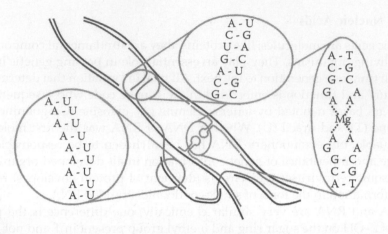

FIGURE 4.4
Structure of bio-inspired polymer from nucleic acids. A line indicates a polymeric molecule. A letter indicates a nitrogen base in the nucleic acid. Mg indicates a magnesium ion.

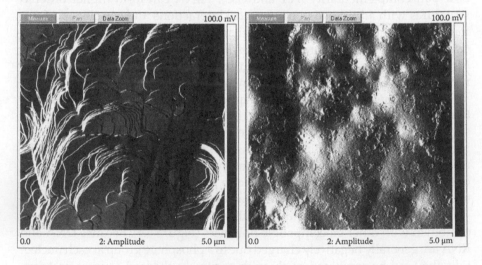

FIGURE 4.5
Morphology at surface of anionic nucleic acid and cationic amphiphile film. Left: DNA film. Right: RNA film. These atomic force microscopy photos show as step-like structure on the DNA film but not on the RNA film. The reason is that the molecular weight of a DNA molecule is 100 times that of a RNA molecule.

Interestingly, a secondary structure of DNA also has a significant effect on the mechanical properties of such biofilms (Figure 4.5). Single-stranded polyadenine and polyuracil films are less elastic and more brittle than complexes of double-stranded polyadenine–uracil.[34] One example of the application that has been medically proven is using nucleic acids in the form of a hydrogel material that can store and release insulin hormone in patients with diabetes.[35]

FIGURE 4.6
The structure of an airplane wing is inspired from the bone structure of a bird.

4.4 Lightweight Materials

Since the dawn of human civilization, people have observed nature, learned from it, and applied it in everyday life. And since the days of mythological Icarus, people have desired to fly. The challenge of flying is using a material that is lightweight, yet relatively strong. Gravity has a significant impact on both natural and synthetic materials. One technique to achieve an optimum balance of weight and strength is through a "sandwich structure" that includes a hard shell exterior and hollow interior.[36]

This physical principle for increasing the strength of hollow tube structures is exemplified in nature in the bones and feathers of birds. Gaps between the hard shells made of materials that resist axial direction are found in plant stems that must withstand strong wind and rain conditions. Plant stems often consist of thin-walled cylindrical structures filled with cellular cores.[37] Their large aspect ratios pose an interesting engineering problem to both nature and technology. Based on lessons learned from nature (bird bones and thin-walled plant stems), scientists found that structures constructed from lightweight materials can be used effectively in building construction and scaffolding, oil platforms, aviation (Figure 4.6), and other fields.

4.5 Building Design

We are inspired by nature from the sky to the deepest part of the ocean. The structures of plants and animals such as sea sponges can inspire new technologies. The structure of a porous multicellular sea sponge is a complex network; the main component is a hollow silica tube about 20 to 25 cm in length and about 2 to 4 cm in diameter, similar to a bird bone.

Enhanced imaging reveals the subtle architecture of these sea sponge tubes. The structure is based on a square type scaffold strengthened by diagonal struts such as those found in the construction of buildings. Additionally, a helical support at approximately a 45-degree angle overlays the entire structure. Although the tubes are architecturally perfect, the silicone structure is brittle and easily damaged.[38,39]

Nevertheless, the sea sponge structures inspired architects to design and engineer some of the world's most beautiful buildings, for example, the Swiss Reinsurance Company building in London, the Hotel de Las Artes in Barcelona, and the Eiffel Tower in Paris.

4.6 Gecko Feet

The ability of geckos and other animals and insects (beetles, spiders, etc.) to walk effortlessly on vertical surfaces is related to the controlled adhesion-anti-adhesion concept that has drawn a lot of attention.[40–42] A gecko's foot is another indication that we can learn a lot from observations of nature.

Electron microscopy revealed that a gecko's foot is covered with a set of tiny hairs (setae) finished at the ends with post (spatulae) type structures about 100 nm in length. Scientists developed a mathematical formula explaining the gecko foot phenomenon with reference to van der Waals forces and Hertzian contact stresses.[43] They found that the calculation and analysis of the stress depend inversely on the radii of the spatulae. The larger the animal, the smaller the radii of the spatulae. For example, a gecko weighs about 100 g and has about 1,000 setae per 100 μm^2; spatula radius is about 0.1 μm. A beetle weighs about 0.01 g and has 1 seta per 100 μm^2 and a spatula radius of about 1 μm.

As a result of this construction, the spatulae covering a gecko's feet can interact with almost any type of surface material via van der Waals interactions between the ends of spatulae atoms with surface atoms. Increasing the number of spatulae results in greater adhesion; this allows a gecko to walk on glossy vertical surfaces. At the same time, the gecko's feet must be able to detach from a surface quickly to allow the animal to move forward. The spatula structure consists of hairs of different sizes.

Each hair has the ability to bend to increase its contact area with a surface. Because this interaction is based on van der Waals forces, the animals can walk on all types of non-polar surfaces. These hydrophobic hairs also make a gecko's feet resilient to dust, dirt, and moisture and capable of repetitive use without damage.

Based on understanding of the gecko foot phenomenon, scientists have tried to develop new artificial materials utilizing brushes of hairs that can adhere to surfaces with a strength exceeding 3 kg/cc.[44] Another example of a technology mimicking a gecko food is the brushed polyethylene bristle.[45] Although these materials lack the advantage of regeneration and lose their adhesion and anti-adhesion properties in the presence or moisture or dust on a surface, the concept still has potential for future applications.

4.7 Natural Glues

Diverse marine species have evolved with clamps that allow them to attach to rocks and other underwater objects and withstand the hostile environment of the ocean. Mussels, for instance, have byssus fibers that work similar to human tendons, but the byssus is 5 times stronger and can be deformed more than 16 times.[46–48] The byssus fibers act like natural glues to adhere to surfaces and are stronger than any human-made adherence compound. The properties of such fibers that we are only beginning to understand can help us develop new types of materials.

Generally, the byssus fiber of a mussel is the only structure that holds the mussel against waves. It is composed mainly of two parts: (i) a rigid distal portion responsible for adherence to surfaces and (ii) a flexible proximal portion involved in vibration adsorption (Figure 4.7).[49,50] The distal part is composed of a histidine-rich collagen fiber that contains a fibroin-like structure.

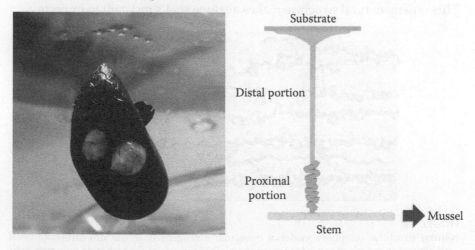

FIGURE 4.7
Left: Byssus fibers from a mussel have good adhesion to surfaces. Right: Structure of byssus fiber from a mussel.

The distal and proximal portions of byssus fibers exhibit different profiles. The proximal portion has an elastic modulus smaller than that of the distal part. As a result, the distal portion is about four times tougher for adhering to surface substrates. The two distinct features of the elastic module and thus the strength of a mussel is still not fully understood. To date, we have not developed an artificial material that can match byssus fiber performance. The fiber is the subject of active investigation.

4.8 Abalone

Although several synthetic materials with required mechanical strengths and controlled rigidities are available, not one is equivalent to the abalone material developed by nature. The material is composed mainly of calcium carbonate ($CaCO_3$) tiles of controlled size and thickness, staggered as laminated layers (Figure 4.8).[51–53] The precise structure and composition of an abalone shell are not known completely, particularly the components of the complex structure of the nano-scale laminated layers and the organic media layer.

However, scientists intrigued by the physical properties of abalone shells studied the structures of the laminates and use the concepts in model systems. They found that the main chemical in an abalone shell is calcium carbonate, with a thickness on the nanometer scale and diameter of 3 to 10 µm. Each nano-sized tile is connected by an adhesive protein and polysaccharides. This orderly natural structure makes abalone shells resistant to impacts.

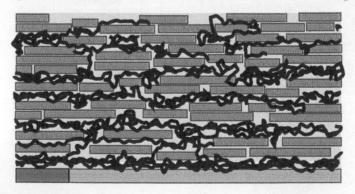

FIGURE 4.8
Natural structural composite model of molecular arrangement of calcium carbonate brick wall aligned in orderly manner for abalone shell. Proteins and polysaccharide polymers help adhere bricks together well.

Scientists have tried to mimic the structure of an abalone shell using materials like tiles or bricks and synthetic polymers as connecting glues. The tiles can be made with microscopic particles of $Mg_3(Si_4O_{10})(OH)_2$ mixed with a polymer in the ratio 10:60%. Using the dispersion technique, charge potential was used to rearrange clay (montmorillonite) with a polymeric component prepared on a silicon wafer.[54] Other examples of thin structures with epitaxial arrangements consist of silicon carbide (SiC) or boron carbide (B_4C) in a hexagonal form covered by a laminated Kevlar thermoset and bonded together by a polymer. This type of structure has the property of absorbing energy and strength similar to its natural abalone precursor. This and other types of bio-inspired materials can be used in large vehicles such as aircraft, ships, and spacecraft that require great structural strength.

4.9 Conclusions

Structural biological materials are complex composites now under extensive investigation by biologists, chemists, materials scientists, and engineers, with the ultimate goal to apply the knowledge gained to develop artificial materials. This chapter summarizes various concepts of bio-inspired materials that mimic natural predecessors. Nature, during thousands of years of evolution, created materials that have inspired humans for generations, but we know that we have revealed only a fraction of nature's mysteries.

Can we successfully reproduce and maybe enhance the properties of some of these biological structures? We want to reach perfection by designing a material with ideal properties. If such a material could be found, would it be useful? It is possible that biological materials are designed to optimize stiffness and toughness simultaneously to provide maximum structural support and flaw tolerance.

Keeping this idea in mind, we formulate optimization problems under the assumptions of appropriate material constitutive models and certain failure criteria. We have shown that within this optimization framework, the staggered microstructures of biological materials may be reproduced successfully at least on a nanometer scale. From an optimization view, bio-mimicking or bio-inspired material design can be formulated as an inverse optimization problem, i.e., finding a functional goal of an optimization problem under the condition that the solution to the problem is known. Using optimization principles to reveal the strategies and mechanisms used by nature to design biomaterials is an interesting research field. Current advancements in this field are just preliminary steps in utilizing nature's material design concepts.

References

1. J.M. Lehn. Supramolecular Chemistry. *Science*, 260 (1993): 1762–1763.
2. J. Ihli, P. Bots, A. Kulak et al. Elucidating Mechanisms of Diffusion-Based Calcium Carbonate Synthesis Leads to Controlled Mesocrystal Formation. *Advanced Functional Materials*, 23 (2013): 1965–1973.
3. E. Dujardin and S. Mann. Bio-Inspired Materials Chemistry. *Advanced Materials*, 14 (2002): 775.
4. A. Chworos, I. Severcan, A.Y. Koyfman et al. Building Programmable Jigsaw Puzzles with RNA. *Science*, 306 (2004): 2068.
5. C.W.P. Foo, J. Huang, and D.L. Kaplan. Lessons from Seashells: Silica Mineralization via Protein Templating. *Trends in Bbiotechnology*, 22 (2004): 577–585.
6. P. Fratzl, I. Burgert, and H.S. Gupta. On the Role of Interface Polymers for the Mechanics of Natural Polymeric Composites. *Physical Chemistry Chemical Physics*, 6 (2004): 5575–5579.
7. R. Chollakup, W. Smitthipong, and A. Chworos. Specific Interaction of DNA-Functionalized Polymer Colloids. *Polymer Chemistry*, 1 (2010): 658–662.
8. L. Jaeger and A. Chworos. The Architectonics of Programmable RNA and DNA Nanostructures. *Current Opinions in Structural Biology*, 16 (2006): 531–543.
9. W. Smitthipong, T. Neumann, A. Chworos et al. Supramolecular Materials Comprising Nucleic Acid Biopolymers. *Macromolecular Symposia*, 264 (2008): 13–17.
10. R.S. Tu and M. Tirrell. Bottom-Up Design of Biomimetic Assemblies. *Advanced Drug Delivery Reviews*, 56 (2004): 1537–1563.
11. L. Ploux, S. Beckendorff, M. Nardin et al. Quantitative and Morphological Analysis of Biofilm Formation on Self-Assembled Monolayers. *Colloids and Surfaces B: Biointerfaces*, 57 (2007): 174–181.
12. R.B. Cook and P. Zioupos. The Fracture Toughness of Cancellous Bone. *Journal of Biomechanics*, 42 (2009): 2054–2060.
13. B. Van Rietbergen, R. Huiskes, F. Eckstein et al. Trabecular Bone Tissue Strains in the Healthy and Osteoporotic Human Femur. *Journal of Bone Mineral Research*, 18 (2003): 1781–1788.
14. M.P. Ferraz, F.J. Monteiro, and C.M. Manuel. Hydroxyapatite Nanoparticles: A Review of Preparation Methodologies. *Journal of Applied Biomaterial Biomechanics*, 2 (2004): 74–80.
15. L. Li, H. Pan, J. Tao et al. Repair of Enamel by Using Hydroxyapatite Nanoparticles as the Building Blocks. *Journal of Materials Chemistry*, 18 (2008): 4079–4084.
16. M. Andiappan, S. Sundaramoorthy, N. Panda et al. Electrospun Eri Silk Fibroin Scaffold Coated with Hydroxyapatite for Bone Tissue Engineering Applications. *Progress in Biomaterials*, 2 (2013): 1–11.
17. M. Black, A. Trent, Y. Kostenko et al. Self-Assembled Peptide Amphiphile Micelles Containing a Cytotoxic T-Cell Epitope Promote a Protective Immune Response in Vivo. *Advanced Materials*, 24 (2012): 3845–3849.
18. A. Harada, R. Kobayashi, Y. Takashima et al. Macroscopic Self-Assembly through Molecular Recognition. *Nature Chemistry*, 3 (2011): 34–37.

19. G.M. Whitesides and M. Boncheva. Beyond Molecules: Self-Assembly of Mesoscopic and Macroscopic Components. *Proceedings of the National Academy of Sciences of the United States of America,* 99 (2002): 4769–4774.

20. C. Branden and J. Tooze. *Introduction to Protein Structure,* 2nd ed. Garland Science, 1999.

21. M. Tirrell, E. Kokkoli, and M. Biesalski. The Role of Surface Science in Bioengineered Materials. *Surface Science,* 500 (2002): 61–83.

22. E. Puklin-Faucher, M. Gao, K. Schulten et al. How the Headpiece Hinge Angle Is Opened: New Insights into the Dynamics of Integrin Activation. *Journal of Cell Biology,* 175 (2006): 349–360.

23. F. Danhier, A.L. Breton, and V. Préat. RGD-Based Strategies to Target Alpha(V) Beta(3) Integrin in Cancer Therapy and Diagnosis. *Molecular Pharmaceutics,* 9 (2012): 2961–2973.

24. P. Berndt, G.B. Fields, and M. Tirrell. Synthetic Lipidation of Peptides and Amino Acids: Monolayer Structure and Properties. *Journal of the American Chemical Society,* 117 (1995): 9515–9522.

25. R.S. Tu, R. Marullo, R. Pynn et al. Cooperative DNA Binding and Assembly by a B-zip Peptide Amphiphile. *Soft Matter,* 6 (2010): 1035–1044.

26. E. Kokkoli, A. Mardilovich, A. Wedekind et al. Self-Assembly and Applications of Biomimetic and Bioactive Peptide Amphiphiles. *Soft Matter,* 2 (2006): 1015–1024.

27. J.D. Hartgerink, E. Beniash, and S.I. Stupp. Self-Assembly and Mineralization of Peptide-Amphiphile Nanofibers. *Science,* 294 (2001): 1684–1688.

28. D. Peters, M. Kastantin, V.R. Kotamraju et al. Targeting Atherosclerosis by Using Modular, Multifunctional Micelles. *Proceedings of the National Academy of Sciences of the United States of America,* 106 (2009): 9815–9819.

29. S.E. Ochsenhirt, E. Kokkoli, J.B. McCarthy et al. Effect of RGD Secondary Structure and the Synergy Site PHSRN on Cell Adhesion, Spreading, and Specific Integrin Engagement. *Biomaterials,* 27 (2006): 3863–3874.

30. M.L.M. Anderson. *Nucleic Acid Hybridization.* Garland Science, 1998.

31. E. Stulz. DNA Architectonics: Toward the Next Generation of Bio-Inspired Materials. *Chemistry,* 18 (2012): 4456–4469.

32. W. Smitthipong, T. Neumann, S. Gajria et al. Noncovalent Self-Assembling Nucleic Acid-Lipid Based Materials. *Biomacromolecules,* 10 (2009): 221–228.

33. R. Chollakup, W. Smitthipong, and A. Chworos. Supramolecular Cooperative-Assembly of Polyelectrolyte Films. *RSC Advances,* 3 (2013): 4745–4749.

34. R. Chollakup and W. Smitthipong. Chemical Structure, Thermal and Mechanical Properties of Poly(Nucleotide)-Cationic Amphiphile Films. *Polymer Chemistry,* 3 (2012): 2350–2354.

35. S.H. Um, J.B. Lee, N. Park et al. Enzyme-Catalysed Assembly of DNA Hydrogel. *Nature Materials,* 5 (2006): 797–801.

36. Y. Seki, B. Kad, D. Benson et al. The Toucan Beak: Structure and Mechanical Response. *Materials Science and Engineering C,* 26 (2006): 1412–1420.

37. G.N. Karam and L.J. Gibson. Biomimicking of Animal Quills and Plant Stems: Natural Cylindrical Shells with Foam Cores. *Materials Science and Engineering C,* 2 (1994): 113–132.

38. J. Aizenberg, J.C. Weaver, M.S. Thanawala et al. Skeleton of Euplectella Sp.: Structural Hierarchy from the Nanoscale to the Macroscale. *Science,* 309 (2005): 275–278.

39. G. Mayer. Rigid Biological Systems as Models for Synthetic Composites. *Science*, 310 (2005): 1144–1147.
40. E. Arzt, S. Gorb, and R. Spolenak. From Micro to Nano Contacts in Biological Attachment Devices. *Proceedings of the National Academy of Sciences of the United States of America*, 100 (2003): 10603–10606.
41. K. Autumn, M. Sitti, Y.A. Liang et al. Evidence for Van der Waals Adhesion in Gecko Setae. *Proceedings of the National Academy of Sciences of the United States of America*, 99 (2002).
42. J. Yu, S. Chary, S. Das et al. Friction and Adhesion of Gecko-Inspired PDMS Flaps on Rough Surfaces. *Langmuir*, 28 (2012): 11527–11534.
43. K.L. Johnson, K. Kendall, and A.D. Roberts. Surface Energy and Contact of Elastic Solids. Paper presented at Proceedings of Royal Society of London, 1971.
44. A.K. Geim, S.V. Dubonos, I.V. Grigorieva et al. Microfabricated Adhesive Mimicking Gecko Foot Hair. *Nature Materials*, 2 (2003): 461–463.
45. M. Sitti and R.S. Fearing. Synthetic Gecko Foot-Hair Micro/Nano-Structures as Dry Adhesives. *Journal of Adhesion Science and Technology*, 17 (2003): 1055–1074.
46. E. Vaccaro and J.H. Waite. Yield and Post-Yield Behavior of Mussel Byssal Thread: A Self-Healing Biomolecular Material. *Biomacromolecules*, 2 (2001): 906–911.
47. J.M. Lucas, E. Vaccaro, and J.H. Waite. A Molecular, Morphometric and Mechanical Comparison of the Structural Elements of Byssus from *Mytilus edulis* and *Mytilus galloprovincialis*. *Journal of Experimental Biology*, 205 (2002): 1807–1817.
48. C.B. Emily and M.G. John. Strategies for Life in Flow: Tenacity, Morphometry, and Probability of Dislodgment of Two Mytilus Species. *Marine Ecology Progress Series*, 159 (1997): 197–208.
49. E. Bell and J. Gosline. Mechanical Design of Mussel Byssus: Material Yield Enhances Attachment Strength. *Journal of Experimental Biology*, 199 (1996): 1005–1017.
50. K. Bertoldi and M.C Boyce. Mechanics of the Hysteretic Large Strain Behavior of Mussel Byssus Threads. *Journal of Materials Science*, 42 (2007): 8943–8956.
51. B. Chen, X. Peng, J.G. Wang et al. Laminated Microstructure of Bivalve Shell and Research of Biomimetic Ceramic–Polymer Composite. *Ceramics International*, 30 (2004): 2011–2014.
52. M.A. Meyers, A.Y.M. Lin, P.Y. Chen et al. Mechanical Strength of Abalone Nacre: Role of the Soft Organic Layer. *Journal of the Mechanical Behavior of Biomedical Materials*, 1 (2008): 76–85.
53. V. Srot, U.G.K. Wegst, U. Salzberger et al. Microstructure, Chemistry, and Electronic Structure of Natural Hybrid Composites in Abalone Shell. *Micron*, 48 (2013): 54–64.
54. N. Almqvist, N.H. Thomson, B.L. Smith et al. Methods for Fabricating and Characterizing a New Generation of Biomimetic Materials. *Materials Science and Engineering C*, 7 (1999): 37–43.

5

Natural Fiber Surface Treatments and Coupling Agents in Bio-Based Composites

Potjanart Suwanruji, Wirasak Smitthipong, and Rungsima Chollakup

CONTENTS

5.1 Introduction

Natural fiber-reinforced composites mainly consist of two natural fiber components that act as fillers or reinforcing materials and polymer matrices. Original sources of natural fibers, their advantages, and comparisons to synthetic reinforcements are discussed in the next section.

Generally, natural fibers present hydrophilic properties while polymer matrices exhibit hydrophobic or hydrophilic properties. To obtain good final mechanical performance or use properties of a resulting reinforced composite, the quality of the interface formed by the filler and matrix is important. Fiber surface treatment and coupling agents are used to improve interfaces by forming a bridge of chemical and/or physical bonds between the two phases.[1] This chapter reviews the classifications of surface treatments and coupling agents used in bio-based composites.

The properties of composite materials based on natural fiber-reinforced polymers should depend on the adhesion at the interface of fiber and polymer. Adhesion is a steady or firm attachment of two bodies and may be characterized as the thermodynamic work of adhesion, i.e., the work required to keep two different bodies in contact with each other under equilibrium conditions.[2] The action of the molecular forces at an interface forms the fundamental basis of the adhesive forces between a substrate and an adhesive.[3] The thermodynamic work of adhesion by polymer composites depends on the ratio of the components in a system and also on the composition and ratio of two phases evolved near the interface. Thus, we can state that adhesion strength may be changed, depending on the conditions of the adhesion joint formation.[4–6]

Initial discussions of the thermodynamic description of adhesion involve the characteristics of two surfaces: the surface tension and interfacial tension at the interface between two bodies in contact.[2,7] In the simplest case of two liquids in contact with each other, the surface tension at the interface (interfacial tension) is always lower than the highest surface tension at the interface with saturated vapor.

The thermodynamic approach to the description of adhesion has more advantages than some other theories. It does not require knowledge of the molecular mechanisms of adhesion and considers only the equilibrium processes at the fiber-polymer interface. In the case of polymer blends with immiscible components, the surface tension (surface energy) of each polymer plays a major role in organizing the polymer position on the surface or on the bulk of the blends to make the thermodynamic system stable.[8,9] In considering adhesion of polymer blends to fibers, the wetting of fibers is important.

Many theories of adhesion have been proposed, and we can describe some features of the formation of adhesive joints and their properties.[2,4–6,10] However, adhesion cannot be recognized as an independent theory; it represents an application of fundamental thermodynamic principles to interfacial interactions. In considering the theories of adhesion, the most useful concept of adhesion stays within the boundaries of the molecular theory and the thermodynamics of interfacial phenomena. At the same time, no single theory can predict the real adhesion between solid and polymer or adhesion joint strength. A large number of theoretical ideas about adhesion do not refer to the phenomena of adhesion but focus on the processes of failures of adhesion joints and their description.

To understand the properties of adhesion joints, we should first consider the mechanisms of their formation[2,7] that follow several stages:

- Spreading of adhesive on the surface of a solid body and its wetting
- Equilibrium establishment (which is not always possible) of the adhesion contact, depending on the viscosity of the adhesive and the processes of adsorption and diffusion
- Formation of the physical or chemical structure of the adhesive by curing, accompanied by the emergence of a surface layer distinguishable from the bulk; includes setting of the adhesive, possible crystallization, and evolution of new phases

The emergence of forces of molecular interactions responsible for adhesion is possible only under conditions by which the molecular contact occurs at the adhesive–substrate interface.[11,12] This process depends on the physical relief characteristics of the surface. The formation of adhesive joints also depends on the kinetics of spreading that in turn depend on the rheological properties of the adhesive.

The strength of adhesion joints determines the main properties of natural fiber-reinforced composites.[13-15] When evaluating adhesion strength, one should take into account many factors, including crack development, distribution of stresses in the system, and the presence of inner stresses.

It is important to distinguish between thermodynamic evaluation of adhesion and adhesion joint strength: the thermodynamic work of adhesion as an equilibrium value that does not depend on test conditions, application of adhesive to a surface, or surface roughness. This value depends only on the thermodynamic characteristics of adhesive and adherent. However, the adhesion strength, similar to the strength of any solid, is a kinetic value affected by conditions of failure, defects in the material structure, weak boundary layers, and other factors.

The enhancement of adhesion at the interface of a polymer and solid is very important for natural fiber-reinforced composite properties.[16-18] It depends on two main factors: the state of the solid surface and possibility of modification of both the substrate and adhesive. The state of the surface determines its wetting. Surface cleanliness is a very important factor, and during the production of a natural fiber-reinforced composite, it is desirable to remove all impurities from the surface.[19] Cleanliness is especially important when working with reinforcing fibers that may have decreased surface tension after treatment with various substances. The degree of roughness of the surface and chemical uniformity also play essential roles.

When considering natural fibers as reinforcing elements in a polymer composite, the heterogeneity of the fiber structure may improve the properties of the composite due to a difference in the surface tension of the fiber on its surface.[1] The surface tension increases as fiber diameter diminishes and the spinning speed increases. The mutual influence of the polymer and organic fiber leads to changes in the structure of the fiber and its orientation. In the surface layers of a fiber, some processes lead to the formation of an intermediate layer. As a result, there is no sharp phase border between the reinforcing fiber and polymer, and in some cases, the result is improved adhesion.

Another solution can be derived from the modification of a filler surface using physical and chemical methods. The purpose of such a modification is to improve the adhesion of filler and polymer.[20-22] Surface modifiers have various functions. First, they should change the adhesive interaction between the two components, i.e., create good contact between adhesive and adherent. If wetting is not complete, some voids may arise at the interface, leading to the formation of weak layers. Coupling agents also improve wetting of a surface by a liquid binder that contributes to improved adhesion between fiber and polymer.[23-25]

5.2 Natural Fibers and Their Properties

Natural fibers originate from plants, animals, or minerals. The main sources of natural cellulosic fibers are plant parts. Cellulosic fibers are usually bound by a natural gummy substance (pectin or hemicellulose) and associated with lignin. We can classify sources of natural cellulosic fibers into various groups such as bast fibers from the bark sections of certain plants, called soft fibers for textile use; leaf fibers, called hard fibers, mainly used in cordage; seed-hair fibers, the most important types used in the textile industry; and brush fibers, from various portions of plants.

Lignocellulosic by-products constitute readily available and low-cost biomass products. Examples are corn stover, wheat, rice, barley straw, sorghum stalks, sugarcane bagasse, pineapple and banana leaves, and oil palm branches. Table 5.1 summarizes types, botanical names, and growing area of commercially important plants. Figure 5.1 shows natural and extracted fibers.

Figure 5.2 shows major sources of natural fibers that could be utilized for composites. The data were extracted from several sources. Estimates and extrapolations for some of the numbers were based on data from the Food and Agriculture Organization of the United Nations[26] and other organizations.[27,28] Thus, the data in this figure should be considered rough relative estimates of world fiber resources.

The fibers that originate from bast and leaf fibers provide strength and support for plant structures. Bast fibers located next to the outer bark of the bast or phloem serve to strengthen plant stems. They are strands that span the

TABLE 5.1

Important Fiber Sources and Biofibers from Agricultural By-Products[27,105]

Commercial Name	Botanical Name	Growing Areas
Bast fibers		
China jute	*Abutilon theophrasti*	China
Cadillo, urena	*Urena lobata*	Zaire, Brazil
Flax	*Linum usitatissimum*	North and south temperate zones
Hemp	*Cannabis sativa*	All temperate zones
Isora	*Helicteres isora*	China and Eastern Europe
Jute	*Corchorus capsularis; Corchorus olitorius*	India
Kenaf	*Hibiscus cannabinus*	India, Iran, Russia and former Soviet republics, South America
Roselle	*Hibiscus sabdarifa*	Brazil, Indonesia
Ramie	*Boehmeira nivea*	China, Japan, U.S.
Sun hemp	*Cortalaria juncea*	India
Leaf fibers		
Abaca	*Musa textilis*	Borneo, Philippines, Sumatra
Banana	*Musa indica*	Borneo, Philippines, Sumatra
Cantala agave	*Agave cantala*	Philippines, Indonesia
Caroa	*Neoglaziovia variegata*	Brazil
Henequen	*Agave fourcroydes*	Australia, Cuba, Mexico
Istle	*Samuela carnerosana*	Mexico
Mauritius hemp	*Furcraea gigantea*	Brazil, Mauritius, Venezuela,
Phormium	*Phormium tenas*	Argentina, Chile, New Zealand
Pineapple	*Ananus comosus*	Philippines, Thailand
Sansevieria (bowstring hemp)	*Sansevieria (entire genus)*	Africa, Asia, South America
Sisal agave	*Agave sisilana*	Haiti, Java, Mexico, South Africa
Seed-hair fibers		
Cotton	*Gossypium species*	U.S., Asia, Africa
Kapok	*Ceiba pentranda*	Tropics
Grass		
Bamboo	>1250 species	China
Other fibers		
Broom root	*Muhlenbergia macroura*	Mexico
Coir (coconut husk fiber)	*Cocos nucifera*	Tropics
Crin vegetal (palm leaf segments)	*Chamaerops humilis*	North Africa
Palmyra palm (palm leaf stems)	*Brossus flabellifera*	India
Piassava (palm leaf bases)	*Attalea funifera*	Brazil
Water hyacinth	*Eichhornia crassipes*	India, Asia, Africa, Australia, New Zealand, Central and South America

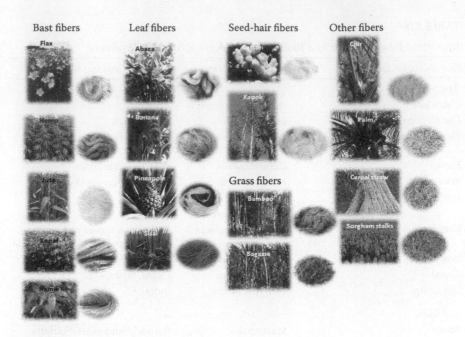

FIGURE 5.1
Groups of natural fibers from plants and extracted fibers.[26]

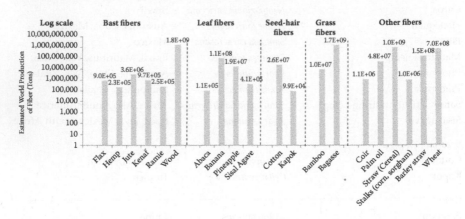

FIGURE 5.2
Estimated world production of important fiber sources and biofibers from agricultural byproducts in tons.

length of the stem or grow between joints. To separate the fibers, the natural gum (pectin and hemicellulose) binding cellulose fibers must be removed. This operation is called retting. Leaf fibers contribute strength to the leaves of non-woody plants. The fibers are separated from the pulp by scraping due to the lack of interaction bonding between fiber and pulp. This operation is called decortication.

Seed-hair fibers are single-celled, and they aid in spreading seeds. A cotton gin applies mechanical force to separate fibers from seeds. New technologies utilize ultrasonic separation and steam explosion decomposition.

The chemical compositions and properties of important natural fibers are presented in Tables 5.2 and 5.3, respectively. The various chemical components of natural fibers (Table 5.2) vary considerably due to origin, age, retting process (mode of extraction of fiber from source), and chemical treatment. Fibers contain cellulose, hemicellulose, pectin, and lignin. The properties of each component contribute to the overall properties of the fiber.

Hemicellulose is charged with biodegradation, moisture absorption, and thermal degradation of fibers because it shows the least resistance. Lignin is thermally stable but is responsible for ultraviolet (UV) degradation. Fibers generally contain 60 to 80% cellulose, 5 to 20% lignin, and up to 20% moisture. Table 5.2 indicates substantial amounts of non-cellulosics, mainly lignin, that contribute to the strength of fibers. Variations in the dimensions of unit cells are major factors in the lack of a good relationship between crystallinity and strength.[28] However, the natural fibers with longer unit cells (fineness) have greater strength, as shown in Table 5.3. Therefore, the amounts of chemical components and dimensions of the unit cells in natural fibers should be considered in designing products from these natural fibers.

5.3 Surface Treatment Techniques for Natural Fibers Used in Composites

Natural fibers have been used as reinforcements in composites in place of conventional synthetic fibers such as glass fiber because of their attractive properties such as low cost, low density, low pollutant emission levels, non-abrasive nature, acceptable specific properties, high aspect ratios, and biodegradability.[29–37] However, the hydrophilic nature of natural fibers due to the presence of hydroxyl and polar groups causes high moisture absorption. Hence, the incorporation of natural fibers in composites leads to a major problem of poor interfacial adhesion between the hydrophilic fibers and hydrophobic matrices.

The incompatibility of the different chemical characteristics of natural fibers and polymer matrices causes ineffective stress transfer, dimensional instability, and debonding throughout interfaces and results in deterioration of the mechanical properties of the resulting composites.[38–42] Thus, surface treatments of reinforced natural fibers are necessary to improve interfacial adhesion of the composites.

TABLE 5.2

Structures and Chemical Compositions of Important Fiber Sources and Biofibers from Agricultural By-Products

Fiber	Average Length (mm)	Average Width (mm)	References	Cellulose %	Hemicellulose %	Lignin %	Pectin %	Moisture %	References
Bast fibers									
Flax	5–60	0.012–0.027	106–110	71–78	18.6–20.6	2.2–2.8	1.8–2.3	10	106–108,110
Hemp	5–55	0.016	106,111	70.2–74.4	17.9–22.4	3.7–5.7	0.9	8.0–10.8	106–109
Jute	2.5	0.02	27	61–71.5	13.6–20.4	12–13	0.2–1.7	12.6	106–108,110
Kenaf	2.6	0.02	27	31–39	21.5	15–19	—		107,112
Ramie	60–250	0.016–0.120	113,114,115	68.6–76.2	13.1–16.7	0.6–0.7	1.9	8	106–109
Leaf fibers									
Abaca	6	0.024	27	70.1	21.8	5.7	0.6		113,114
Banana	0.9–4.0	0.08–0.25	28	63–64	19	5			116
Pineapple	3–9	0.02–0.08	28	70–82	18	5–12	—	11.8	117
Sisal agave	3.3	0.05–0.20	106,108,111,118	67–78	10.0–14.2	8–11	10	11	106–108,119

Fiber			[Ref.]						[Ref.]
Seed-hair fibers									
Cotton	1.5–5.6	0.02	120	82.7		5.7			121
Kapok	15–30	0.01–0.03	113,114	64	13	23	23		113,114
Grass									
Bamboo	2.7	0.014	27	78.8	10.2	12.5			110
Bagasse	1.7	0.02	27	32–48	23–32	19–24	0.4	8.8	122
Other fibers									
Coir (coconut husk fiber)	0.7	0.1–45	106,111	36–43	41–45	0.2–0.3	3–4	8	106,119,123
Oil palm	17.5	0.15–0.30	124	60–65	11–19				116
Cereal straw	1.5	0.023	27	31–45	16–19				27
Corn straw	1.5	0.018	27	32–35	16–27				27
Wheat straw	1.4	0.015	27	33–39	16–23			10	27,28
Rice straw	1.4	0.008	27	28–36	12–16	23–28			27
Barley straw	0.7–3.1	0.007–0.024	77	31–45	14–19	27–38		8–12	77
Sorghum stalks	0.8–1.2	0.03–0.08	122	27	11	25		8–12	122

TABLE 5.3

Properties of Important Fiber Sources and Biofibers from Agricultural By-Products

Fiber	Density (g/cm³)	Tensile Strength (MPa)	Young's Modulus (GPa)	Elongation %	References
Bast fibers					
Flax	1.5	345–1100	27.6	2.7–3.2	106,108,118,125
Hemp	1.52	690–920	46	1.6	106,108
Jute	0.6–1.4	393–773	13–26.5	1.16–1.5	106,108,111,118
Kenaf	1.19	240–600	14–38	2.7	114,126
Ramie	1.5	400–938	61.4–128	1.2–3.8	106,108,125
Leaf fibers					
Abaca		320–690		2–4.5	114,126
Banana	1.35	600	17.9	3.36	127
Pineapple	1.56	413–1627	34.5–82.5	1.6	111
Sisal agave	1.45	468–640	9.4–22	3–7	106,108,111,118,125
Seed-hair fibers					
Cotton	1.5	287–800	5.5–12.6	7–8	128
Kapok		160–300	13	1.2	114,126
Grasses					
Bamboo	0.8–1.4	391–1000		48–49	129,130
Other fibers					
Coir (coconut husk fiber)	1.15–1.25	131–175	4–6	15–40	106,111
Oil palm	1.03	50–400	0.57–9.0	2.5–1.8	131–134

5.3.1 Alkali Treatment

Alkali treatment or mercerization has been widely used to improve natural fiber properties prior to composite fabrication. Important modifications of natural fibers accomplished with alkali treatment have been reported.

First, waxy materials and impurities on fiber surface were removed. As a result, the surface tension and wetting ability of alkali-treated fiber increased. A rougher surface morphology was produced to increase interfacial adhesion through a mechanical interlock between fiber and matrix. Examples of such composites are alfa fiber–polyester,[43] palm leaf stalk fiber–polyester,[37] jute–epoxy,[44] and sisal low density polyethylene (LDPE) composites.[45] After the impurities were removed, an increase in nucleating ability of the alkali-treated hemp fiber was reported and a dense trans-crystalline layer of poly(lactic acid) (PLA) could be induced.[42,46]

Moreover, an increase in fiber hydrophilicity due to the removal of waxy materials led to a higher number of available hydroxyl groups on the fiber surface. Therefore, a better interfacial interaction with the hydrophilic resin was obtained as reported in a curaua fiber–corn starch-based resin,[43] and hemp fiber–PLA composite.[42] In coir fiber–poly(butylene succinate) (PBS) composites, the removal of cuticle layers exposed lignin on the fiber surface, leading to the enhancement of chemical compatibility with PBS resin.[36]

Second, the dissolution of hemicellulose and lignin in the amorphous region caused fibrillation (the breakdown of the fiber bundle into smaller microfibrils). The effective surface area available for contact with the resin increased from the fibrillation. The destruction of the cellular structure due to the removal of the cementing hemicellulose allowed the cellulose chains to rearrange themselves in a closer configuration, increasing the crystallinity index. The molecular orientation also increased from a decrease in the spiral angle. More homogeneous fibers were obtained as a result of microvoid elimination.[30,43,44] In addition, the reduction in fiber diameter by the dissolution of the hemicellulose enhanced the aspect ratio.[34,36,45]

Third, the microfibrils became more flexible due to the loss of crystallinity of cellulose from the partial conversion of cellulose I into cellulose II.[43,44,47]

In general, alkali-treated fiber reinforced composites possess superior mechanical properties compared to untreated fiber reinforced composites. The alkali concentration and treatment time significantly affect fiber morphology; consequently, the properties of treated fiber-reinforced composites are different.

For example, the tensile properties of alkali-treated coir fiber–PBS composites were markedly higher than those of composites with untreated fibers. The composites reinforced with coir fibers pretreated in 5% NaOH at room temperature for 72 hr exhibited the best mechanical properties.[36] In the case of bagasse fiber–aliphatic polyester composites, 1% NaOH-treated bagasse-reinforced composites showed maximum improvement compared with untreated fiber composites—13, 14, and 30% increases in tensile strength, flexural strength, and impact strength, respectively.[30] The 10% NaOH-treated banana fiber–polyurethane composites exhibited better tensile and dynamic mechanical properties than the untreated fiber composites. The scanning electron micrographs showed no evidence of fiber pull-out for the treated banana composites, indicating good interfacial adhesion.[48]

5.3.2 Acetylation Treatment

During acetylation treatment, an esterification reaction occurs between the cell wall hydroxyl groups of natural fibers and acetyl groups (CH_3CO-) from acetylating reagents such as acetic acid and acetic anhydride together with a catalyst or cosolvent.[49–52] The hydroxyl groups of lignin, hemicellulose, and amorphous cellulose mainly react with acetylating reagents because they are more easily accessible than the closely packed hydroxyl groups of crystalline cellulose.[40,50] As a result, the hydrophilicity of natural fibers decreases based on the acetyl content as indicated by the reduction of moisture absorption.[40,52]

When the degree of acetylation of flax fibers increases from 12 to 34%, moisture absorption compared with the untreated flax fibers decreases from 18 to

42%, respectively.[51] Acetylation treatment also removes surface microroughness and produces a smoother fiber surface.[40,50] The dispersive free energy and the polar ability of the acetylated fibers are also enhanced.[50] Acetylation helps improve composite properties such as dimensional stability,[29,40] tensile strength,[51,53,54] thermal resistance,[51,55] and biological resistance.[29]

5.3.3 Benzoylation Treatment

The hydroxyl groups of natural fibers are commonly esterified with benzoyl chloride, yielding benzyl groups (C_6H_5COO-) at the fiber surface.[56] The alkali pretreatment before benzoylation causes a reduction in fiber diameter and hence the aspect ratio increases due to the removal of alkali-soluble materials such as waxes, lignin, and hemicellulose.[56,57]

The rougher surface from small voids produced during the treatment promotes mechanical interlock between the fiber and matrix. The presence of phenyl groups in benzoylated fiber makes it more hydrophobic and also improves compatibility with the resin in composites such as sisal fiber–polyester,[56] banana fiber–polypropylene,[57] and sisal fiber–polystyrene.[58] Improvements in composites resulting from benzoylated fiber reinforcement were noted in tensile properties,[58] thermal conductivity,[57] and melt viscosity.[56]

5.3.4 Fatty Acid Derivative Treatment

Fatty acid derivatives can react with the hydroxyl groups of natural fibers to form ester groups such as dodecanoyl and octadecanoyl chlorides with bagasse[59] and oleoyl chloride with jute fibers (Figure 5.3).[60] The long carbon chains of the fatty acid derivatives introduce hydrophobicity to the fiber surface, resulting in a higher dispersion level and improvements of wettability

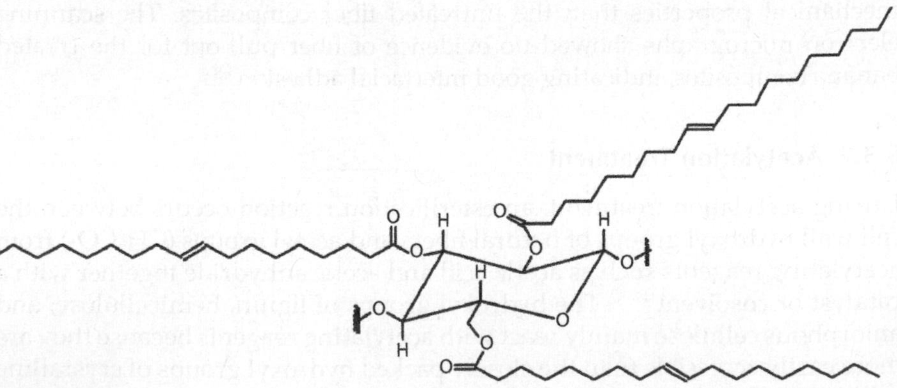

FIGURE 5.3
Chemical structure of oleoyl chloride-treated cellulose.

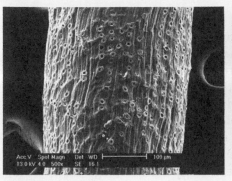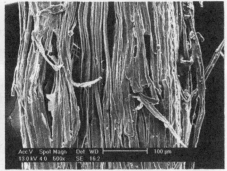

FIGURE 5.4
Scanning electron microscope micrographs of coir fibers before (left) and after (right) peroxide treatment.

and interfacial adhesion with non-polar matrices such as polyethylene[59,61] and poly(ethylene sebacate).[62] The composites exhibited enhanced homogeneity, mechanical properties, and water resistance.[59–63] The degree of substitution and fatty chain length significantly affect the performances of composites.[61]

5.3.5 Peroxide Treatment

Benzoyl peroxide (BP) and dicumyl peroxide (DCP) have been used to treat natural fibers such as sisal[64,65] and jute.[66] The peroxide initiates free radical addition between the polymer matrix and the fibers (Figure 5.4). DCP is reported to be more effective than BP due to a lower decomposition rate.[45] Peroxide treatment significantly improves the tensile properties of composites.[45,67] However, a decrease of thermal stability of treated fiber composites is attributed to peroxide treatment, as evident from the decreases of the melting temperatures of sisal–LDPE[64,65] and jute–polylactide[66] composites with DCP and BP treatments, respectively.

5.3.6 Permanganate Treatment

Natural fibers treated with potassium permanganate ($KMnO_4$) result in the formation of a highly reactive permanganate that is responsible for the induced grafting of polymer onto natural fibers.[45,56,57,66] The permanganate concentration is a critical factor affecting the tensile properties of treated fiber-reinforced composites since the cellulosic fibers may be degraded at high permanganate concentrations.[45] As compared to untreated composites, the permanganate treatment improved the flexural strength of sisal fiber–polyester composites,[56] the tensile strength of sisal fiber–LDPE composites,[45] and the thermal diffusivity of banana fiber–polypropylene composites.[57]

5.3.7 Fungal Treatment

The use of fungi for fiber surface treatment is considered an environmentally friendly method and a possible alternative to chemical treatment. White rot fungi *Schizophyllum commune* (*S. com*) from the Basidomycetes group is capable of separating hemp fiber bundles, removing non-cellulosic compounds such as lignin, wax, and pectin and increasing fiber surface roughness. White rot fungi produces enzymes such as lignin peroxidases, manganese peroxidases, and laccases to degrade non-cellulosic compounds selectively and produce fine holes on fiber surfaces. This fungal treatment improves the tensile strength of hemp–polypropylene composites through the enhancement of interfacial interactions.[68–71]

5.3.8 Enzymatic Treatment

Enzymatic treatment was also introduced as an environmentally friendly surface modification to improve fiber–matrix interfaces. Laccase has a high selectivity to remove lignin efficiently with little damage to fiber surfaces. The removal of lignin enhances the interfacial interactions of sisal fiber–phenolic resin composites, resulting in a significant increase in tensile and internal bonding strengths.[42]

Abaca fibers modified with Fungamix, a commercial enzyme, exhibited smoother surfaces and fibrillation was also observed. The tensile strength of enzyme-treated abaca fiber–polypropylene (PP) composites increased 45% as compared to unmodified composites.[72] Likewise, the improvement of tensile and flexural characteristics of pectinase-treated hemp–PP composites was reported.[73] Novamix, a mixture of lipase, protease, and amylase–xylanase, was used to separate microfibers from wheat husks and remove unwanted materials on the fiber surfaces without causing damage. Enzyme modification facilitated improvements in modulus, failure strains, and strength properties of the treated microfiber–PP and –PLA composites.[74]

5.4 Coupling Agents

In addition to fiber treatments, coupling agents are used to form chemical and/or physical bonds between fiber and matrix phases to improve interfacial adhesion of composites. Coupling agents facilitate functional groups, especially hydroxyl groups, at one end to react with the fiber surface and another functional group at the other end to bond with the polymer matrix.[75–77] Examples of coupling agents used for natural fiber–polymer composites are described below.

5.4.1 Silanes

Silanes can form a chemical link between a fiber surface and a polymer matrix through their bifunctional structures. Silane chemical structure can be represented simply as R-Si-X$_3$, where R is an organofunctional group and X is a hydrolyzable group that can form silanols in the presence of moisture.[75,76] Silanols and hydroxyl groups of natural fibers can subsequently form covalent bonds of –Si-O-C.[78–80] Various silanes have been used as coupling agents in composites as shown in Table 5.4.

The functionality of the R group plays a major role on the reactivity of silane toward a polymer matrix. Alkyl groups are designed to create better compatibility with the hydrocarbon backbones of matrices. Vinyl, amino, mercapto, and methacryl groups are capable of forming covalent bonds with the reactive groups of a matrix.

Hexadecyltrimethoxy silane (HDS) adheres to polyethylene through van der Waals forces and the formation of a cage-like interpenetrating polymer network using its long aliphatic hydrocarbon chain. A marginal enhancement of the properties of HDS-treated-fiber-reinforced polyethylene composites was observed in the work of Abdelmouleh et al.[81] Vinyl and methacryl silanes covalently bond onto the polymer backbone by a free radical grafting reaction in the presence of a peroxide initiator.

A better interfacial bonding from vinyltrimethoxy silane (VTS) coupling agents is evident from an enhancement of tensile strength and flexural

TABLE 5.4

Chemical Structures of Silanes and Their Polymer Matrices

Name	R	X	Matrix	References
Hexadecyltrimethoxy silane	-(CH$_2$)$_{15}$CH$_3$	-OCH$_3$	Epoxy Polyethylene Natural rubber	78,81
Vinyltrimethoxy silane	-CH = CH$_2$	-OCH$_2$	Polyethylene	83
Vinyltri(2-methoxy ethoxy silane	-CH = CH$_2$	-OCH$_2$CH$_2$-OCH$_3$	Polyethylene Polystyrene	135,136
γ-Aminopropyl triethoxy silane	-(CH$_2$)$_3$NH$_2$	-OCH$_2$CH$_3$	Polyethylene Poly(lactic acid) Epoxy	79,84,85,137
γ-Mercaptopropyl trimethoxy silane	-(CH$_2$)$_3$SH	-OCH$_3$	Polyethylene Epoxy PMMA Natural rubber	78,81
γ-Methacryloxypropyl trimethoxy silane	$-(CH_2)_3 - O - \overset{O}{\overset{\|}{C}} - \underset{\underset{CH_3}{\|}}{C} = CH_2$	-OCH$_3$	Polyethylene Polystyrene PMMA Natural rubber	78,81,135

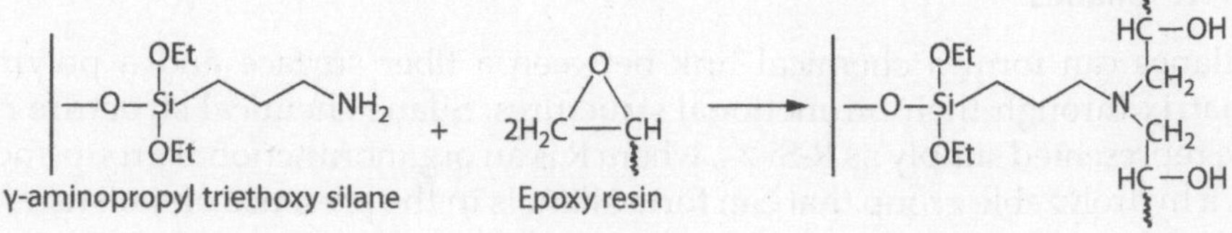

γ-aminopropyl triethoxy silane   Epoxy resin

FIGURE 5.5
Chemical reaction of γ-aminopropyltriethoxy silane and epoxy resin.

strength of VTS-treated henequen fiber–HDPE composites,[82] an increase of flexural toughness of VTS-treated wood–polyethylene composites,[83] and a reduction in water absorption of VTS-treated pineapple-leaf fiber–LDPE composites,[38] for instance. Similarly, short cellulose fibers treated with γ-methacryloxypropyltrimethoxy silane (MCS) were reported to display good mechanical properties in the reinforcement of polyethylene composites.[81]

The amine groups of the γ-aminopropyltriethoxy silane (APS) can form hydrogen bonds with the hydrolyzed PLA backbone. The compatibility of kenaf fibers and PLA was enhanced by APS coupling agents, leading to a significant improvement of mechanical and thermal properties of the composites.[84] APS can also react with epoxy resins, as shown in Figure 5.5, resulting in both tensile and flexural strength of sisal–epoxy composites to be slightly higher than those of untreated composites.[85] Moreover, the interfacial adhesion of APS and PVC was improved through acid–base interaction.[86] Abdelmouleh et al. reported the use of γ-mercaptopropyl trimethoxy silane (MPS) in cellulose fibers and polyethylene matrices. Covalent bonds were generated between the fiber and matrix through a transfer reaction with the mercapto moiety.[81]

5.4.2 Isocyanates

Isocyanates (R–N = C = O) consist of –N = C = O functional groups that can react with hydroxyl groups of cellulosic fibers and form covalent bonds between them through a urethane linkage (Figure 5.6).[22,87] Isocyanates are very susceptible to hydrolysis in the presence of moisture; blocked isocyanates are thus used to prevent the hydrolysis reaction prior to the reaction with fiber hydroxyl groups.[88] At the other ends of isocyanates, different R groups are applied for better compatibility and dispersibility

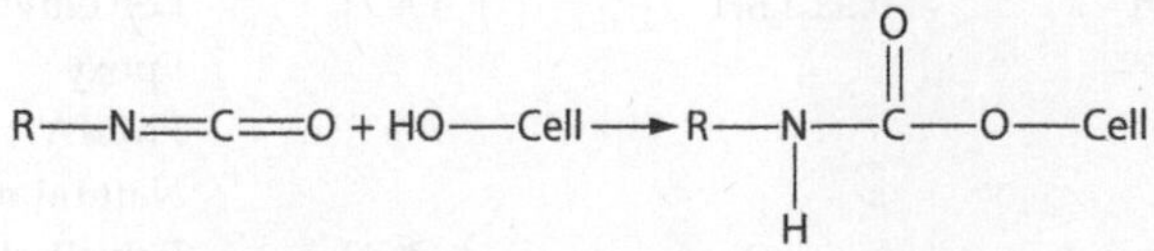

FIGURE 5.6
Chemical reaction of isocyanate and cellulose.

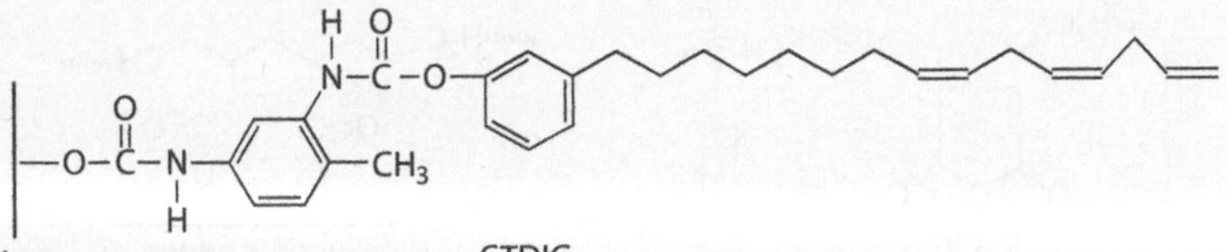

FIGURE 5.7
CTDIC-treated sisal fiber.

of the fiber in the matrix. Examples are alkyl (e.g., hexamethylene), alkenyl (e.g., 3-isopropenyl-α,α'-dimethylbenzyl), aromatics (e.g., toluene and polymethylene polyphenyl), amino acids (e.g., lysine) and plant extracts (e.g., cardanol).

The long and flexible hydrocarbon chains can improve the interfacial adhesion between the matrix and treated fibers through an entanglement interlock and van der Waals interaction. Examples are composites of LDPE with poly(methylene) poly(phenyl) isocyanate (PMPPIC)-treated pineapple-leaf fiber (PALF)[89] and a cardanol derivative of toluene diisocyanate (CTDIC)-treated sisal fiber (Figure 5.7).[45,90]

The alkenyl moiety of 3-isopropenyl-α,α'-dimethylbenzyl isocyanate and 2-isocyanoethyl methacrylate can copolymerize with matrix monomers such as styrene and methyl methacrylate.[91] The delocalized π-electrons in the benzene rings of toluene 2,4-diisocyanate (TDIC) and methylenediphenyl diisocyanate (MDIC) also promote stronger interactions with polystyrene matrices.[88,92]

In corn starch–PLA composites, isocyanate groups can react with terminal hydroxyl or carbonyl groups of PLA along with the hydroxyl groups of corn starch to form urethane linkages.[93] The use of isocyanate coupling agents results in improvement of mechanical properties and reduction of water uptake of composites such as MDIC-treated wheat starch–PLA,[93,94] MDIC-treated pine fiber–polystyrene,[88] TDIC-treated sisal fiber–polystyrene,[92] CTDIC-treated sisal fiber–LDPE,[45,90] and hexamethylene diisocyanate (HMDIC)-treated cellulose fiber–PP composites.[95] Lysine diisocyanate (LDIC)-treated bamboo fiber-reinforced PLA and PBS composites exhibited higher thermal degradation and delayed enzymatic degradation.[31]

5.4.3 Maleated Coupling Agents

Maleated coupling agents are prepared by grafting maleic anhydrides onto polymers to form copolymers such as maleated polyethylene (MAPE), maleated polypropylene (MAPP), and maleated polystyrene (MAPS). During grafting, the anhydride groups of maleated copolymers can react with hydroxyl groups of natural fibers through esterification, coating natural fibers with maleated copolymers as shown in Figure 5.8.[75] The similar

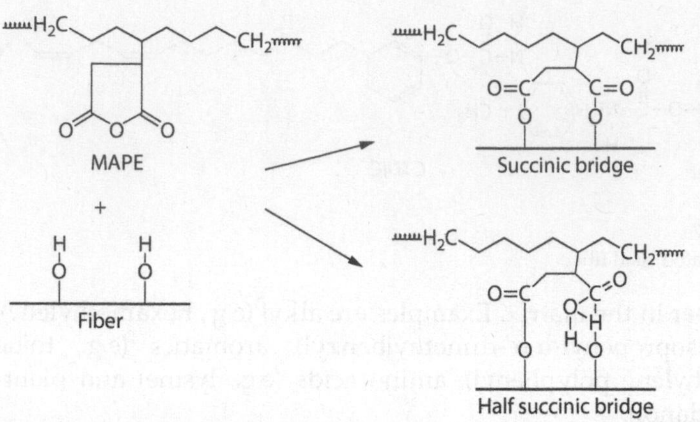

FIGURE 5.8
Chemical reaction of MAPE and cellulose.

polarities between the polymer matrix and the treated fiber facilitate a better entanglement interlock at the interface.[87,96]

The incorporation of maleated coupling agents in fiber–polymer composites leads to the improvement of the mechanical properties of the composites. The enhancement of strength properties of composites through the addition of MAPP was reported, for example, for PALF–PP,[77] jute fiber–PP,[97] and wood fiber–PP composites.[98]

Improvements in the abrasive wear resistance of jute-reinforced PP composites in the presence of MAPP were noted by Chand and Dwivedi.[32] The influence of MAPE on the mechanical and dynamic mechanical properties of jute-reinforced HDPE composites was studied by Mohanty et al. At 30% jute fiber loading, the composites with 1% MAPP exhibited significant increases in tensile, flexural, and impact strengths (38, 45, and 67%, respectively) compared with results from untreated composites. The MAPE-treated composites also showed increases in storage modulus and thermal stability.[99] The inclusion of MAPS in sisal fiber-reinforced polystyrene composites showed better mechanical properties and dimensional stability but a slight increase of water absorption.[92]

5.4.4 Triazine Derivatives

Triazine derivatives such as dichlorotriazines can function as chemical bridges between fibers and polymer matrices. The reactive chlorine atoms on the heterocyclic ring are able to form a covalent bond with the hydroxyl groups of natural fibers through an ether linkage. The hydrogen bonds also possibly link the nitrogen atoms in the triazine ring with the hydroxyl groups of the fibers. Zedorecki et al. used triazine derivatives as coupling agents to treat cellulose fibers and then reinforce them with polyester resin. The improvement in interfacial adhesion also improved the environmental aging behaviors of the resulting cellulose–polyester composites.[100,101]

Moreover, biopolymers such as chitin, chitosan, and zein have also been used as coupling agents in fiber–polymer composites in an effort to produce green composites. Chitin and chitosan are natural polysaccharide polymers of N-acetyl glucosamine and glucosamine, respectively. Chitin and chitosan coupling agents enhanced the mechanical properties of wood flour–PVC composites in comparison to composites without the coupling agents. The stronger interfacial adhesion was attributed to the acid–base interaction between the chlorine atoms of PVC and the amino groups of chitin and chitosan on the wood surface.[102]

Zein—a natural protein extracted from corn—possesses many functional groups on its molecular chain that can act as coupling agents for fiber–polymer composites. Zein is composed of polar groups such as carboxyl and amino groups that can react with the hydroxyl groups of fibers through hydrogen bonds. The alkyl and aryl groups can react with a polymer matrix through hydrophobic interactions.[103] In the presence of zein, the storage moduli of flax–PP composites improved.[23,104] Similarly, improvements of the mechanical and viscoelastic properties of kenaf–PP composites by the addition of zein coupling agents were also reported.[103]

5.5 Conclusions

Natural fibers are considered potential replacements for man-made fibers in composite materials. Although natural fibers have the advantages of low cost and low density, they are not totally free of problems.

A serious issue with natural fibers is their strong polar character that creates incompatibilities with most polymer matrices. Surface treatments, despite their negative impacts on production economics, are potentially able to overcome the incompatibility problem. Chemical treatments can increase the interface adhesion between the fiber and matrix and decrease the water absorption of fibers. Therefore, chemical treatments can be considered for modifying the properties of natural fibers.

Coupling agents can be used to improve the adhesion between natural fibers and polymer matrix. Fiber modification methods discussed in this chapter exhibited different efficacy levels in aiding adhesion of matrix and fiber. Both chemical treatments and coupling agents have achieved various levels of success in improving fiber strength, fiber fitness, and fiber–matrix adhesion in natural fiber-reinforced composites.

References

1. R. Chollakup, W. Smitthipong, and P. Suwanruji. Environmentally Friendly Coupling Agents for Natural Fibre Composites. In *Natural Polymers. Volume 1: Composites*, M.J. John and S. Thomas, Eds. Cambridge: Royal Society of Chemistry, 2012, pp. 161–182.
2. J. Schultz and M. Nardin. Determination of the Surface Energy of Solids by the Two-Liquid-Phase Method. In *Modern Approaches to Wettability: Theory and Applications*, M.E. Schrader and G. Loeb, Eds. New York: Plenum Press, 1992.
3. W. Smitthipong, M. Nardin, J. Schultz et al. Study of Tack Properties of Uncrosslinked Natural Rubber. *Journal of Adhesion Science and Technology*, 18 (2004): 1449–1463.
4. A.J. Kinloch. *Adhesion and Adhesives: Science and Technology*. London: Chapman & Hall, 1987.
5. L.H. Lee. *Fundamentals of Adhesion*. New York: Plenum Press, 1991.
6. K.L. Mittal and A. Pizzi, Eds. *Adhesion Promotion Techniques: Technological Applications*. Boca Raton, FL: CRC Press, 1999.
7. M. Nardin and E. Papirer. *Powders and Fibers: Interfacial Science and Applications*. Boca Raton, FL: CRC Press, 2007.
8. W. Smitthipong, R. Gadiou, L. Vidal et al. Raman Images of Rubber Blends (IR–HNBR). *Vibrational Spectroscopy*, 46 (2008): 8–13.
9. W. Smitthipong, M. Nardin, J. Schultz et al. Adhesion and Self-Adhesion of Immiscible Rubber Blends. *International Journal of Adhesion and Adhesives*, 29 (2009): 253–258.
10. S.S. Voyutskii. *Autohesion and Adhesion of High Polymers*. New York: Interscience, 1963.
11. W. Smitthipong, M. Nardin, J. Schultz, et al. Adhesion and Self-Adhesion of Rubbers Crosslinked by Electron Beam Irradiation. *International Journal of Adhesion and Adhesives*, 27 (2007): 352–357.
12. A.D. Roberts and A.B. Othman. Rubber Adhesion and the Dwell-Time Effect. *Wear*, 42 (1977): 119–133.
13. B. Bhushan. Adhesion and Stiction: Mechanisms, Measurement Techniques, and Methods for Reduction. *Journal of Vacuum Science & Technology B*, 21 (2003): 2262–2296.
14. N. Venkateshwaran and A. Elayaperumal. Banana Fiber-Reinforced Polymer Composites: A Review. *Journal of Reinforced Plastics and Composites*, 29 (2010): 2387–2396.
15. R. Malkapuram, V. Kumar, and Y.S. Negi. Recent Developments in Natural Fiber Reinforced Polypropylene Composites. *Journal of Reinforced Plastics and Composites*, 28 (2009): 1169–1189.
16. S. Zhan-Ying, H. Hai-Shan, and D. Gan-Ce. Mechanical Properties of Injection-Molded Natural Fiber-Reinforced Polypropylene Composites: Formulation and Compounding Processes. *Journal of Reinforced Plastics and Composites*, 29 (2010): 637–50.
17. T. Huber, U. Biedermann, and J. Müssig. Enhancing the Fibre Matrix Adhesion of Natural Fibre Reinforced Polypropylene by Electron Radiation Analyzed with the Single Fibre Fragmentation Test. *Composite Interfaces*, 17 (2010): 371–381.

18. J.N. Israelachvili. *Intermolecular and Surface Forces.* New York: Academic Press, 2011.
19. E. Papirer, E. Brendle, H. Balard et al. Inverse Gas Chromatography Investigation of the Surface Properties of Cellulose. *Journal of Adhesion Science and Technology,* 14 (2000): 321–337.
20. R. Chollakup, W. Smitthipong, W. Kongtud, et al. Polyethylene Green Composites Reinforced with Cellulose Fibers (Coir and Palm Fibers): Effect of Fiber Surface Treatment and Fiber Content. *Journal of Adhesion Science and Technology,* 27 (2012): 1290–1300.
21. A. Valadez-Gonzalez, J.M. Cervantes-Uc, R. Olayo et al. Effect of Fiber Surface Treatment on the Fiber–Matrix Bond Strength of Natural Fiber-Reinforced Composites. *Composites Part B,* 30 (1999): 309–320.
22. X. Li, L.G. Tabil, and S. Panigrahi. Chemical Treatments of Natural Fiber for Use in Natural Fiber-Reinforced Composites: A Review. *Journal of Polymers and the Environment,* 15 (2007): 25–33.
23. R. Kumar, M.K. Yakabu, and R.D. Anandjiwala. Effect of Montmorillonite Clay on Flax Fabric Reinforced Poly Lactic Acid Composites with Amphiphilic Additives. *Composites Part A,* 41 (2010): 1620–1627.
24. M. Khan, S. Ghoshal, R. Khan et al. Preparation and Characterization of Jute Fiber-Reinforced Shellac Biocomposites: Effect of Additive. *Journal Chemistry & Chemical Technology,* 2 (2008).
25. T. Stuart, Q. Liu, M. Hughes et al. Structural Biocomposites from Flax. Part I: Effect of Bio-Technical Fibre Modification on Composite Properties. *Composites Part A,* 37 (2006): 393–404.
26. Food and Agricultural Organization of United Nations. AOSTAT. Food and Agricultural Commodities Production. http://faostat.fao.org/site/339/default.aspx
27. R.M. Rowell. The State of Art and Future Development of Bio-Based Composite Science and Technology Towards the 21st Century. Paper presented at Fourth Pacific Rim Bio-Based Composites Symposium, Indonesia, November 2–5, 1998.
28. N. Reddy and Y. Yang. Biofibers from Agricultural Byproducts for Industrial Applications. *Trends in Biotechnology,* 23 (2005): 22–27.
29. H.P.S. Abdul Khalil and H. Ismail. Effect of Acetylation and Coupling Agent Treatments upon Biological Degradation of Plant Fibre Reinforced Polyester Composites. *Polymer Testing,* 20 (2000): 65–75.
30. Y. Cao, S. Shibata, and I. Fukumoto. Mechanical Properties of Biodegradable Composites Reinforced with Bagasse Fibre before and after Alkali Treatments. *Composites Part A,* 37 (2006): 423–429.
31. S.H. Lee and S. Wang. Biodegradable Polymer–Bamboo Fiber Biocomposite with Bio-Based Coupling Agent. *Composites Part A,* 37 (2006): 80–91.
32. N. Chand and U.K. Dwivedi. Effect of Coupling Agent on Abrasive Wear Behaviour of Chopped Jute Fibre-Reinforced Polypropylene Composites. *Wear,* 261 (2006): 1057–1063.
33. N. Sgriccia, M.C. Hawley, and M. Misra. Characterization of Natural Fiber Surfaces and Natural Fiber Composites. *Composites Part A,* 39 (2008): 1632–1637.
34. X. Peng, L. Zhong, J. Ren et al. Laccase and Alkali Treatments of Cellulose Fibre: Surface Lignin and Its Influences on Fibre Surface Properties and Interfacial Behaviour of Sisal Fibre–Phenolic Resin Composites. *Composites Part A,* 41 (2010): 1848–1856.

35. K. Mylsamy and I. Rajendran. The Mechanical Properties, Deformation and Thermomechanical Properties of Alkali Treated and Untreated Agave Continuous Fibre Reinforced Epoxy Composites. *Materials & Design*, 32 (2011): 3076–3084.

36. T.H. Nam, S. Ogihara, N.H. Tung et al. Effect of Alkali Treatment on Interfacial and Mechanical Properties of Coir Fiber Reinforced Poly(Butylene Succinate) Biodegradable Composites. *Composites Part B*, 42 (2011): 1648–1656.

37. D. Shanmugam and M. Thiruchitrambalam. Static and Dynamic Mechanical Properties of Alkali-Treated Unidirectional Continuous Palmyra Palm Leaf Stalk Fiber–Jute Fiber-Reinforced Hybrid Polyester Composites. *Materials & Design*, 50 (2013): 533–542.

38. J. George, S.S. Bhagawan, and S. Thomas. Effects of Environment on the Properties of Low-Density Polyethylene Composites Reinforced with Pineapple-Leaf Fibre. *Composites Science and Technology*, 58 (1998): 1471–1485.

39. A.K. Mohanty, L.T. Drzal, and M. Misra. Novel Hybrid Coupling Agent as an Adhesion Promoter in Natural Fiber-Reinforced Powder Polypropylene Composites. *Journal of Materials Science Letters*, 21 (2002): 1885–1888.

40. V. Tserki, N.E. Zafeiropoulos, F. Simon et al. A Study of the Effect of Acetylation and Propionylation Surface Treatments on Natural Fibres. *Composites Part A*, 36 (2005): 1110–1118.

41. E. Franco-Marquès, J.A. Méndez, M.A. Pèlach et al. Influence of Coupling Agents in the Preparation of Polypropylene Composites Reinforced with Recycled Fibers. *Chemical Engineering Journal*, 166 (2011): 1170–1178.

42. M.A. Sawpan, K.L. Pickering, and A. Fernyhough. Effect of Fibre Treatments on Interfacial Shear Strength of Hemp Fibre-Reinforced Polylactide and Unsaturated Polyester Composites. *Composites Part A*, 42 (2011): 1189–1196.

43. A. Gomes, T. Matsuo, K. Goda et al. Development and Effect of Alkali Treatment on Tensile Properties of Curaua Fiber Green Composites. *Composites Part A*, 38 (2007): 1811–1820.

44. J. Gassan and A.K. Bledzki. Possibilities for Improving the Mechanical Properties of Jute–Epoxy Composites by Alkali Treatment of Fibres. *Composites Science and Technology*, 59 (1999): 1303–1309.

45. K. Joseph, S. Thomas, and C. Pavithran. Effect of Chemical Treatment on the Tensile Properties of Short Sisal Fibre-Reinforced Polyethylene Composites. *Polymer*, 37 (1996): 5139–5149.

46. M.S. Islam, K.L. Pickering, and N.J. Foreman. Influence of Alkali Treatment on the Interfacial and Physico-Mechanical Properties of Industrial Hemp Fibre-Reinforced Polylactic Acid Composites. *Composites Part A*, 41 (2010): 596–603.

47. E.T.N. Bisanda. The Effect of Alkali Treatment on the Adhesion Characteristics of Sisal Fibres.

48. A.C. Merlini, V. Soldi, and G.M.O. Barra. Influence of Fiber Surface Treatment and Length on Physico-Chemical Properties of Short Random Banana Fiber-Reinforced Castor Oil Polyurethane Composites. *Polymer Testing*, 30 (2011): 833–840.

49. S.M. Luz, J. Del Tio, G.J.M. Rocha et al. Cellulose and Cellulignin from Sugarcane Bagasse-Reinforced Polypropylene Composites: Effect of Acetylation on Mechanical and Thermal Properties. *Composites Part A*, 39 (2008): 1362–1369.

50. N.E. Zafeiropoulos, D.R. Williams, C.A. Baillie et al. Engineering and Characterisation of the Interface in Flax Fibre–Polypropylene Composite Materials. Part I: Development and Investigation of Surface Treatments. *Composites Part A,* 33 (2002): 1083–1093.

51. A.K. Bledzki, A.A. Mamun, M. Lucka-Gabor et al. The Effects of Acetylation on Properties of Flax Fibre and Its Polypropylene Composites. *Express Polymer Letters,* 2, (2008): 413–422.

52. M.M. Kabir, H. Wang, K.T. Lau, and F. Cardona. Effects of Chemical Treatments on Hemp Fibre Structure. *Applied Surface Science,* 276 (2013): 13–23.

53. S. Joseph, P. Koshy, and S. Thomas. The Role of Interfacial Interactions on the Mechanical Properties of Banana Fibre-Reinforced Phenol Formaldehyde Composites. *Composite Interfaces,* 12 (2005): 581–600.

54. H.P.S.A. Khalil, H. Ismail, H.D. Rozman et al. The Effect of Acetylation on Interfacial Shear Strength between Plant Fibres and Various Matrices. *European Polymer Journal,* 37 (2001): 1037–1045.

55. M. Nogi, K. Abe, K. Handa et al. Property Enhancement of Optically Transparent Bionanofiber Composites by Acetylation. *Applied Physics Letters,* 89 (2006): 1–3.

56. P.A. Sreekumar, S.P. Thomas, J.M. Saiter et al. Effect of Fiber Surface Modification on the Mechanical and Water Absorption Characteristics of Sisal–Polyester Composites Fabricated by Resin Transfer Molding. *Composites Part A,* 40 (2009): 1777–1784.

57. S.A. Paul, A. Boudenne, L. Ibos et al. Effect of Fiber Loading and Chemical Treatments on Thermophysical Properties of Banana Fiber–Polypropylene Commingled Composite Materials. *Composites Part A,* 39 (2008): 1582–1588.

58. K.C.M. Nair, S.M. Diwan, and S. Thomas. Tensile Properties of Short Sisal Fiber-Reinforced Polystyrene Composites. *Journal of Applied Polymer Science,* 60 (1996): 1483–1497.

59. D. Pasquini, E. De M. Teixeira, A.A. Da Silva-Curvelo et al. Surface Esterification of Cellulose Fibres: Processing and Characterisation of Low-Density Polyethylene–Cellulose Fibre Composites. *Composites Science and Technology,* 68 (2008): 193–201.

60. F. Corrales, F. Vilaseca, M. Llop et al. Chemical Modification of Jute Fibers for the Production of Green Composites. *Journal of Hazardous Materials,* 144 (2007): 730–735.

61. C.S.R. Freire, A.J.D. Silvestre, C.P. Neto et al. Composites Based on Acylated Cellulose Fibers and Low-Density Polyethylene: Effect of Fiber Content, Degree of Substitution, and Fatty Acid Chain Length on Final Properties. *Composites Science and Technology,* 68 (2008): 33583364.

62. T.F. Fernandes, E. Trovatti, C.S.R. Freire et al. Preparation and Characterization of Novel Biodegradable Composites Based on Acylated Cellulose Fibers and Poly(Ethylene Sebacate). *Composites Science and Technology,* 71, (2011): 1908–1913.

63. D. Pasquini, M.N. Belgacem, A. Gandini et al. Surface Esterification of Cellulose Fibers: Characterization by Drift and Contact Angle Measurements. *Journal of Colloid and Interface Science,* 295 (2006): 79–83.

64. A.S. Luyt and M.E. Malunka. Composites of Low-Density Polyethylene and Short Sisal Fibres: Effect of Wax Addition and Peroxide Treatment on Thermal Properties. *Thermochimica Acta,* 426, (2005): 101–107.

65. E.E.M. Ahmad and A.S. Luyt. Effects of Organic Peroxide and Polymer Chain Structure on Morphology and Thermal Properties of Sisal Fibre-Reinforced Polyethylene Composites. *Composites Part A,* 43 (2012): 703–710.

66. B.K. Goriparthi, K.N.S. Suman, and N. Mohan Rao. Effect of Fiber Surface Treatments on Mechanical and Abrasive Wear Performance of Polylactide–Jute Composites. *Composites Part A,* 43 (2012): 1800–1808.
67. S. Sapieha, P. Allard, and Y.H. Zang. Dicumyl Peroxide-Modified Cellulose– LLDPE Composites. *Journal of Applied Polymer Science,* 41 (1990): 2039–2048.
68. Y. Li and K.L. Pickering. The Effect of Chelator and White Rot Fungi Treatments on Long Hemp Fibre-Reinforced Composites. *Composites Science and Technology,* 69 (2009): 1265–1270.
69. Y. Li, K.L. Pickering, and R.L. Farrell. Analysis of Green Hemp Fibre-Reinforced Composites Using Bag Retting and White Rot Fungal Treatments. *Industrial Crops and Products,* 29 (2009): 420–426.
70. Y. Li, K.L. Pickering, and R.L. Farrell. Determination of Interfacial Shear Strength of White Rot Fungi-Treated Hemp Fibre Reinforced Polypropylene. *Composites Science and Technology,* 69 (2009): 1165–1171.
71. K.L. Pickering, Y. Li, R.L. Farrell et al. Interfacial Modification of Hemp Fiber-Reinforced Composites Using Fungal and Alkali Treatment. *Journal of Biobased Materials and Bioenergy,* 1 (2007): 109–117.
72. A.K. Bledzki, A.A. Mamun, A. Jaszkiewicz et al. Polypropylene Composites with Enzyme-Modified Abaca Fibre. *Composites Science and Technology,* 70 (2010): 854–860.
73. Z. Saleem, H. Rennebaum, F. Pudel et al. Treating Bast Fibres with Pectinase Improves Mechanical Characteristics of Reinforced Thermoplastic Composites. *Composites Science and Technology,* 68 (2008): 471–476.
74. A.A. Mamun and A.K. Bledzki. Micro Fibre-Reinforced PLA and PP Composites: Enzyme Modification, Mechanical and Thermal Properties. *Composites Science and Technology,* 78 (2013): 10–17.
75. J.Z. Lu, Q. Wu, and H.S. McNabb. Chemical Coupling in Wood Fiber and Polymer Composites: A Review of Coupling Agents and Treatments. *Wood and Fiber Science,* 32 (2000): 88–104.
76. Y. Xie, C.A.S. Hill, Z. Xiao et al. Silane Coupling Agents Used for Natural Fiber–Polymer Composites: A Review. *Composites Part A,* 41 (2010): 806–819.
77. D.N. Saheb and J.P. Jog. Natural Fiber Polymer Composites: A Review. *Advances in Polymer Technology,* 18 (1999): 351–363.
78. M. Abdelmouleh, S. Boufi, M.N. Belgacem et al. Modification of Cellulosic Fibres with Functionalised Silanes: Development of Surface Properties. *International Journal of Adhesion and Adhesives,* 24 (2004): 43–54.
79. M.C.B. Salon, M. Abdelmouleh, S. Boufi et al. Silane Adsorption onto Cellulose Fibers: Hydrolysis and Condensation Reactions. *Journal of Colloid and Interface Science,* 289 (2005): 249–261.
80. B.D. Park, S.G. Wi, K.H. Lee et al. X-Ray Photoelectron Spectroscopy of Rice Husk Surface Modified with Maleated Polypropylene and Silane. *Biomass and Bioenergy,* 27 (2004): 353–363.
81. M. Abdelmouleh, S. Boufi, M. N. Belgacem et al. Short Natural-Fibre Reinforced Polyethylene and Natural Rubber Composites: Effect of Silane Coupling Agents and Fibres Loading. *Composites Science and Technology,* 67 (2007): 1627–1639.
82. P.J. Herrera-Franco and A. Valadez-González. A Study of the Mechanical Properties of Short Natural Fiber-Reinforced Composites. *Composites Part B,* 36 (2005): 597–608.

83. M. Bengtsson, N.M. Stark, and K. Oksman. Durability and Mechanical Properties of Silane Cross-Linked Wood Thermoplastic Composites. *Composites Science and Technology*, 67 (2007): 2728–2738.

84. M.S. Huda, L.T. Drzal, A.K. Mohanty et al. Effect of Fiber Surface Treatments on Properties of Laminated Biocomposites from Poly(Lactic Acid) (PLA) and Kenaf Fibers. *Composites Science and Technology*, 68 (2008): 424–432.

85. M.Z. Rong, M.Q. Zhang, Y. Liu et al. Effect of Fiber Treatment on the Mechanical Properties of Unidirectional Sisal-Reinforced Epoxy Composites. *Composites Science and Technology*, 61 (2001): 1437–1447.

86. L.M. Matuana, R.T. Woodhams, J.J. Balatinecz et al. Influence of Interfacial Interactions on the Properties of PVC–Cellulosic Fiber Composites. *Polymer Composites*, 19 (1998): 446–455.

87. J. George, M. S. Sreekala, and S. Thomas. A Review on Interface Modification and Characterization of Natural Fiber Reinforced Plastic Composites. *Polymer Engineering & Science*, 41 (2001): 1471–1485.

88. J. Gironès, M.T.B. Pimenta, F. Vilaseca et al. Blocked Isocyanates as Coupling Agents for Cellulose-Based Composites. *Carbohydrate Polymers*, 68 (2007): 537–543.

89. J. George, S. Thomas, and S.S. Bhagawan. Effect of Strain Rate and Temperature on the Tensile Failure of Pineapple Fiber-Reinforced Polyethylene Composites. *Journal of Thermoplastic Composite Materials*, 12 (1999): 443–464.

90. A. Paul, K. Joseph, and S. Thomas. Effect of Surface Treatments on the Electrical Properties of Low-Density Polyethylene Composites Reinforced with Short Sisal Fibers. *Composites Science and Technology*, 57 (1997): 67–79.

91. V. Botaro and A. Gandini. Chemical Modification of the Surface of Cellulosic Fibres. 2. Introduction of Alkenyl Moieties via Condensation Reactions Involving Isocyanate Functions. *Cellulose*, 5 (1998): 65–78.

92. K.C.M. Nair and S. Thomas. Effect of Aging on the Mechanical Properties of Short Sisal Fibre-Reinforced Polystyrene Composites. *Journal of Thermoplastic Composite Materials*, 16 (2003): 249–271.

93. T. Ohkita and S.H. Lee. Effect of Aliphatic Isocyanates (HDI and LDI) as Coupling Agents on the Properties of Eco-Composites from Biodegradable Polymers and Corn Starch. *Journal of Adhesion Science and Technology*, 18 (2004): 905–924.

94. H. Wang, X. Sun, and P. Seib. Mechanical Properties of Poly(Lactic Acid) and Wheat Starch Blends with Methylenediphenyl Diisocyanate. *Journal of Applied Polymer Science*, 84 (2002): 1257–1262.

95. W. Qiu, F. Zhang, T. Endo et al. Isocyanate as a Compatibilizing Agent on the Properties of Highly Crystalline Cellulose–Polypropylene Composites. *Journal of Materials Science*, 40 (2005): 3607–3614.

96. J.M. Felix and P. Gatenholm. Nature of Adhesion in Composites of Modified Cellulose Fibers and Polypropylene. *Journal of Applied Polymer Science*, 42 (1991): 609–620.

97. S. Mohanty, S.K. Nayak, S.K. Verma et al. Effect of MAPP as a Coupling Agent on the Performance of Jute–PP Composites. *Journal of Reinforced Plastics and Composites*, 23 (2004): 625–637.

98. M. Kazayawoko, J.J. Balatinecz, R.T. Woodhams et al. Effect of Ester Linkages on the Mechanical Properties of Wood Fiber–Polypropylene Composites. *Journal of Reinforced Plastics and Composites*, 16 (1997): 1383–1406.

99. S. Mohanty, S.K. Verma, and S.K. Nayak. Dynamic Mechanical and Thermal Properties of MAPE-Treated Jute–HDPE Composites. *Composites Science and Technology*, 66 (2006): 538–547.

100. P. Zadorecki and P. Flodin. Surface Modification of Cellulose Fibers. II. Effect of Cellulose Fiber Treatment on the Performance of Cellulose–Polyester Composites. *Journal of Applied Polymer Science*, 30 (1985): 3971–3983.

101. P. Zadorecki and T. Rönnhult. An ESCA Study of Chemical Reactions on the Surfaces of Cellulose Fibers. *Journal of Polymer Science Part A*, 24 (1986): 737–745.

102. B.L. Shah, L.M. Matuana, and P.A. Heiden. Novel Coupling Agents for PVC–Wood Flour Composites. *Journal of Vinyl and Additive Technology*, 11 (2005): 160–165.

103. M.J. John, C. Bellmann, and R.D. Anandjiwala. Kenaf–Polypropylene Composites: Effect of Amphiphilic Coupling Agent on Surface Properties of Fibres and Composites. *Carbohydrate Polymers*, 82 (2010): 549–554.

104. M.J. John and R.D. Anandjiwala. Chemical Modification of Flax-Reinforced Polypropylene Composites. *Composites Part A*, 40 (2009): 442–448.

105. M.J. John and S. Thomas. Biofibres and Biocomposites. *Carbohydrate Polymers*, 71 (2008): 343–364.

106. A.K. Bledzki, S. Reihmane, and J. Gassan. Properties and Modification Methods for Vegetable Fibers for Natural Fiber Composites. *Journal of Applied Polymer Science*, 59 (1996): 1329–1336.

107. D.N.S. Hon. Chemical Modification of Lignocellulosic Materials: Old Chemistry, New Approaches. *Polymer News*, 17 (1992): 102–107.

108. S.C.O. Ugbolue. Structure–Property Relationships in Textile Fibres. *Textile Progress*, 20 (1990): 1–43.

109. E.W. Wuppertal. *Die Textilen Rohstoffe (Natur und Chemiefasern)*. Frankfurt: Dr. Spohr/Deutscher Fachverlag, 1981.

110. Y.P. Wang, G. Wang, and H.T. Cheng. Structures of Bamboo Fiber for Textiles. *Textile Research Journal*, 84 (2010): 334–343.

111. N. Chand and P.K. Rohatgi. *Natural Fibres and Their Composites*. Delhi: Periodical Experts Publishers, 1994.

112. R.M. Rowell. Opportunities for Lignocellulosic Materials and Composites. In *Emerging Technologies for Materials and Chemicals*, R.M. Rowell and R. Narayan, Eds. Washington, DC: American Chemical Society, 1992, p. 476.

113. M. Harris. *Handbook of Textile Fibers*. Washington, DC: Harris Research, 1954.

114. T. Zylinski. *Fiber Science*. Warsaw, Poland: National Science Foundation, 1964.

115. W. Von Bergen and Walter. Krauss. *Textile Fiber Atlas: A Collection of Photomicrographs of Common Textile Fibers*. New York: American Wool Handbook Co., 1942.

116. M.S. Sreekala, N.R. Neelakantan, and S. Thomas. Utilization of Short Oil PalmEmpty Fruit Bunch Fiber (OPEFB) as Reinforcement in Phenol–Formaldehyde Resins: Studies of Mechanical Properties. *Journal of Polymer Engineering*, 16 (1996): 265–294.

117. P.S. Mukherjee and K.G. Satyanarayana. Structure and Properties of Some Vegetable Fibers. Part 2: Pineapple Fiber. *Journal of Materials Science*, 21 (1986): 51–56.

118. E.T.N. Bisanda and M.P. Ansell. Properties of Sisal–CNSL Composites. *Journal of Materials Science*, 27 (1992): 1690–1700.

119. D.S. Varma, M. Varma, and I.K. Varma. Coir Fibers. Part 1: Effect of Physical and Chemical Treatment on Properties. *Textile Research Institute,* 54 (1984): 821–832.

120. K. Joseph, L.H.C. Mattoso, R.D. Toledo et al. Natural Fiber Reinforced Composites. In *Natural Polymers and Agrofibers,* E. Frallini, A.L. Leao, and L.H.C. Mattoso, Eds. San Carlos, Brazil: Embrapa, 2000.

121. A.K. Bledzki and J. Gassan. In *Handbook of Engineering Polymeric Materials,* N.P. Cheremisinoff, Ed. New York: Marcel Dekker, 1997.

122. R.M. Rowell. Chemical Modification of Agro-Resources for Property Enhancement. Paper presented at Paper and Composites from Agro-Based Resources Conference, Boca Raton, FL, 1997.

123. D.S. Varma., M. Varma, and I.D. Varma. Thermal Properties of Coir Fibres. *Thermochimica Acta,* 108 (1986): 199–210.

124. R. Zulkifli, M.J.M. Nor, A.R. Ismail et al. Comparison of Acoustic Properties of Coir Fibre and Oil Palm Fibre. *European Journal of Scientific Research,* 33 (2009): 144–152.

125. A.K. Bledzki and J. Gassan. Composites Reinforced with Cellulose-Based Fibres. *Progress in Polymer Science,* 24 (1999): 221–274.

126. W.E. Mortan and J.W.S. Hearle. *Physical Properties of Textile Fibers,* 2nd ed. New York: John Wiley & Sons, 1975.

127. M.R.K. Murali and R.K. Mohan. Extraction and Tensile Properties of Natural Fibres: Vakka, Date, and Bamboo. *Composite Structures,* 77 (2007): 288–295.

128. K. Joseph, L.H.C. Mattoso, R.D. Toledo et al. Natural Polymers and Agrofiber Composites. In *Natural Fiber-Reinforced Composites,* E. Frallini, A.L. Leao, and L.H.C. Mattoso, Eds. San Carlos, Brazil, Embrapa, 2000.

129. J. Holbery and Houston D. Natural Fiber-Reinforced Polymer Composites in Automotive Applications. *Journal of Mechanics,* 58 (2006): 80–86.

130. B.C. Suddell and W.J. Evans. Natura Fibers, Biopolymers, and Biocomposites. In *Natural Fiber Composites in Automotive Applications,* A.K. Mohanty, M. Misra, and L.T. Drzal, Eds. Boca Raton, FL: Taylor & Francis, 2005.

131. M. S. Sreekala, M. G. Kumaran, M. L. Geethakumariamma et al. Environmental Effects in Oil Palm Fiber-Reinforced Phenol–Formaldehyde Composites: Studies on Thermal, Biological, Moisture, and High Energy Radiation Effects. *Advanced Composite Materials,* 13 (2004): 171–197.

132. A. Bismarck, S. Mishra, and T. Lampke. Plant Fibres as Reinforcement for Green Composites. *In Natural Fibres Biopolymers and Biocomposites,* A.K. Mohanthy, M. Misra, and L.T. Drzal, Eds. Boca Raton, FL: Taylor & Francis, 2005.

133. A. Kalam, B.B. Sahari, Y.A. Khalid et al. Fatigue Behaviour of Oil Palm Fruit Bunch Fibre–Epoxy and Carbon Fibre–Epoxy Composites. *Composite Structures,* 71 (2005): 34–44.

134. A.A. Bakar, A. Hassan, and A.F.M. Yusof. The Effect of Oil Extraction of the Oil Palm Empty Fruit Bunch on the Processability, Impact, and Flexural Properties of PVC-U Composites. *International Journal of Polymeric Materials,* 55 (2006): 627–41.

135. D. Maldas, B.V. Kokta, and C. Daneault. Influence of Coupling Agents and Treatments on the Mechanical Properties of Cellulose Fiber–Polystyrene Composites. *Journal of Applied Polymer Science,* 37, (1989): 751–775.

136. A. Valadez-Gonzalez, J.M. Cervantes-Uc, R. Olayo et al. Chemical Modification of Henequén Fibers with an Organosilane Coupling Agent. *Composites Part B*, 30 (1999): 321–331.

137. K.L. Pickering, A. Abdalla, C. Ji et al. The Effect of Silane Coupling Agents on Radiata Pine Fibre for Use in Thermoplastic Matrix Composites. *Composites Part A*, 34 (2003): 915–26.

6

Reinforcement of Polymers by Flax Fibers: Role of Interfaces

Christophe Baley, Antoine Le Duigou, Alain Bourmaud,
Peter Davies, Michel Nardin, and Claudine Morvan

CONTENTS

6.1 Introduction

The use of natural fibers to reinforce composite materials is justified based on sustainable development that enables local resources to be used. Use of these fibers can act as a vector for socioeconomic development in non-industrial countries.

Plant fibers are noble materials that have been produced and recycled naturally for millions of years (e.g., renewal of biodegradable materials via plant growth). Their use as composite reinforcements produces environmental benefits that can be shown by life cycle analyses.[1-4] Components made from biocomposites with thermoplastic matrices can be recycled,[5-7] and if a matrix is a biodegradable biopolymer, it can be composted after grinding.[8,9]

Elementary flax fibers (*Linum usitatissimum*) have a complex structure and composition related to their role within the plants. Their structure is comparable to that of a composite material and thus the interfaces should be identified and studied. Flax fibers[10-15] show high quasi-static tensile properties (E = 60.4 GPa, σ = 1262 MPa, and ε = 2.2% are mean values based on authors' tests over the past 10 years) comparable (specific properties) to those for glass fibers. This justifies their use as reinforcements for composite materials.

These mechanical properties vary and simple explanations for the variabilities are (1) genetic factors that depend on the variety, (2) agronomic practices, and (3) environment. It is possible to correlate via micromechanics expressions the mean tensile properties of elementary flax fibers with the properties of unidirectional plies[16] by accounting for the fiber and matrix properties, the quality of the fiber-matrix adherence, the fiber volume fraction, and the separation and distribution of fibers in the matrix.

Several parameters influence the properties of composite materials. Obviously, the intrinsic mechanical properties of the matrix and fibers play a key role. The fiber–matrix interface strength is also crucial in ensuring load transfer from the matrix to the reinforcement. This property is governed mainly the quality of the fiber–matrix adhesion and the aspect ratio of the fiber reinforcement.[17]

This chapter is devoted to interfaces in composite materials reinforced by flax fibers. First, the structure and organization of flax stem and fibers will be presented, and the influences of these constitutive elements on the mechanical and thermal behaviors of the fibers will be described. An analysis of the surfaces of flax fibers and adherence capacities of the fibers with different polymers will follow. The results of various treatments and the characterization of adherence will also be described. Finally, at the composite scale, the influence of dispersion of reinforcements, the appearance of defects, and the choice of manufacturing route will be discussed.

6.2 Analysis of Flax Stem Structure

6.2.1 Organization of Stem

The fibers within a stem are supporting (reinforcing) elements located around the exterior (Figure 6.1) and assembled in groups in the form of bundles (about 40 bundles of 30 fibers). The mean diameter of a flax fiber is 15 to 20 μm and length is between 5 and 80 mm. An individual flax plant produces 15,000 to 20,000 fibers, yielding 0.3 to 0.5 g of dry fiber.

Examination of the structure of a flax stem provides much information. The stem has the structure of a composite material reinforced by continuous vessels (wood part) and discontinuous fibers (in both cortex and wood). The wind subjects plants to flexural and torsion loads. A bending moment results in tensile and compression strains in stems. Behavior in compression is the weak point of plant support elements. The presence of cohesive bundles of fibers regularly distributed around the outside of a stem (Figure 6.1) limits the risk of micro-buckling in compression zones. Load transfer between fibers is performed by shared lamella between fibers. To extract the fibers, it is necessary to damage the lamella during retting and scutching operations (see Section 6.2.3).

In order to use biomimetics, we must understand the natures of the elements that provide good cohesion (efficient interfaces) to fiber bundles in flax plants. These features will be described in the next section.

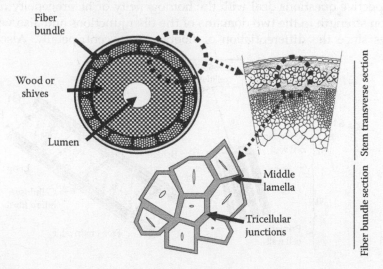

FIGURE 6.1
Stem fiber section and details of elementary fiber bundle.

6.2.2 Role Cell Junctions in Fiber Cohesion within Bundles

In flax stems, fibers tied together by their pectic junctions within a bundle structure constitute a type of composite material (Figure 6.1). The stress is transferred to the reinforcing fibers through their junctions that may be considered pectic polymers. The junctions consisting of two main types of pectins (homogalacturonan and rhamnogalacturonan[18]) are present in two morphological domains: (1) the middle lamella between the primary cell walls of two neighboring fibers and (2) the tricellular junctions of three fibers in the corners.

The pectic compositions of fiber junctions vary based on space and time dimensions during stem growth and adapt to the differentiation stages of the fibers (elongation, expansion, and cell wall thickening). Junction remodeling may be catalyzed by enzymes (e.g., de-esterification of pectins and pectin cross-linking) or occur by apposition of newly synthesized pectic polysaccharides. The strength of the junctions generally increases during the expansion and thickening of the fibers, but the molecular mechanisms depend on the culture environment.

The environment may be the normal sun-and-rain alternation, abiotic stress caused by drought,[19] or high levels of pollution.[20] In normal conditions, homogalacturonans are de-esterified, thus increasing their affinity for calcium ions, leading to strong calcium bridges. The rhamnogalacturonans, especially in the middle lamella, react chemically with those of the primary cell walls (Figure 6.2). During a stress event, phenolics may be polymerized as lignin points throughout the junctions.[21]

Prospective questions deal with the homogeneity or heterogeneity of the cohesion strength in the two domains of the fiber junctions and also within bundles since the differentiation of fibers occurs centripetally. Also, for

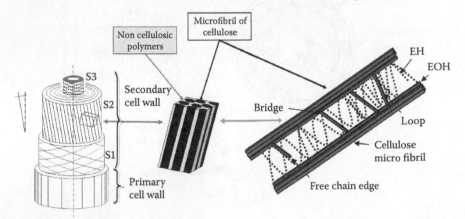

FIGURE 6.2
Flax fiber showing helix arrangement of cellulose fibrils and chemical structures in S2 layer. EH = matrix non-cellulosic polymer. EOH = structural non-cellulosic polymer.

the facilitation of the fiber extraction from stems, questions about cohesion strength at the boundaries between fiber bundles and the neighboring tissues (cortical on one side and phloem on the other; see next section) must be considered. Transmission electronic microscopy analyses coupled with immunocytogold labeling indicate similar pectic components in the various domains but with different proportions and linkages. Most data were collected at the University of Rouen in France. Local mechanical data are being collected using nanoindentation approaches.[22]

6.2.3 Impact of Retting on Fiber Cohesion

Retting—known as dew retting in Northern Europe—consists of enzyme-catalyzed degradation of cell junctions and cell walls caused by the combined actions of moisture and heat.[23] Retting may be divided into two spatio-temporal steps. First, the epidermis and cortical tissues of stems layered on soil are colonized by fungi that release several enzymes that degrade pectins and polysaccharides.[24,25] As a result, the epidermis and cortical parenchyma are degraded partially and the fiber bundles split.[26] Calcium pectates are considered resistant to retting and are thus limiting factors.[27]

In a second step, the fungi colonize the fiber bundles where the complex pectins present in the fiber junctions may be degraded partially.[28,29] The two-step retting process facilitates fiber decortication.[30,31]

On the other hand, if continuous rains at the latter retting stage prevent harvest, the fungi continue their development within the elementary fibers and initiate rotting. Indeed, the fungi undergo the degradation of the cellulose microfibrils in the successive cell wall layers, due to the secretion of glucanase and cellulose enzymes.[32] This step is called over-retting. For practical and economical reasons, it is hardly possible to harvest and store wet flax straw[33] because of mold development during storage. Therefore, retting is a critical agricultural process that determines the divisions of the bundles and the surface quality of the so-called technical fibers.

There are alternative methods to field dew retting. Water retting is no longer an option due to pollution of the water by substances created by the anaerobic degradation of the plants. Various enzyme retting processes have been developed.[34–36] They are very efficient in avoiding fiber rot and maintain the mechanical properties of the elementary fibers; However, these alternative methods are not used because they are costly.

Some variation in degrees of retting can be reduced by (1) modifying the subsequent mechanical treatments of the straw (breaking, scutching, and hackling) that separate the bast fibers from woody cores and (2) degumming treatment during wet spinning.[37] Partially degraded bundles constitute the so-called technical fibers containing a mixture of 1 to 10 fibers with loose cohesion between themselves and the debris of cortical tissues.

6.2.4 Impacts of Retting on Physicochemical Properties of Technical and Elementary Fibers

Physicochemical properties of technical fibers obviously depend on the retting, scutching, hackling, degumming, and possibly surfacting treatments. Chemical composition, swelling, vapor water sorption, surface properties, and thermal stability have been reported to vary with the degree of retting and the subsequent mechanical and/or chemical treatments.[31,35,38-40] Also, the mechanical properties of single fibers are influenced by the degree of retting.[35,41] Greater retting can result in higher Young's modulus and strength.[35,40,41] There is little variation in failure strains. The scatter in values tends to decrease with increasing retting.[41] Conversely, over-retting and most chemical treatments negatively impact tensile properties.[39,42]

All these properties along with the surface aspect (clean or containing cortical residues; smooth or rough) and morphology (ratio of perimeter to surface) impact the establishment of interfaces when manufacturing composites with polymers.

6.3 Composition and Structure of Single Flax Fiber

The mechanical properties of composites depend on both the characteristics of reinforcing fibers and the quality of the interface with the matrix resin. This section focuses on the role of interfaces and thus on the surface conditions of the fibers. Properties depend on the agroindustrial processes of extraction of technical fibers (see above) and also on the development history of the fibers within plant stems. The next section describes the development of fibers—the composition of the interfiber connections, the primary cell walls that constitute the surface state of the fiber, and the synthesis of the secondary walls (particularly the S2 layer). This discussion serves as a model for the construction of a composite material with a matrix and reinforcement of similar compositions and optimized interfaces in terms of quality and quantity, resulting in excellent fiber properties.

6.3.1 Development of Fibers

Throughout the growth of a plant, fibers are formed around the exterior in a coordinated manner with the criblo-vascular bundles (conducting tissues) of the stem. This occurs in four distinct steps: (1) fiber cell multiplication at the top of the stem,[43] (2) elongation in the top 3 to 5 cm of the stem, (3) fiber expansion and thickening of the walls below the snap point zone [44] in which stem stiffness increases significantly, and (4) structuring of the wall leading in some cases to a reduction in fiber diameter[28,45] after seed formation in certain varieties and/or as a function of the environment.

During the elongation phase that lasts 3 to 5 days per fiber,[46] the cell extends by 5 to 20 mm per day [47] and may reach up to 100 mm[48] mainly by so-called intrusive growth. This enables the fibers to penetrate through the shared lamella between neighboring cells.[47] If the intrusive elongation starts with a tip growth leading to a spindle-like shape of the fiber, the main phase of elongation occurs by diffuse symplasmic growth along the entire cell.[46,47,49] This means that the final length of the cells is not obtained by classical division (multiplication of chromosomes and cellular division by construction of a septum, then laying down cell walls on both sides of the septum). Final length is achieved uniquely by multiplication of the nuclei and then by cellular elongation. At the end, each fiber can possess several tens of nuclei.[47] The length of nuclei is generally included from 200 to 500 nm.[46] The number of nuclei is a function of the elementary fiber length, which is not constant.

This extension of the fibers stops progressively at the snap point.[48] The fiber, like all cells, is bounded along its length by a continuous primary cell wall whose cellulose mesofibrils* are aligned parallel to the growth direction. The thickness of the primary wall is 200 to 500 nm[50] and usually consists of one third pectins, one third hemicelluloses (definition below), and one third cellulose.[51] The length of nuclei is generally included from 200 to 500 nm.[46] The number of nuclei is a function of the elementary fiber length which is not a constant.

The interfiber junctions are acidified and reinforced by calcium links and by the establishment of covalent bonds between RG-I of the shared lamella and primary cell wall (see above). The surfaces of the fibers after retting are particularly rich in these RG-I whose linear density of negative charges is on average 1 nm^{-1}. The third phase corresponds to the synthesis of a secondary cell wall that fills the cellulosic fibers of the flax layer by layer. A study of the synthesis of this first layer illustrates the strategy of fibers in producing a composite material.

6.3.2 S1 to S3 Layers in Secondary Wall

Gorshova et al.[44] have shown that the initiation of the first deposit is preceded by a strong secretion of non-cellulosic polysaccharides (NCPSs), glycoproteins rich in galactose and with high molecular weights (>1 M). It has been proposed that the environment of such an extracellular matrix in which the cellulose mesofibrils are excreted plays an important role in determining the final size and orientation of the mesofibrils.[45]

---

* The terminology used to describe cell wall reinforcement geometry is confusing. Based on a diameter of 2 to 4 nm, Burgert et al.,[52] Altaner and Jarvis,[53] and Peterlin and Ingram[54] used the *cellulose microfibril* term. Fraztl[55] cited *cellulose fibrils*. The *macrofibril* term is used for structures with diameters between 100 and 200 nm by Altaner and Jarvis[53] and Peterlin and Ingram.[54] Bos et al.[56] in their work on the compression behaviors of fibers use the *mesofibril* term for assembly of microfibrils of dimensions around 200 nm. We will use this latter designation.

The use in transmission electron microscopy of a gold-labeled cellobio-hydrolase enzyme specific to crystalline cellulose with Patag's reactive and a marker of non-crystalline cellulose enabled His et al.[57] to show that this first layer gradually becomes more structured and bound to the primary cell walls. The cellulose of the first layer appears highly crystalline at the moment when the second layer is deposited. Thus, this first layer progressively brings a certain stiffness to the fiber.[58]

The molecular processes involved in this stiffening were reviewed in previous works.[45,59] The enzymes secreted at the same time as the non-cellulosic polymers catalyze the remodeling of the latter, allowing the cellulosic nanofibrils (2 to 4 nm sections for ~36 molecules) to group and form microfibrils (20 to 40 nm sections). These microfibrils are surfaced by hemicelluloses and encrusted within a remodeled pectin matrix.

Whatever the process, the formation of mesofibrils and their association with the non-cellulosic polymers is complex and results in multilayer fiber-reinforced composite wall structures composed of three main layers S1 (~0.5 to 2 µm), S2 (~5 to 10 µm), and S3 (~0.5 to 1 µm). S2, the main structural layer, is responsible for most of the mechanical performance and the physical properties of the fibers (Figure 6.2). S3 surrounds a cavity known as a lumen whose volume is inversely proportional to that of the secondary cell wall.

6.3.3 Influence of Microstructure on Tensile Behaviors of Fibers

As described above (Figure 6.2), the stiffness of the secondary cell walls arises from the presence of highly oriented cellulose mesofibrils within the S2 layer. These mesofibrils represent more than 80% of the weight, and they form a high performance composite with the NCPSs in which they are inserted. Various theories based on the work of mechanical engineers and biologists enable the arrangements of the different cell wall constituents to be explained.

Hearle[60] showed that the cellulose mesofibrils are closely linked to the amorphous polysaccharides and form a non-covalent network with the hemicelluloses (principally glucomananes and galactanes according to Gorshkova and Morvan[61]). The hemicelluloses are bonded to the cellulose mesofibrils by hydrogen bonds that, because of their large numbers, create strong associations with the cellulose. Some of the hemicelluloses can form intercellulose bridges and/or entanglements (and untangle) with pectic matrices. Beyond a certain shear stress threshold, these networks can fail due to breakage of hydrogen bonds between the constituents.

Other studies of wood have revealed a stick–slip (Velcro®)-type mechanism in the secondary cell walls.[62-64] According to Burgert[65] and Keckes et al.,[62] the mechanism is partly reversible after loading and other hydrogen bonds between cellulose and hemicellulose will be built elsewhere. Altaner and Jarvis[64] proposed a slightly different model in which, in addition to hydrogen bonds with cellulose, the hemicellulose chains can make bridges or loops with cellulose mesofibrils.

Comparative work on the fiber structures of different varieties of flax[66] have shown stiffness differences related to cellulose contents in cell walls. The varieties with the highest cellulose contents were represented with cellulose fibrils and less well-developed matrices between fibrils. The bridges between glucomananes chains have been proposed to explain the stiffness differences observed.[67]

In flax fibers, other components such as arabinogalactan proteins rich in glycine (GRP) and β-1-4 galactans have also been proposed to play the role of "hemicelluloses" to achieve compatibility between the celluloses mesofibrils and the pectic chains (RG-I) to which they are linked. This represents a true interphase between the matrix and the cellulose and guaranteeing the cohesion of the system.[45,67,68] The configuration can be reinforced by homogalacturonans that will consolidate the interphase of the galactans and the pectic matrix.[66,69] This analysis was based on models of in vitro interactions studied by Zykwinska et al.[70] and primary cell wall models elaborated by Cosgrove.[71]

A schematic synthesis of the structure of the S2 cell wall is proposed in Figure 6.2. Recent studies[12] have shown the prime importance of the organization of the plant cell wall components and their structures in mechanical performance. More important than sufficient quantities of cellulose to ensure good performance, the ratio between the structural polysaccharides and matrix within the S2 layer presents a high level of correlation with the stiffness of flax fibers.

In conclusion, cellulose mesofibrils (> 80%) within cellulosic flax fibers arranged in spirals[72] within an amorphous polysaccharide matrix consisting of hemicelluloses and pectins rich in galactose[68] and oriented at an angle of around 8 to 10 degrees[73] with respect to the fiber axis constitute a composite material structure, with many levels of interfaces. The multilayer growth of this material involving several remodeling mechanisms results in good mechanical behavior of the individual fibers and may provide a source of inspiration for developing biomaterial composites. Open questions concern the detailed compositions of composites throughout the successive layers and the strengths of their interfaces.

6.3.4 Influence of Wall Composition on Thermal Behaviors of Fibers

The nature of the constituents of cell walls, the temperature, and water content all exert strong influences on the mechanical properties of plant fibers. Hearle[60] underlined the importance of water as early as 1963 and proposed that the Young's moduli of plant fibers will decrease as water content increases; water will not penetrate into the crystalline zones. He attributed the drop in properties to a modification of interfibrillar constituents.

According to Altaner et al.,[64] water enters the polysaccharide network and causes the cellulose fibrils to separate. Thus, the water acts to plasticize the cell walls[74] that will transform the polysaccharides, particularly the water-soluble glycans, into a quasi-fluid gel. The carboxylic groups of pectins are mainly responsible for water absorption by the fibers. Water content is thus conditioned by the degree of fiber retting.

Baley et al.[75] demonstrated that water causes losses of interactions (and hydrogen bonds) between cell wall components. Water, by plasticizing the non-crystalline polysaccharides of the cell walls and absorption by the fibers, generally leads to an increase in failure strain. After heating for 14 hr at 105°C,[76] despite retention of part of the water linked to the polysaccharide constituents of the S2 layer, the tensile strength of the fibers was strongly affected by this drying cycle. In addition, early appearance of a damage threshold was noted on modulus-versus-strain plots. The water removal embrittled the constituents,[75] especially the hydrated gel network formed by the polysaccharide matrix.[77] These phenomena will affect the interactions of mesofibrils and pectin matrices.

It is possible to examine the thermal behaviors of fibers using thermo-gravimetric analysis (TGA). Van de Velde et al.[40] used the technique to show that weight losses can be related to loss of water and degradation of the cellu-lose and its non-cellulosic components (mainly hemicelluloses and pectins). Gassan and Bledzki[78] have shown that the mechanical properties of jute and flax fibers start to be affected by the temperature at around 170°C.

Placet[79] performed dynamic mechanical analyses (DMA) on hemp fiber bundles at different temperatures. A drop in storage modulus and increase in loss factor were revealed between 20 and 200°C. These changes may be caused by transition phenomena within the polymers that constitute plant cell walls (cellulose, hemicelluloses, and pectins). Their rheological behavior is controlled by the intrinsic responses of these three main constituents.

Bundles have also been subjected to cycling for several hours under iso-thermal conditions. An accommodation phenomenon revealed as an increase in stiffness was noted. In addition, the fatigue behavior was strongly affected; bundles can support more than 60,000 cycles at 20°C compared to only 6,500 at 220°C. The accommodation phenomenon may be explained by the progressive decrease in the microfibril angle observed by various authors using x-ray diffraction during tensile tests.[62,65]

In general, an increase in temperature is harmful to the mechanical perfor-mance of a plant fiber and causes decreases in short- and long-term proper-ties. Thermal treatments inspired by wood technologies such as Duralin® can be applied to plant fibers[80] to reduce their hydrophilic behavior and improve thermal resistance.

6.4 Surface of Flax Fiber

The surface of a material is generally composed of the first planes of atoms that interact with the neighboring environment. The multilayer structure of

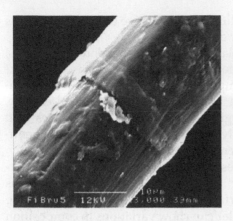 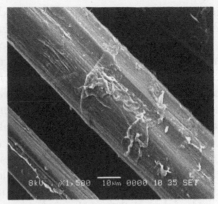

FIGURE 6.3
Scanning electron microscopy images of flax fibers showing different surface states according to degree of retting.

flax fibers[81] implies that the outer layer is the fiber surface and corresponds to the primary cell wall whose thickness varies between 200 and 500 nm.[82] This structure was described and its composition is heterogeneous because it includes cellulose, hemicelluloses, and pectins. The compositions of fiber surfaces also vary according to degrees of retting.[82–84]

Certain residues exert strong influences on surfaces and their roughness levels (Figure 6.3).[85] Le Duigou et al.[86] used atomic force microscopy (AFM) to evaluate correctly retted fibers (Hermès variety) and measured roughness varying from 10.8 to 14.3 nm (compared to a few nanometers for a glass fiber) over an area of 1 μm^2. When a larger surface area (25 μm^2) was analyzed, the roughness reached 35 nm.[87] Cleaning in water for 72 hr or applying a thermal cycle can reduce this roughness to ~5 nm.

In general, these residues adhere poorly to the surfaces of fibers and cannot serve as mechanical anchor points for polymers. They can play an important role in energy dissipation during interface failures, but in the present case are more likely to be zones of low cohesion and failure initiation.

Among the surface properties of flax fibers, the surface energy γ_S is a thermodynamic property that indicates the aptitude of a surface to create interactions (van der Waals and/or acid–base) with a matrix. Wetting of fibers by a liquid matrix—an essential step in the manufacture of composites—is controlled mainly by the surface energies of the two components. This wetting is a necessary but not sufficient condition to achieve good fiber–matrix adherence.

For plant fibers, the heterogeneity of the surface composition, roughness, porosity, adsorption of liquids, geometrical variations, and very small dimensions of samples make the characterization of surfaces with the usual techniques (goniometer,[88,89] Wilhelmy balance,[90] Washburn tube,[91,92] and inverse gas chromatography[93]) very difficult. Innovative techniques such as AFM with a colloidal probe are under development.[94]

Surface energies, involving non-polar components (London interactions) and polar components (Debye interactions, Keesom, and acid–base) are influenced strongly by flax fiber surface composition. Van Hazendonk et al.[95] demonstrated by selective extraction that each constituent (cellulose, hemicellulose, pectin) plays a role in the surface energies of fibers. For example, waxes with low surface energies (~25 mJ/m²) globally reduce γs, whereas pectins and hemicelluloses show intermediate surface energies (around 40 mJ/m²).

The surface energy of a fiber increases as a function of the cellulose proportion (~60 mJ/m²). These values should be treated with caution, however, as the polysaccharides present on the surfaces of flax fibers, even if they belong to the same family (pectins, for example), can show variations in composition (methylesterification, polymerization, and crystallinity).[68,96,97] Thus, the application of thermodynamic wetting theory should be performed carefully.[98]

To obtain high performance and durable composites, surface treatments can also be envisaged for composites in addition to mechanical treatments during fiber extraction (scutching and hackling). Surface treatments have been the subject of recent reviews.[99–101] Fiber treatments can be divided into three distinct categories: (1) intended to retain the properties of the plant cell walls but with different objectives; (2) individualization of the fibers; and (3) cleaning and/or modification of surfaces. These treatments are discussed in other chapters.

6.5 Adherence of Flax Fiber and Polymer

6.5.1 Choice of Micromechanics Test to Characterize Adherence

Several micromechanics tests are available for studying the shear strength of a fiber–matrix interface. The most common tests are fragmentation, compression, microindentation, pull-out, and microdroplet debonding.[102] The method described here is the debonding of a microdroplet of matrix from a single fiber[103] (Figure 6.4). It allows the evaluation of fiber–matrix adherence and the mechanisms of debonding and wetting, the latter via contact angle measurement.

The first step is to extract a single fiber from a bundle and place a microdroplet on it. The method is straightforward for a thermoset matrix that is liquid at room temperature. For a thermoplastic that is solid at room temperature, the simplest approach is to use a thin polymer thread to make a knot around the fiber, then apply a heating cycle to melt the polymer and form a droplet.

Before characterization, each specimen is examined under an optical microscope to measure the geometry (fiber diameter, droplet diameter, and

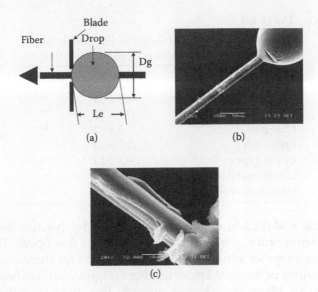

FIGURE 6.4
Fiber–matrix interface characterization using microdrop debonding (A and B) and peeling (C).

bonded length), check the symmetry of the droplet, and search for defects. For the debonding force to be proportional to the bonded area, the bonded length should be short (fewer than 250 μm).[104] In reality, a length shorter than 150 μm is generally used to reduce the likelihood of fiber breakage, particularly during long thermal treatments. When the specimens are placed in the test machine, the knives on the micrometric plateau (Figure 6.4) are adjusted under a microscope. The debonding of the droplets is observed by microscope and the force and displacement are recorded continuously. The tensile loading rate is 0.1 mm/min.

Various authors analyzed stress distribution in microdroplets by finite element methods,[102,105–108] photoelasticity,[102] and Raman spectroscopy.[109,110] These studies have shown that the loading is complex (shear stresses are not constant along the fiber–matrix interface) and residual thermal stresses are not negligible. The simplest analysis is based on the determination of apparent interfacial shear stress (IFSS). A uniform shear stress distribution along the bonded length is assumed. This simplifying assumption allows comparisons of different fiber–matrix combinations.

6.5.2 Flax–Polymer Bonding: Experimental Examples

The interfacial strength between plant fibers and polymers is often considered poor in the literature. Many authors consider that adherence of a hydrophilic fiber to a hydrophobic polymer cannot give a good result. This section presents interface test results obtained from flax and glass fibers with three different polymer matrices (Table 6.1). For the three polymers studied

TABLE 6.1

Examples of Apparent Interfacial Shear Strength
(IFSS) Determined by Microdroplet Debonding

Material	IFSS (MPa)	Reference
PLLA (Naturworks®)–flax	16.4 ± 3.8	86
PLLA (Biomer®)–flax	15.3 ± 3.3	132
PLA–glass	7.5 ± 2.3	86
Epoxy–flax	22.7 ± 0.8	88
Epoxy–glass	29.3 ± 2.4	133
Unsaturated polyester–flax	14.2 ± 0.4	88
Unsaturated polyester–glass	16.1 ± 0.5	88

(poly(L)-lactic acid [PLLA], epoxy, and unsaturated polyester), the adherence was of the same order of magnitude for glass and flax fibers. These values are high; flax–polymer adherence is not a weakness for these materials.

Partial peeling of internal layers of fibers (Figure 6.4) has been observed after drying of fibers or aging in water.[86] This type of damage can be explained by an evolution of the fiber microstructure. The peeling also affects post-debonding friction behavior, particularly for flax–PLA.[86] Thus, for a composite material reinforced by flax fibers, even though the fiber–matrix interface is of prime importance, it is also absolutely essential to consider the internal interfaces between layers within the fibers.

6.5.3 Influences of Various Treatments on Flax–PLA System

The treatments presented in this section were applied to investigate the role of the fiber–matrix interface. The apparent interfacial shear stress (IFSS) of a flax–PLA system is controlled by the many interactions between the fiber surface and matrix, e.g., van der Waals forces, acid–base reactions, and hydration.[111,112] The residual stresses induced during a manufacturing cycle arising from the differences in thermal expansion coefficients and changes in crystallinity will affect the IFSS.[86]

Cleaning a fiber surface with water will dissolve certain components[11] and in combination with reduced surface roughness[86] results in a small increase in IFSS (Table 6.2). A high level of roughness of a fiber is therefore not an explanation for good interfacial properties, as has been proposed. An increase in the amount of peeling during debonding was noted after this treatment.

A drying cycle at 105°C implies a loss of water of around 6% and a reduction in surface roughness. Here the IFSS drops by around 20% (Table 6.2). The migration of low molecular weight components, the modification of these components, and surface porosity may explain this change.[113–117] Again an increase in fiber peeling was noted. Adding a debonding agent to the fiber surface reduced the IFSS as expected.[86]

TABLE 6.2

Examples of Apparent Interfacial Shear Stresses of Flax–PLA Systems Determined by Microdroplet Debonding

Material	Treatment	IFSS (MPa)	Reference
Treatments on fibers	Untreated fibers	15.3 ± 3.3; 16.4 ± 3.8	Table 6.1
	Water (72 hr)	18.8 ± 3.2	86
	Heat (105°C at14 hr)	13.6 ± 3.3	
	Release agent	4.8 ± 1.3	
Treatments on composites	Air cooling	15.3 ± 3.3	132
	10°C/min cooling	18.2 ± 1.8	
	1°C/min cooling	22.2 ± 3.4	
	Annealing (50°C at 72 hr)	9.9 ± 1.5	

The cooling rate of PLA after molding plays an important role. The slower the cooling, the higher the IFSS. This change may be caused by residual stresses, but the matrix morphology near the fiber also changes. An annealing step below the glass transition temperature allows a relaxation of residual stresses, underlining their role in interfacial adherence.

6.6 Flax–Polymer Interfaces and Composite Materials

6.6.1 Damage Mechanisms

In a composite material reinforced by glass fibers or polyacrylonitrile [PAN] carbon, observation of fracture surfaces reveals four damage mechanisms: axial fiber breakage, matrix and fiber–matrix interface cracking, and fiber debonding. For a polymer reinforced by flax fibers, additional mechanisms can be observed: fiber breakage in the transverse direction[118] and in shear between and within secondary cell walls.[76]

If a polymer is reinforced by fiber bundles, cracks can also appear in the shared lamella.[118] Stress concentrations related to the presence of kink bands[119] and modifying load transfers between fibers and matrix can also appear in composites. These defects can develop when a fibrillar structure is loaded in compression,[10] for example, during operations to separate the fibers from plants.

6.6.2 Influence of Fiber Separation on Mechanical Properties of Composites

Fibers within plants are grouped around the outside of the stems in the form of bundles (Figure 6.1). The bundles consist of tens of fibers linked together by shared lamella. In spite of the retting and mechanical separation operations the cohesive nature of these bundles results in the retention of a certain number through to the final composite. The flow conditions during manufacture can also cause the formation of clusters of fibers. The bundles may cause a significant drop in the reinforcement aspect ratio and hence a drop in their reinforcing capacity.

Bos[120] has shown how flax fiber bundles can affect the mechanical properties of composites. Rask et al.[121] studied the damage processes of unidirectional flax–polypropylene (PP) via x-ray diffraction and concluded that well-separated fibers are recommended for composite reinforcements. Andersons and Joffe[122] announced similar conclusions by demonstrating that a probabilistic model, assuming perfect separation and regular spacing of fibers, yields an upper limit of strength for unidirectional (UD) flax composites.

The influence of the dispersion of single fibers on the mechanical properties of flax–PP composites has also been shown.[16] Morphological analyses highlighted the importance of a hackling step for fiber dispersion. This process reduces the number of bundles in the final composite. Fibers subjected to combing have been used to manufacture composites with separated fiber contents similar to those found with glass fibers.

The study of damage in PP composites reinforced by injected flax fibers[123] has shown, as for traditional composites, a significant skin–core effect.[124,125] Shearing effects close to the mold result in high orientation of the fibers in the flow direction in these regions. In the core, as a result of divergent flow and lower shear, the orientation is more isotropic. This results in more fiber bundles in the core that may cause premature failure in this region and drops in composite strength and failure strain. Tensile tests performed inside a scanning electron microscope (Figure 6.5) confirmed this effect, showing crack propagation within clusters of fibers.

Fiber clusters and bundles promote damage initiation and fracture propagation; to improve composite quality and performance, it is necessary to improve the separation and dispersion of fibers by optimizing both the extraction procedures and manufacturing conditions.

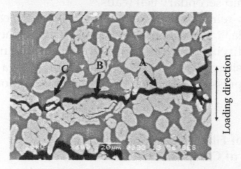

FIGURE 6.5
Development of crack in UD polyester–flax within bundle of flax fiber (A), at the fiber–matrix interface (B), and failure within a single fiber (C) [1].

6.6.3 Choice of Manufacturing Process

Composite materials reinforced by natural fibers can be manufactured using classic processes; References 8 and 126 show various fiber–matrix combinations. During processing, the reinforcing fibers are subjected to high temperatures that affect their properties.

Van de Velde et al.[40] showed that water removal and degradation of waxes occurred from 120°C while pectin degradation started around 180°C. For Gassan et al.,[78] the degradation of flax fibers and a reduction in their toughness appeared from 170°C due to depolymerization and chain breakage. Nanoindentation tests performed in situ on flax fiber cross-sections before and after manufacture[38] revealed a strong drop in fiber stiffness after injection molding. A transformation by film stacking was shown to be less severe for the plant walls due to lower shear levels associated with this process. Similar conclusions about process–mechanical property relationships were reached by Le Duigou et al.[7] for flax–PLA biocomposites.

The use of aggressive processes involving high shear stresses strongly influences fiber morphology,[127] causing a reduction in the lengths and diameters of the fiber bundles. The reduction in diameter is interesting for the aspect ratio, but the drop in lengths must be limited. Thus high rotation speeds induce improved fiber dispersion but lower aspect ratios. Technologies such as Buss co-mixers can be used to reduce fiber changes[128] thanks to adapted profiles and delayed fiber introduction.

Residence time is an important parameter. Too short a duration results in poor dispersion while long residence can favor matrix and fiber degradation.[129] Increasing the rotation speed of the extrusion screw reduces residence time but can in some cases induce drops in modulus and flexural and tensile strengths of composites.[127] Finally, the temperatures of extruders and injection processes are important. A high temperature favors fiber dispersion through higher matrix fluidity.[130] However, as shown by Li et al.[131] with flax–polyethylene composites, a compromise must be found as high temperatures can cause permanent damage to plant fibers.

6.7 Conclusions

As a result of their role as a support framework for plant stems, flax fibers fulfill mechanical functions in nature. A first interface is observed when they are assembled in bundles with shared lamella. These interfaces are damaged during retting. This step thus directly influences surface properties (composition, roughness, etc.) and volume (composition, mechanical properties, etc.) of the fibers. The subsequent mechanical treatments (scutching and hackling) do not cause a complete separation of the fibers.

Additional operations can be performed to separate the fibers more completely, clean the surface, or add a function to optimize the interfacial area and homogenize the composite material. A second interface can be defined at this scale.

A flax fiber is a multilayer structure consisting of a primary wall on a secondary wall reinforced by cellulose mesofibrils revealing another level of interface. The S2 layer of the secondary wall controls the mechanical properties of the fibers at a nanometric level. The load transfer between fibrils is ensured by constituents acting as an interphase. This is the final interface zone that affects composite properties.

Water also plays a key role, acting as a plasticizer. However, composite manufacture requires a thermal cycle that causes a loss of water. Thus, manufacturing can embrittle a material. This chapter presented a detailed description of the flax fiber structure, fiber growth, and reinforcement capacity when combined with polymers. The existence of interfacial zones within cell walls, between fibers within bundles, and between fibers and matrix polymers reveals the complexity of the damage mechanisms present in natural fiber composites. An approach different from conventional composite production is required because the regions that are most sensitive to damage are not necessarily the fiber–matrix interfaces but may be within the fibers.

References

1. C. Baley. Fibres Naturelles De Renfort Pour Matériaux Composites. In *Techniques de l'Ingénieur*, 2013.
2. A. Le Duigou, P. Davies, and C. Baley. Environmental Impact Analysis of the Production of Flax Fibres to Be Used as Composite Material Reinforcement. *Journal of Biobased Material and Bioenergy*, 5 (2011): 1–13.
3. A. Le Duigou, P. Davies, and C. Baley. Replacement of Glass–Unsaturated Polyester Composites by Flax–PLLA Biocomposites: Is It Justified? *Journal of Biobased Material and Bioenergy*, 5 (2012): 466–482.
4. A. Le Duigou, J.M. Deux, P. Davies et al. PLLA–Flax Mat–Balsa Bio-Sandwich: Environmental Impact and Simplified Life Cycle Analysis. In *Applied Composite Materials*. Heidelberg: Springer, 2012, pp. 1–5.
5. A. Bourmaud and C. Baley. Investigations on the Recycling of Hemp and Sisal Fibre Reinforced Polypropylene Composites. *Polymer Degradation and Stability*, 92 (2007): 1034–1045.
6. A. Bourmaud and C. Baley. Rigidity Analysis of Polypropylene–Vegetal Fibre Composites after Recycling. *Polymer Degradation and Stability*, 94 (2009): 297–305.
7. A. Le Duigou, I. Pillin, A. Bourmaud et al. Effect of Recycling on Mechanical Behaviour of Biocompostable Flax/Poly(L-Lactide) Composites. *Composites Part A*, 39 (2008): 1471–1478.

8. C. Baley, Y. Grohens, and I. Pillin. Etat de L'Art Sur Les Matériaux Composites Biodégradables. *Revue des Composites et des Matériaux Avancés*, 14 (2005): 135–166.

9. E. Bodros, I. Pillin, N. Montrelay et al. Could Biopolymers Reinforced by Randomly Scattered Flax Fibre Be Used in Structural Applications? *Composites Science and Technology*, 67 (2007): 462–470.

10. C. Baley. Analysis of the Flax Fibre Tensile Behaviour and Analysis of the Tensile Stiffness Increase. *Composites Part A*, 33 (2002): 939–948.

11. A. Bourmaud, C. Morvan, and C. Baley. Importance of Fiber Preparation to Optimize the Surface and Mechanical Properties of Unitary Flax Fiber. *Industrial Crops and Products*, 32 (2010): 662–667.

12. A. Lefeuvre, A. Bourmaud, L. Lebrun et al. Study of the Yearly Reproducibility of Flax Fiber Tensile Properties. *Industrial Crops and Products*, 50 (2013): 400–407.

13. K. Charlet, C. Baley, C. Morvan et al. Characteristics of Hermès Flax Fibres as a Function of Their Location in the Stem and Properties of the Derived Unidirectional Composites. *Composites Part A*, 38 (2007): 1912–1921.

14. K. Charlet, J.P. Jernot, M. Gomina et al. Influence of an Agatha Flax Fibre Location in a Stem on Its Mechanical, Chemical and Morphological Properties. *Composites Science and Technology*, 69 (2009): 1399–1403.

15. I. Pillin, A. Kervoelen, A. Bourmaud et al. Could Oleaginous Flax Fibers Be Used as Reinforcement for Polymers? *Industrial Crops and Products*, 34 (2011): 1556–1563.

16. G. Coroller, A. Lefeuvre, A. Le Duigou et al. Effect of Flax Fibre Individualisation on Tensile Failure of Flax–Epoxy Unidirectional. *Composites Part A*, 51 (2013): 62–70.

17. A. Kelly and W.R. Tyson. Tensile Properties of Fibre-Reinforced Metals: Copper–Tungsten and Copper–Molybdenum. *Journal of the Mechanics and Physics of Solids*, 13 (1965): 329–338.

18. C. Andème-Onzighi, R. Girault, I. His et al. Immunocytochemical Characterization of Early-Developing Flax Fiber Cell Walls. *Protoplasma*, 213 (2000): 235–245.

19. S.B. Chemikosova, N.V. Pavlencheva, O.P. Gur'yanov et al. Effect of Soil Drought on the Phloem Fiber Development in Long-Fiber Flax. *Russian Journal of Plant Physiology*, 53 (2006): 656–662.

20. O. Douchiche, A. Driouich, and C. Morvan. Spatial Regulation of Cell Wall Structure in Response to Heavy Metal Stress: Cadmium-Induced Alteration of Methylesterification Pattern of Homogalacturonan. *Annals of Botany*, 105 (2010): 481–491.

21. A. Day, K. Ruel, G. Neutelings et al. Lignification in the Flax Stem: Evidence for an Unusual Lignin in Bast Fibers. *Planta*, 222 (2005): 234–245.

22. A. Bourmaud and C. Baley. Nanoindentation Contribution to Mechanical Characterization of Vegetal Fibers. *Composite Part B*, 43 (2012): 2861–2866.

23. A.E. Brown, H.S.S. Sharma, and D.L.R. Black. Relationship between Pectin Content of Stems of Flax Cultivars, Fungal Cell Wall-Degrading Enzymes and Pre-Harvest Retting. *Annals of Applied Biology*, 109 (1986): 345–351.

24. F.P. De Franca, J.A. Rosemberg, and A.M. De Jesus. Retting of Flax by *Aspergillus niger*. *Applied Microbiology*, 17 (1969): 7–9.

25. W.J.M. Meijer, N. Vertregt, B. Rutgers et al. Pectin Content as a Measure of the Retting and Rettability of Flax. *Industrial Crops and Products*, 4 (1995): 273–284.

26. D.E. Akin, G.R. Gamble, W.H. Morrison et al. Chemical and Structural Analysis of Fibre and Core Tissues from Flax. *Journal of the Science of Food and Agriculture*, 72 (1996): 155–165.
27. H.S.S. Sharma. Chemical Retting of Flax Using Chelating Compounds. *Annals of Applied Biology*, 113 (1988): 159–165.
28. C. Andème-Onzighi, R. Girault, I. His et al. Immunocytochemical Characterization of Early-Developing Flax Fiber Cells. *Protoplasma*, 213 (2000): 235–245.
29. A. Jauneau, C. Morvan, F. Lefebvre et al. Differential Extractability of Calcium and Pectic Substances in Different Wall Regions of Epicotyl Cells in Young Flax Plants. *Journal of Histochemistry and Cytochemistry*, 40 (1992): 1183–1189.
30. R.W. Kessler, U. Becker, R. Kohler et al. Steam Explosion of Flax: A Superior Technique for Upgrading Fibre Value. *Biomass and Bioenergy*, 14 (1998): 237–249.
31. H.S.S. Sharma, G. Faughey, and G. Lyons. Comparison of Physical, Chemical, and Thermal Characteristics of Water-, Dew-, and Enzyme-Retted Flax Fibers. *Journal of Applied Polymer Science*, 74 (1999): 139–143.
32. A.E. Brown and H.S.S. Sharma. Spoilage of Flax Straw by *Ceraceomyces sublaevis*. *Transactions of the British Mycological Society*, 86 (1986): 173–175.
33. B.E. Pallesen. Quality of Combine-Harvested Fibre Flax for Industrial Purposes Depends on the Degree of Retting. *Industrial Crops and Products*, 5 (1996): 65–78.
34. D.E. Akin. Linen Most Useful: Perspectives on Structure, Chemistry, and Enzymes for Retting Flax. *ISRN Biotechnology*, 2013 (2013): 23.
35. S. Alix, L. Lebrun, S. Marais et al. Pectinase Treatments on Technical Fibres of Flax: Effects on Water Sorption and Mechanical Properties. *Carbohydrate Polymers*, 87 (2012): 177–185.
36. A. Jauneau, F.C. Rihouey, and C. Morvan. Les Traitements Biologiques Du Lin. *Biofutur*, 167 (1997): 34–37.
37. M. Fausten. Méthodes De Préparation Et De Qualification Des Fibres De Lin. In *Journées d'échanges Franco-Allemandes sur le lin*. Rouen, 1992.
38. A. Bourmaud and C. Baley. Effects of Thermomechanical Processing on the Mechanical Properties of Biocomposite Flax Fibers Evaluated by Nanoindentation. *Polymer Degradation and Stability*, 95 (2010): 1488–1494.
39. C. Morvan, A. Jauneau, A. Flaman et al. Degradation of Flax Polysaccharides with Purified Endo-Polygalacturonase. *Carbohydrate Polymers*, 13 (1990): 149–163.
40. K. Van De Velde and E. Baetens. Thermal and Mechanical Properties of Flax Fibres as Potential Composite Reinforcement. *Macromolecular Materials and Engineering*, 286 (2001): 342–349.
41. N. Martin, N. Mouret, P. Davies et al. Influence of the Degree of Retting of Flax Fibers on the Tensile Properties of Single Fibers and Short Fiber–Polypropylene Composites. *Industrial Crops and Products*, 49 (2013): 755–767.
42. C. Baley, F. Busnel, Y. Grohens et al. Influence of Chemical Treatments on Surface Properties and Adhesion of Flax Fibre–Polyester Resin. *Composites Part A*, 37 (2006): 1626–1637.
43. K. Esau. *Anatomy of Seed Plants*. New York: John Wiley & Sons, 1977.
44. T.A. Gorshkova, S.B. Chemikosova, V.V. Lozovaya et al. Turnover of Galactans and Other Cell Wall Polysaccharides during Development of Flax Plants. *Plant Physiology*, 114 (1997): 723–729.
45. T. Gorshkova and C. Morvan. Secondary Cell Wall Assembly in Flax Phloem Fibres: Role of Galactans. *Planta*, 223 (2006): 149–158.

46. A.V. Snegireva, M.V. Ageeva, S.I. Amenitskii et al. Intrusive Growth of Sclerenchyma Fibers. *Russian Journal of Plant Physiology*, 57 (2010): 342–355.
47. M.V. Ageeva, B. Petrovská, H. Kieft et al. Intrusive Growth of Flax Phloem Fibers Is of Intercalary Type. *Planta*, 222 (2005): 565–574.
48. T.A Gorshkova, V.V. Sal'nikov, S.B Chemikosova et al. Snap Point: A Transition Point in *Linum usitatissimum* Bast Fiber Development. *Industrial Crops and Products*, 18 (2003): 213–221.
49. A.V. Snegireva, M.V. Ageeva, V.N. Vorob'ev et al. Plant Fiber Intrusive Growth Characterized by NMR Method. *Russian Journal of Plant Physiology*, 53 (2006): 163–168.
50. H. Bos. *Potential of Flax Fibres as Reinforcement for Composite Materials*. Technische Universiteit Eindhoven, 2004.
51. T.A. Gorshkova, V.V. Salnikov, N.M. Pogodina et al. Composition and Distribution of Cell Wall Phenolic Compounds in Flax (*Linum usitatissimum* L.) Stem Tissues. *Annals of Botany*, 85 (2000): 477–486.
52. I. Burgert. Exploring the Micromechanical Design of Plant Cell Walls. *American Journal of Botany*, 93 (2006): 1391–1401.
53. C.M. Altaner and M.C. Jarvis. Modelling Polymer Interactions of the Molecular Velcro Type in Wood under Mechanical Stress. *Journal of Theoretical Biology*, 253 (2008): 434–445.
54. A. Peterlin and P. Ingram. Morphology of Secondary Wall Fibrils in Cotton. *Textile Research Journal*, 40 (1970): 345–354.
55. P. Fratzl, I. Burgert, and H.S. Gupta. On the Role of Interface Polymers for the Mechanics of Natural Polymeric Composites. *Physical Chemistry, Chemical Physics*. 6 (2004): 5575–5579.
56. H.L. Bos, M.J.A. Van Den Oever, and O.C.J.J. Peters. Tensile and Compressive Properties of Flax Fibres for Natural Fibre Reinforced Composites. *Journal of Materials Science*, 37 (2002): 1683 1692.
57. I. His, C. Andème-Onzighi, C. Morvan et al. Microscopic Studies on Mature Flax Fibers Embedded in LR White: Immunogold Localization of Cell Wall Matrix Polysaccharides. *Journal of Histochemistry & Cytochemistry*, 49 (2001): 1525–1535.
58. U. Kutschera. The Growing Outer Epidermal Wall: Design and Physiological Role of a Composite Structure. *Annals of Botany*, 101 (2008): 615–621.
59. T. Gorshkova, N. Brutch, B. Chabbert et al. Plant Fiber Formation: State of the Art, Recent and Expected Progress, and Open Questions. *Critical Reviews in Plant Sciences*, 31 (2012): 201–228.
60. J.W.S Hearle. The Fine Structures of Fibers and Crystalline Polymers. III: Interpretation of the Mechanical Properties of Fibers. *Journal of Applied Polymer Science*, 7 (1963): 1207–1223.
61. T. Gorshkova and C. Morvan. Secondary Cell Wall Assembly in Flax Phloem Fibres: Role of Galactans. *Planta*, 223 (2006): 149–158.
62. J. Keckes, I. Burgert, K. Fruhmann et al. Cell Wall Recovery after Irreversible Deformation of Wood. *Nature Materials*, 2 (2003): 810–884.
63. P. Fratzl, I. Burgert, and H.S. Gupta. On the Role of Interface Polymers for the Mechanics of Natural Polymeric Composites. *Physical Chemistry Chemical Physics*, 6 (2004): 5575–5579.

64. C.M. Altaner and M.C. Jarvis. Modelling Polymer Interactions of the Molecular Velcro Type in Wood under Mechanical Stress. *Journal of Theoretical Biology,* 253 (2008): 434–445.

65. I. Burgert. Exploring the Micromechanical Design of Plant Cell Walls. *American Journal of Botany,* 93 (2006): 1391–1401.

66. S. Alix, E. Philippe, C. Morvan, and C. Baley. Putative Role of Pectins in the Tensile Properties of Flax Fibres: A Comparison between Linseed and Flax Fibers Varieties. In *Pectins and Pectinases.* Wageningen, 2008.

67. S. Alix, E. Philippe, C. Morvan et al. Putative Role of Pectins in the Tensile Properties of Flax Fibres: A Comparison between Linseed and Flax Fibres Varieties. In *Pectins and Pectinases.* Wageningen, Netherlands, 2008.

68. C. Morvan, C. Andème-Onzighi, R. Girault et al. Building Flax Fibres: More Than One Brick in the Walls. *Plant Physiology and Biochemistry,* 41 (2003): 935–944.

69. C. Morvan. Le Lin, Les Fibres Cellulosiques: Leurs Parois: Variabilité Et Défauts. In *Ecole d'été.* Lorient, 2010.

70. A. Zykwinska, J.F. Thibault, and M.C. Ralet. Organization of Pectic Arabinan and Galactan Side Chains in Association with Cellulose Microfibrils in Primary Cell Walls and Related Models Envisaged. *Journal of Experimental Botany,* 58 (2007): 1795–1802.

71. D.J. Cosgrove. Growth of the Plant Cell Wall. *Nature Reviews,* 6 (2005): 850–861.

72. T. Gorshkova, O. Gurjanov, P. Mikshina et al. Specific Type of Secondary Cell Wall Formed by Plant Fibers. *Russian Journal of Plant Physiology,* 57 (2010): 328–341.

73. A.K. Bledzki and J. Gassan. Composites Reinforced with Cellulose Based Fibres. *Progress in Polymer Science,* 24 (1999): 221–274.

74. J.F. Vincent. From Cellulose to Cell. *Journal of Experimental Biology,* 202 (1999): 3263b–3268b.

75. C. Baley, C. Morvan, and Y. Grohens. Influence of the Absorbed Water on the Tensile Strength of Flax Fibers. *Macromolecular Symposia,* 222 (2005): 195–202.

76. C. Baley, A. Le Duigou, A. Bourmaud et al. Influence of Drying on the Mechanical Behaviour of Flax Fibres and Their Unidirectional Composites. *Composites Part A,* 43, (2012): 1226–1233.

77. R. Girault, F. Bert, C. Rihouey et al. Galactans and Cellulose in Flax Fibres: Putative Contributions to Tensile Strength. *International Journal of Biological Macromolecules,* 21 (1997): 179–188.

78. J. Gassan and A.K. Bledzki. Thermal Degradation of Flax and Jute Fibers. *Journal of Applied Polymer Science,* 82 (2001): 1417–1422.

79. V. Placet. Characterization of the Thermomechanical Behaviour of Hemp Fibres Intended for the Manufacturing of High Performance Composites. *Composites Part A,* 40 (2009): 1111–1118.

80. N.E. Zafeiropoulos, C.A. Baillie, and F.L. Matthews. A Study of Transcrystallinity and Its Effect on the Interface in Flax Fibre-Reinforced Composite Materials. *Composites Part A,* 32 (2001): 525–543.

81. J.W.S. Hearle. The Fine Structure of Fibers and Crystalline Polymers. III: Interpretation of the Mechanical Properties of Fibers. *Journal of Applied Polymer Science,* 7 (1963): 1207–1223.

82. H.L. Bos. Potential of Flax Fibres as Reinforcement for Composite Materials. Thesis, Eindhoven (2004).

83. N.E. Zafeiropoulos, C.A. Baillie, and J.M. Hodgkinson. Engineering and Characterisation of the Interface in Flax Fibre–Polypropylene Composite Materials. II: Effect of Surface Treatments on the Interface. *Composite Part A,* 33 (2002): 1185–1190.

84. N.E. Zafeiropoulos, P.E. Vickers, C.A. Baillie et al. Experimental Investigation of Modified and Unmodified Flax Fibres with XPS, ToF-SIMS, and ATR-FTIR. *Journal of Materials Science,* 38 (2003): 3903–3914.

85. A. Bismarck, I. Aranberri-Askargorta, J. Springer et al. Surface Characterization of Flax, Hemp and Celullose Fibers: Surface Properties and Water Uptake Behavior. *Polymer Composites,* 23 (2002): 872–894.

86. A. Le Duigou, A. Bourmaud, E. Balnois et al. Improving the Interfacial Properties between Flax Fibres and PLLA by a Water Fibre Treatment and Drying Cycle. *Industrial Crops and Products,* 39 (2012): 31–39.

87. F. Busnel, E. Balnois, C. Baley et al. Influence Des Traitements Chimiques Sur Les Propriétés De Surface Des Fibres De Lin: Approche Nano Et Macroscopiques. *Revue des composites et matériaux avancés-renforcemnt des polymères par des fibres végétales,* 16 (2006): 75–87.

88. C. Baley, F. Busnel, Y. Grohens et al. Influence of Chemical Treatments on Surface Properties and Adhesion of Flax Fibre–Polyester Resin. *Composite Part A,* 37 (2006): 1626–1637.

89. Y.Y.J. Heng, F.D. Pearse, F. Thielmann et al. Methods to Determine Surface Energies of Natural Fibres: A Review. *Composite Interface,* 14 (2007): 581–604.

90. K. Van De Welde and P. Kiekens. Wettability of Natural Fibres Used as Reinforcement for Composites. *Die Angewandte Makromolekulare Chemie,* 271 (1999): 87–93.

91. G. Cantero, A. Arbelaiz, R. Llano-Ponte et al. Effects of Fibre Treatment on Wettability and Mechanical Behaviour of Flax–Polypropylene Composites. *Composite Science and Technology,* 63 (2003): 1247–1254.

92. I. Aranberri-Askargorta, T. Lampke, and A. Bismarck. Wetting Behavior of Flax Fibers as Reinforcement for Polypropylene. *Journal of Colloid and Interface Science,* 263 (2003): 580–589.

93. N. Cordeiro, C. Gouveia, and M.J. John. Investigation of Surface Properties of Physico-Chemically Modified Natural Fibres Using Inverse Gas Chromatography. *Industrial Crops and Products,* 33 (2011): 108–115.

94. G. Raj, E. Balnois, C. Baley et al. Probing Cellulose–Polylactic Acid Interactions in Model Biocomposite by Colloidal Force Microscopy. *Colloids and Surfaces A,* 352 (2009): 47–55.

95. J.M. Van Hazendonk, J.C. Van Der Putten, J.T.F. Keurentjes et al. Simple Experimental Method for the Measurement of the Surface Tension of Cellulosic Fibres and Its Relation with Chemical Composition. *Colloids and Surfaces A,* 81 (1993): 251–261.

96. A. Zykwinska, J.F. Thibault, and M.C. Ralet. Competitive Binding of Pectin and Xyloglucan with Primary Cell Wall Cellulose. *Carbohydrate Polymers,* 74 (2008): 957–961.

97. G.C. Davies, and D.M. Bruce. Effect of Environmental Relative Humidity and Damage on the Tensile Properties of Flax and Nettle Fibers. *Textile Research Journal*, 68 (1998): 623–629.

98. S. Barsberg and L.G. Thygesen. Nonequilibrium Phenomena Influencing the Wetting Behavior of Plant Fibers. *Journal of Colloid and Interface Science*, 234 (2001): 59–67.

99. M.M. Kabir, H. Wang, K. T. Lau et al. Chemical Treatments on Plant-Based Natural Fibre-Reinforced Polymer Composites: An Overview. *Composites Part B*, 43 (2012): 2883–2892.

100. O. Faruk, A.K. Bledzki, H.P. Fink et al. Biocomposites Reinforced with Natural Fibers: 2000–2010. *Progress in Polymer Science*, 37 (2012): 1552–1596.

101. X. Li, L.G. Tabil, and S. Panigrahi. Chemical Treatments of Natural Fiber for Use in Natural Fiber-Reinforced Composites: A Review. *Journal of Polymers and the Environment*, 15 (2007): 25–33.

102. P.J. Herrera-Franco and L.T. Drzal. Comparison of Methods for Measurement of Fibre–Matrix Adhesion in Composites. *Composite Science and Technology*, 23 (1992): 2–27.

103. B. Miller. Microbond Method for Determination for Determination of the Shear Strength of a Fiber–Resin Interface. *Composite Science and Technology*, 28 (1987): 17–32.

104. C.H. Liu and J.A. Nairn. Analytical and Experimental Methods for a Fracture Mechanics Interpretation of the Microbond Test Including the Effects of Friction and Thermal Stresses. *International Journal of Adhesion and Adhesives*, 19 (1999): 59–70.

105. J.T. Ash, W.M. Cross, D. Svalstad et al. Finite Element Evaluation of the Microbond Test: Meniscus Effect, Interphase Region, and Vise Angle. *Composite Science and Technology*, 63 (2003): 641–651.

106. A. Hodzic, S. Kalyanasundaram, J.K. Kim et al. Application of Nano-Indentation, Nano-Scratch, and Single Fibre Tests in Investigation of Interphases in Composite Materials. *Micron*, 32 (2001): 765–775.

107. H. Kessler, T. Schüller, W. Beckert et al. Fracture-Mechanics Model of the Microbond Test with Interface Friction. *Composite Science and Technology*, 59 (1999): 2231–2242.

108. E. Pisanova, S. Zhandarov, E. Mäder et al. Three Techniques of Interfacial Bond Strength Estimation from Direct Observation of Crack Initiation and Propagation in Polymer–Fibre Systems. *Composites Part A*, 32 (2001): 435–443.

109. R.J. Day and J.V. Cauich-Rodrigez. Investigation of the Micromechanics of the Microbond Test. *Composite Science and Technology*, 58 (1997): 907–914.

110. S.J. Eichhorn and R.J. Young. Deformation Micromechanics of Natural Cellulose Fibre Networks and Composites. *Composite Science and Technology*, 63 (2003): 1225–1230.

111. G. Raj, E. Balnois, C. Baley et al. Role of Polysaccharides on Mechanical and Adhesion Properties of Flax Fibres in Flax–PLA Biocomposite. *International Journal of Polymer Science* 20 (2011).

112. G. Raj, E. Balnois, C. Baley et al. Colloid Force Measurements between Cellulose and Polylactic Acid (PLA). *RCMA*, 18 (2008): 177–183.

113. J. Tissaoui. Effects of Long-Term Creep on the Integrity of Modern Wood Structures. Blacksburg: Virginia Polytechnic Institute and State University, 1996.

114. P. Sernek. Comparative Analysis of Inactivated Wood Surface. Blacksburg: Virginia Polytechnic Institute and State University, 2002.

115. A.W. Christiansen. How Overdrying Reduces Bonding to Phenol Formlaldehyde Adhesives: A Critical Review of the Literature. I: Physical Responses. *Wood and Fiber Science*, 22 (1990): 441–449.

116. A.W. Christyiansen. How Overdrying Reduces Bonding to Phenol Formaldehyde Adhesives: A Critical Review of the Literature. II: Chemical Reactions. *Wood and Fiber Science*, 23 (1991): 69–84.

117. P. Gérardin, M. Petric, M. Petrissans et al. Evolution of Wood Surface Free Energy after Heat Treatment. *Polymer Degradation and Stability*, 92 (2007): 653–657.

118. C. Baley, Y. Perrot, F. Busnel et al. Transverse Tensile Behaviour of Unidirectional Plies Reinforced with Flax Fibres. *Materials Letters*, 60 (2006): 2984–2987.

119. M. Hugues, G. Sèbe, J. Hague et al. Investigation into the Effects of Micro-Compressive Defects on Interphase Behaviour in Hemp–Epoxy Composites Using Half-Fringe Photoelasticity. *Composite Interfaces*, 7 (2000): 13–29.

120. H. Bos, M. Van Den Oever, and O. Peters. Tensile and Compressive Properties of Flax Fibres for Natural Fibre Reinforced Composites. *Journal of Material Science*, 37 (2002): 1683–1692.

121. M. Rask, B. Madsen, B.F. Sørensen et al. In Situ Observations of Microscale Damage Evolution in Unidirectional Natural Fibre Composites. *Composites Part A*, 10 (2012): 1639–1649.

122. J. Andersons and R. Joffe. Estimation of the Tensile Strength of an Oriented Flax Fiber-Reinforced Polymer Composite. *Composites Part A*, 42 (2011): 1229–1235.

123. A. Bourmaud, G. Ausias, G. Lebrun et al. Observation of the Structure of a Composite Polypropylene–Flax and Damage Mechanisms under Stress. *Industrial Crops and Products*, 43 (2013): 225–236.

124. R.S. Bay and C.L. Tucker. Fiber Orientation in Simple Injection Moldings. I: Theory and Numerical Methods. *Polymer Composites*, 13 (1992): 317–331.

125. R.S. Bay and C.L. Tucker. Fiber Orientation in Simple Injection Moldings. II: Experimental Results. *Polymer Composites*, 13 (1992): 332–341.

126. Q. Zhou, L. Greffe, M. Baumann et al. Use of Xyloglucan as a Molecular Anchor for the Elaboration of Polymers from Cellulose Surfaces: General Route for the Design of Biocomposites. *Macromolecules*, 38 (2005): 3547–3549.

127. B. Mano, J.R. Araújo, M.A.S. Spinacé et al. Polyolefin Composites with Curaua Fibres: Effect of the Processing Conditions on Mechanical Properties, Morphology, and Fibre Dimensions. *Composite Science and Technology*, 70 (2010): 29–35.

128. K. Shon and J.L. White. Comparative Study of Fiber Breakage in Compounding Glass Fiber-Reinforced Thermoplastics in a Buss Kneader, Modular Co-Rotating and Counter-Rotating Twin Screw Extruders. *Polymer Engineering & Science*, 39 (1999): 1757–1768.

129. P.V. Joseph, K. Joseph, and S. Thomas. Effect of Processing Variables on the Mechanical Properties of Sisal Fiber-Reinforced Polypropylene Composites. *Composite Science and Technology*, 59 (1999): 1625–1640.

130. B. Siaotong, L. Tabil, S. Panigrahi et al. Determination of Optimum Extrusion Parameters in Compounding Flax Fibre-Reinforced Polyethylene Composites. *American Society of Agricultural and Biological Engineers,* 2006, 1–12.

131. X. Li, L. Tabil, S. Panigrahi et al. Study on Flax Fibre-Reinforced Polyethylene Biocomposites by Injection Moulding. *Applied Engineering in Agriculture,* 25 (2009): 525–531.

132. A. Le Duigou, P. Davies, and C. Baley. Interfacial Bonding of Flax–Poly(L-Lactide) Biocomposites. *Composite Science and Technology,* 70 (2010): 231–239.

133. C. Baley, C. Morvan, and Y. Grohens. Influence of Absorbed Water on the Tensile Strength of Flax Fibers. In *Polymer–Solvent Complexes and Intercalates V,* Y. Grohens, Ed. Weinheim: Wiley-VCH, 2004.

7

Effects of Reinforcing Fillers and Coupling Agents on Performances of Wood–Polymer Composites

Diène Ndiaye, Mamadou Gueye, Ansou Malang Badji, Coumba Thiandoume, Anicet Dasylva, and Adams Tidjani

CONTENTS

7.1 Introduction

The incorporation of various types of fillers into polymer matrices is an interesting route for producing polymer composites with different properties. Considerable research has focused extensively on the use of natural fibers as reinforcement materials in thermoplastic matrices. The utilization of vegetable fibers is driven by growing market trends in terms of environmental impact. The most common composites with natural fibers are made with polyolefin (polyethylene [PP] and polypropylene [PE]) and polyvinyl chloride (PVC) matrices and wood fibers as reinforcements.

In recent years, the utilization of fibers and powders derived from agricultural sources has attracted the attention of many researchers mainly due to their low densities, low costs, non-abrasiveness, high filling levels, renewable and non-toxic organic fibers (compared to glass and carbon fibers), biodegradability, and, most importantly, availability from renewable sources.[1-5]

The WPC acronym indicating wood–polymer composites covers a very wide range of composite materials utilizing plastics ranging from PP to PVC and binders and fillers ranging from wood flour to natural fibers (e.g., flax).[6] WPCs are experiencing a growing market demand. For this reason, it is logical to study ways to enhance their performance attributes.

The use of various reinforcing fillers in composites and their effects should be classified to provide the industry with better understanding of properties and enable delivery of better products. Extensive research and product development has focused on reinforcing polyolefins and other non-biodegradable plastics, but research on reinforcing biodegradable polymers has been because of the incompatibility of the two entities. The application of biodegradable polymers has focused on the medical, agricultural, and consumer packaging industries.[7,8] The lignocelluloses fibers used in polyolefins include cellulose, wood, flax, *Cannabis sativa* (hemp), jute, pine, sisal, rice husk, sawdust, wheat straw paper, mud, coir, kenaf, cotton, pineapple leaf, bamboo, and palm.[9]

Reinforcements such as wood have been successful in improving the mechanical properties of thermoplastic composites. Extensive efforts to develop biodegradable composites using renewable resources are continuing in an attempt to replace the non-biodegradable synthetic polymers commonly used for composites.[10] Composites made from blends of thermoplastics and natural fibers have gained popularity in a variety of applications because they combine the desirable durability of plastics with the cost effectiveness of natural fibers as fillers or reinforcing agents.[11] The product has the aesthetic appearance of wood and the processing capability of thermoplastics. It also exhibits good performance in humid conditions.

Another attraction is the easy availability of these materials from natural wastes. For this reason, WPCs are used in a variety of innovative applications in the automotive (door panels, trims, and trunk liners), construction (decking, fencing, siding, windows, door frames, interior paneling, and decorative trim), and packaging industries. Natural fibers possess excellent sound absorbing efficiency, resist shattering, and have better energy management characteristics than glass fiber-reinforced composites. The incorporation of natural fibers into polycaprolactone (PCL) has been shown to enhance the biodegradability of the resulting composites.[12] However, the biodegradability of the resulting product is limited if all polymers are not biodegradable.[13]

Properties of WPCs depend on the characteristics of fillers and matrices, chemical interactions of wood fibers and polymers, humidity, absorption, and processing conditions. Wood fibers also demonstrate some disadvantages such as fiber–polymer incompatibility and low temperature of thermal degradation due to the presence of cellulose and hemicellulose. The inclusion of hemicellulose, lignin, and other impurities in organic reinforcements causes a lack of adhesion between fibers and polymers. The low thermal degradation limits the allowed processing temperature to less than 200°C. The compatibility of wood fibers and polymer matrices constitutes one important factor in the production of WPCs with improved mechanical properties.[14–16]

These disadvantages of wood led some researchers to use other materials as reinforcements. In recent years, natural fillers such as jute, kenaf, hemp, sisal, pineapple, and rice husk have been used successfully to improve the mechanical properties of thermoplastic composites.[17] The hydrophobic nature of PP poses a potential problem in achieving good fiber–matrix adhesion in these systems. However, cellulose is an inherently hydrophilic polymer because of the numerous hydroxyl groups it contains. To alleviate this obstacle, chemical compatibilizers or coupling agents have been developed to alter the surfaces of hydrophilic cellulose fibers to improve the dispersion and interfacial adhesion between fiber and matrix.

The most popular coupling agents used by many researchers are maleic anhydride-grafted polyolefins such as polyethylene (MAPE) and polypropylene (MAPP).[15,18] Many in-depth studies have elucidated the mechanisms of adhesion between MAPP-treated wood fibers and a PP matrix that cause improvement by forming linkages between the OH groups of wood and maleic anhydride.[14,15]

Coupling agents can form chemical bonds on the surfaces of wood and the interfaces between wood and polymer and can easily infiltrate wood surfaces, leading to lower surface tension of the wood material.[19–21] One such compatibilizer is maleic anhydride-grafted polypropylene (MAPP), a waxy polymer system that has proven useful in the production of cellulose-reinforced PP composites.[21,22] MAPP is formed by reacting maleic anhydride (MA) with PP in the presence of an initiator to produce PP chains with pendant MA groups.

The PP fraction of MAPP can entangle and co-crystallize with the unmodified PP, while the maleic anhydride groups can bond to the hydroxyl (-OH) groups on the fibers. When mixed with cellulose, the hydroxyl group breaks one of the C-O bonds in the MA group and forms a new bond between one of the carbons from the MAPP group and the oxygen from the cellulose. The resulting chemical bond between the oxygen of the cellulose and the carbon of the MA group enhances the bond of fiber and matrix.[23,24] MAPP can also compensate for insufficient break-up forces such as low shear stress by reducing the interfacial tension between cellulose and PP during processing. This leads to finer dispersion of fibers throughout the system.[25]

The addition of MAPP also reduces water absorption (> 20%), making these materials more suitable for use in damp environments.

Some studies focused on other coupling agents such as silanes[26,27] and isocyanates.[28,29] All cases of use of coupling agents produced significant improvement of the mechanical properties of the final composites. Our study was inspired by the principle of ecological replacement of inorganic polluting substances by agricultural products such as wood fibers, rice straws, and dried distiller grains with soluble (DDGS) as reinforcements in polymer matrices. DDGS are co-products of a dry grind corn process that produces fuel ethanol from corn. Considerable efforts have been made to utilize the co-products obtained during the processing of corn, wheat, soybeans, wheat gluten, and soy proteins for composite applications.[30]

The exponential growth of the ethanol industry in recent years has created a problem because the supplies of DDGS have increased dramatically. Currently, the ethanol industry's only use for DDGS is in animal feeds.[32,33] The huge production rate of DDGS exceeds the rate of use in animal feeds. Most research publications about DDGS focus on its feed applications.[34] A successful separated DDGS fiber-based WPC would benefit dry plants, wood composite manufacturers, and the rural economies that produce them by increasing revenues. As technology develops in the area of utilization of natural by-products, classical wood fibers have shown strong potential as reinforcements in polymer matrix composites.

Composites were prepared by extruding DDGS with polypropylene and phenolic resin 38. The need for materials with environmentally friendly characteristics has increased due to limited natural resources and increasing environmental regulation.[34-36] It is interesting to note that natural fibers such as jute, rice husk, DDGS, banana, and sisal are abundantly available in developing countries in Africa but are not optimally utilized.

Risk husk (RH) is a coating or protective layer formed during the growth of rice grains. These shells are removed during the refining of rice. They have low commercial value because the SiO_2 and the fibers they contain have poor nutritional value and are used minimally in animal feeds. RH has been a problem for rice farmers due to its resistance to decomposition in the ground, difficult digestion, and low nutritional value for animals.[31] Therefore, the development of new polymer composites filled with RH is a very interesting approach.

In this study, polypropylene was reinforced with pine wood, rice husk, and DDGS by-products of ethanol production. The objectives of this study were to (1) develop composites with PP matrices and pinewood, rice husk, and DDGS reinforcing fillers and (2) explore the performances and limitations of these reinforcing fillers on the mechanical, thermal, and morphological properties of the composites. The results are discussed below. We also explain how to provide competitive alternative materials to natural wood that is becoming increasingly expensive and diminishing in supply.

7.2 Experimental Details

7.2.1 Materials

Polypropylene (PP) was selected as a matrix material due to its versatility to accept numerous types of fillers and reinforcements. PP in the form of pellets donated by Eastman Chemical Co. (Kingsport, TN) was used as the matrix. It had a melt flow index of 5.2 g/10 min (at 190°C and a 2.16 kg load) and a density of 0.910 g/cm³. To increase interface adhesion, PP grafted with maleic anhydride was added as a coupling agent to all the composites studied. MAPP (G-2010) was supplied by Eastman Chemical Co. Wood filler particles of approximately 425 mm (40-mesh size) were donated by American Wood fibers (Schofield, WI) and consisted predominantly of Ponderosa pine, maple, oak, and spruce hard woods.

Rice husk flour (RHF) was used as a natural fiber reinforcement. The fibers were hand chopped to an average length of 30 mm. The fibrous material was then ground into flour with a Thomas-Wiley mill (particles exited through a 60-mesh screen). Rice husk was treated first with dilute hydrochloric acid (10% for 60 min) to hydrolyze and remove the low molecular weight hemicellulose. After that, the rice husk was alkali-treated with 0.5 N NaOH solution (described below) to remove silica and lignin from the rice husks. NaOH is known to be an effective reagent for reducing the protein content of organic fibers. The rice husk flour was oven dried at 100°C for 24 hr to adjust the moisture content to less than 2 wt% and then stored over desiccant before compounding.

DDGS was obtained as a commercial animal feed pellet product. It was milled with a Thomas-Wiley grinder with the particles exiting through a 2-mm-diameter stainless screen and collected into a 1.81-liter Mason jar. It is well known that the presence of oil in DDGS acts as a plasticizer to lubricate relative molecular motion, hence lowering the modulus. We extracted DDGS with hexane (to remove oils), then with dichloromethane (to remove polar extractibles) employing a Soxhlet extractor. This treatment removes certain amounts of oils and other impurities, making the surface rougher. The residue protein components in the separated DDGS fibers degraded at the melting temperature of the thermoplastic polymers.

To start alkali treatment, the rice husks and DDGS were soaked in a 0.5 N NaOH solution at room temperature, maintaining a ratio of 500 ml alkali solution to 50 g reinforcing filler. Reinforcements were kept immersed in the alkali solution for 2 hr. The fibers were washed several times with distilled water to remove any NaOH on the fiber surfaces, and then they were neutralized with dilute acetic acid and again washed with distilled water. Figure 7.1 shows the different reinforcements used in our study.

Alkali treatment is well known to increase mechanical properties. The process alters the chemical contents of crude fibers by removing lignin, pectin,

FIGURE 7.1
Photos of reinforcements: (a) pine wood, (b) rice husk, and (c) DDGS.

and hemicelluloses and also changing the material state from hydrophilic to hydrophobic. The large amount of hemicellulose lost made the fibers lose their cementing capacity, causing them to separate from each other and making them finer.[10,37,38]

7.2.2 Compounding and Processing

Before compounding, wood flour, rice husks, and DDGS were dried in an oven for at least 48 hr at 105°C to expel moisture before blending with PP and then stored in polyethylene bags. First, the PP was put into a high-intensity Papenmeier mixer (TGAHK20), and then reinforcement was added after the PP reached its melting temperature. The mixing process took 10 min on average. After blending, the compounded materials were stored in a sealed plastic container. Several formulations were produced with amounts of PP, wood flour, rice husks, DDGS, and 5% of MAPP in all the samples (Table 7.1). For the extraction of volatile and harmful gases, the hood was open.

For the mechanical property experiments, test specimens were molded in a Cincinnati Milacron 33 reciprocating screw-injection molder (Batavia, OH). The nozzle temperature was set to 204°C. The extrudate in the form of strands was cooled in the air and pelletized. The resulting pellets were dried at 105°C for 24 hr before they were injection-molded into ASTM test specimens for tensile (Type I, ASTM D 638) and Izod impact strength testing. The dimensions of the specimens for the tests were $120 \times 3 \times 12$ mm^3 (length × thickness × width).

TABLE 7.1

Compositions of Reinforcement–Polypropylene Composites

Sample	PP (wt%)	Reinforcement (50%)	MAPP (wt%)
PP	100	—	—
PM	95	—	5
PWM	45	Wood	5
PRM	45	Rice husk	5
PDM	45	DDGS	5

7.2.3 Property Evaluations

7.2.3.1 Scanning Electron Microscopy (SEM)

The state of dispersion of the wood flour in the polymeric matrix was analyzed with SEM. A FEI Quanta 400 microscope (Hillsboro, Oregon) working at 30 kV was used to obtain microphotographs of the fractured surfaces of the composites. Samples were cut in liquid nitrogen to avoid deformation of the surfaces.

7.2.3.2 Differential Scanning Calorimetry (DSC)

DSC is widely used to characterize the thermal properties of WPCs. It can measure important thermoplastic properties, including melting temperature (T_m), heat of melting, degree of crystallinity, crystallization, and the presence, composition, and compatibility of recyclates, nucleating agents, plasticizers, and polymer blends. Thermal analysis of the WPC samples was carried out on a PerkinElmer calorimeter (Shelton, Connecticut) with the temperature calibrated with indium.

All DSC measurements were performed with samples of about 9.5 ± 0.1 mg under a nitrogen atmosphere with a flow rate of 20 ml/min. Three replicates were run for each specimen. All samples were subjected to the same thermal history with the following thermal protocol that was slightly modified from the one reported by Valentini et al.[39]

1. The samples were heated from 40 to 180°C at a rate of 20°C/min to remove any previous thermal history.

2. The samples were cooled from 180 to 40°C at a rate of 10°C/min to detect the crystallization temperature (T_c).

3. Finally, the samples were heated from 40 to 180°C at a rate of 10°C/min to determine T_m. The heat of fusion (ΔH_m) was calculated from the thermograms obtained during the second heating and normalized on the basis of the weight fraction of PP present in a sample. The values of ΔH_m were used to estimate χ, which was adjusted for each sample in χ_{cor} (%) based on the percentage of PP in the composite. Crystallinity (χ_{cor}) was estimated according to the following equation:

$$\chi_{cor}(\%) = \frac{\Delta H_m}{\Delta H_0 \cdot X_{PP}} \tag{7.1}$$

ΔH_m and ΔH_0 are the heats (J/g) of melting of composite and 100% crystalline PP, respectively, taken as 207.1 J/g[39] and X_{PP} is the PP fraction in the composite.

7.2.3.3 Mechanical Properties

Tensile strength (TS) and flexural modulus (FM) were examined using an Instron 5585H testing machine (Norwood, MA) with crosshead rates of 12.5 and 1.35 mm/min according to the procedures in ASTM standards D 638 and D 790, respectively. The measured flexural moduli of samples were obtained from a three-point bending test. Eight replicates were run to obtain an average value for each formulation.

Before each test, the films were conditioned in a 50% relative humidity chamber at 23°C for 48 hr. Notched Izod impact resistance was tested according to ASTM standard D 256 by a Zwick 5101 testing pendulum at room temperature. Each mean value represented an average of eight tests. Impact strength (IS) was defined as the ability of a material to resist fracture under stress applied at high speed. The impact properties of composite materials are directly related to their overall toughness. In the Izod standard test, the only measured variable was the total energy required to break a notched sample.

7.3 Results and Discussions

7.3.1 Characterization of Composite Morphologies

Analyzing the scanning electron micrographs provided helpful information about the distributions and compatibilities of the various phases in the composites. The Figure 7.2 micrographs (×100 magnification) show the typical states of adhesion between the reinforcing fillers and PP matrices of the composites.

Figures 7.2b, c, and d indicate that the PP matrix melted and had a non-uniform surface. The reinforcement embedded in the polymer matrix suggested that the polymer had been plasticized. Fiber pull-out was observed on all PP composites with organic reinforcing filler fracture surfaces

FIGURE 7.2
Scanning electron micrographs of fractured surfaces of composites: (a) PM, (b) PWM, (c) PRM, and (d) PDM.

examined. These results indicated a lack of complete adhesion between PP and the reinforcing filler.

However, Figure 7.2a (PP with MAPP) exhibits finer morphology and a smoother appearance. The higher miscibility of that composite created the morphology of the polymer blend in the co-continuous structure. However, in all the composites with reinforcements, the micrographs taken from the fractured surfaces of the specimens showed different organizations of the fibers (Figure 7.2b, c, and d). Reinforcement particles were not spread uniformly throughout the polymer matrices and occurred randomly or even in clumps.

Reinforcement particles are of various sizes and irregular shapes. The incorporation of filler into polymer matrices disrupted the homogeneity of the matrices. However, more rice clumps were observed on the fractured surfaces of the PP with rice and maleic anhydride-grafted PP (Figure 7.2c) with rice reinforcement than in the blend of PP with maleic anhydride-grafted PP and PP with DDGS and maleic anhydride-grafted polypropylene.

The very rough external surface of the PRM is visible and shows many aligned lumps. This observation suggests that rice was less favorable for better adherence with polymer matrices. Further, when large aggregates of particles of rice appeared, they followed large cracks between the reinforcement and the polymer matrix (Figure 7.2c).

In Figure 7.2d, the surface of the sample shows characteristic even-sized rectangular nodules, creating a crocodile skin effect. A more detailed analysis of the micrograph at higher magnifications revealed a very rough surface formed by rounded nodules that would facilitate adhesion between the particles and the matrix.[40,41] The dark voids visible in Figures 7.2b, c, and d arise from the pull-out of reinforcement agglomerates during deformation and indicate weak filler–matrix adhesion.

Large plastic deformation and fibrillation can be seen clearly, indicating the ductile mode fractures of the former composites. Poor fiber–matrix adhesion is confirmed by the presence of gaps between the fibers and the matrix and fiber pull-outs. The coupled composite displayed a rough morphology with voids between the filler particles and the polymer matrix, clearly indicating weak interaction between them. Higher void contents usually mean lower fatigue resistance, greater susceptibility to water penetration and weathering, and increased variation or scatter in strength properties.

Figure 7.2a shows no voids; the surface is homogeneous, confirming the effect on promoting adhesion played by MAPP in the interfacial region in the composite. We note that the use of coupling agents or additives to provide good binding between PP matrix and the fibers leads to composites with good properties.

The micrographs show that the level of dispersion of the reinforcement in a matrix is not as good as MAPP with PP. The SEM morphological study shows that the composites of PP with reinforcements have less fiber–matrix adhesion and wettability than PP with MAPP. Aggregates and net porosities of the reinforcing particles are seen clearly even in the presence of MAPP.

7.3.2 Thermal Properties of Composites

The thermal properties of the composite blends containing various reinforcements (rice, wood, and DDGS) measured by DSC are shown in Table 7.2. The thermograms of the second heating of the flour–PP reinforcement blends subjected to the same rate flow are listed. Only one endothermic peak corresponding to PP can be observed.

All the composites invariably exhibited slightly higher T_m values compared to the T_m of neat PP. The DSC curves corresponding to the cooling scan for PP and its composites are illustrated in Figure 7.3. They all show exothermic peaks corresponding to the crystallization of the polymeric matrix. A shift of T_c toward higher temperatures in the presence of MAPP and reinforcement was observed in the composites. This indicated that the phenomenon of crystallization during cooling occurred more rapidly in composites containing MAPP than in pure PP.

The effect of rising crystallization rates was clear for all of the composites containing MAPP. The results imply that MAPP acted as a precursor and increased crystallization. The presence of reinforcement (wood, rice, or DDGS) decreased the thermal stability. Heat in turn caused scissions of chains and all these phenomena generated early fusion.

All curves in Figure 7.3 show exothermic peaks corresponding to the crystallization of the polymeric matrix. Every biofiller affects the thermal properties of a composite differently.[42,43] The cooling characteristics have shown very interesting behavior. The temperatures corresponding to the onset of crystallization and peak crystallization increased due to the presence of filler reinforcement. These temperatures further increased due to chemical treatment of rice husks and DDGS. Thus, it seems that the addition of reinforcement causes early crystallization of PP.

This indicates that reinforcement can influence the degree of super cooling of PP. MAPP does not significantly modify the crystallization temperature but leads to an increase in crystallinity. It is recognized that wood and MAPP

TABLE 7.2

Thermal Properties of Composites of Polypropylene Matrices with Reinforcements

Sample	T_m (°C)	ΔH_m (J/g)	T_c (°C)	χ_{corr} (%)
PP	160.8	79.5	120.5	38.9
PM	160.5	82.8	121.0	39.0
PWM	164.7	84.7	125.1	42.5
PRM	163.6	83.9	124.0	40.7
PDM	163.2	84.0	124.3	41.0

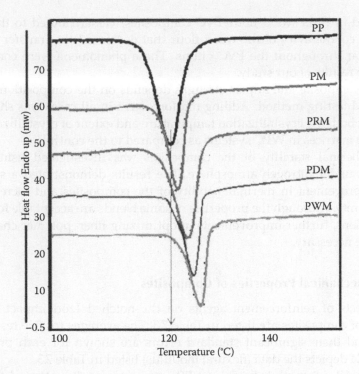

FIGURE 7.3
DSC curves (cooling) of neat PP and composites with various reinforcements.

act as nucleating agents.[16,44] The presence of these two elements generates the formation of more crystals.

T_m, T_c, and ΔH_m are reported in Table 7.2 for composites of PP matrices and different reinforcements of wood, rice, and DDGS. An increase in T_m was observed when reinforcement was loaded into the polymer matrix. The addition of reinforcement had the effect of shifting T_m to higher values. This increase was accompanied by an increase of the degree of crystallinity which was corrected (as percent) by taking into account the reinforcement concentration.[45,46]

These results suggest that crystallization occurred earlier with the incorporation of a reinforcement playing the role of a nucleating agent. Reinforcement provided sites for heterogeneous nucleation; this induced crystallization of the polymeric matrix that was ascribed to the poor thermal conductivity of the reinforcement. In the composites, reinforcement acted as an insulating material, hindering heat conductivity. As a result, the composite compounds needed more heat to melt.

Similar findings were previously reported by Matuana and Kim[47] for PVC-based wood–plastic composites. They found that the addition of wood flour to a PVC resin caused significant increases in the temperature and energy at which fusion of the particles started. The delayed fusion time observed in rigid wood flour–PVC composites was attributed to the poor thermal conductivity of the wood flour that decreased the transfer of heat and shear throughout the PVC grains. These phenomena were consistent with the results of our study.

For a composite, the impact strength depends on the composition, structure, and testing method. Adding reinforcement in all cases was shown to increase both the crystallization temperature and extent of crystallization of polymer matrices in WPC systems as compared to the controls.

The thermal stability of the composites was investigated using DSC analysis under nitrogen atmosphere. The results demonstrated a spectacular improvement in thermal stability of the composites and increases of crystallinity. Although the properties of some blends are acceptable for some applications, further improvements in optimizing fiber–polymer characteristics are necessary.

7.3.3 Mechanical Properties of Composites

The effects of reinforcement agents on the notched Izod impact energy testing of composites are listed in Table 7.3. The averages for five test specimens and their significant standard errors are shown for each property. Figure 7.4 depicts the data (flexural modulus) listed in Table 7.3.

The tensile strength behaviors of PP and its composites showed that the addition of reinforcement weakened the matrices. This result was observed for all matrix reinforcement combinations, although the rate in reduction of the tensile strength and Izod impact strength varied from case to case, depending on the reinforcement.

Most plant fibers are hydrophilic and exhibit high moisture content due to the cellulose in cell structures. All these organic reinforcements generally have high aspect ratios. Thus, the efficiency of transmitting stress from matrix to these types of agents is very poor. On the other hand, lignin increased the hard segments of the composite films, making the films less elastic and more

TABLE 7.3

Tensile Strength (TS), Flexural Modulus (FM) and Notched Izod Impact Strength (IS) of Polypropylene and Its Composites with Various Reinforcements

Sample	TS (MPa)	FM (GPa)	IS (J/m)
PP	27.5	1.40	29.0
PM	29.5	1.85	27.5
PWM	15.4	4.85	20.6
PRM	17.1	4.87	21.5
PDM	17.5	4.88	22.3

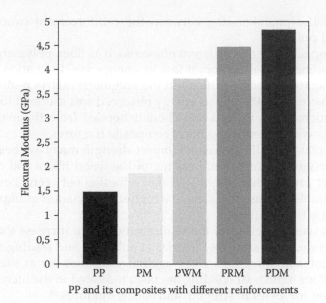

FIGURE 7.4
Flexural modulus of PP and its composites with various reinforcements.

brittle, which led to decreased impact strength. This explains why impact strength decreased as the filler content reached 40 wt%.

Poor interfacial bonding caused partial separation of the filler and matrix polymer, decreasing tensile strength and increasing the brittleness of the composite. However, a compatibilizing agent can partially solve this problem and justifies the use of a coupling agent such as MAPP in all formulations.

As the reinforcement loading (50%) is higher, filler–filler agglomeration occurs and the weak interface regions between reinforcement particles and matrix increase and lead to deterioration in tensile strength. As the degree of agglomeration increases, the filler–matrix interaction becomes poor, leading to a decrement in tensile strength. Incorporation of all three types of fillers (rice, wood, and DDGS) generates a reduction in tensile strength of PP and its composites.

As shown in Table 7.3, notched impact strength decreased in all the composites. This may be explained by the fact that the presence of reinforcing filler ends within the body of a composite can initiate cracks and lead to subsequent failure. The ends of reinforcing fillers act as notches and generate considerable stress concentrations that may initiate microcracks in a ductile PP matrix.

The impact test machine used in this study did not provide enough energy to break the neat PP because of the high flexibility of the PP matrix. By contrast, all specimens were broken completely into two pieces. Introducing reinforcements into the composites led to increased stress concentrations because of the poor bonding between the reinforcement (wood, rice, or DDGS) and polymer.

As the impact wave met different phases such as fiber, polymer, and voids in the cross-machine direction, it lost its energy via dissipation. Although crack propagation became difficult in the polymeric matrix reinforced with filler, the decrease in the impact energy observed was ascribed to fiber ends at which microcracks formed and fibers debonded from the matrix. These microcracks were potential points of composite fractures.

Another reason for the decreased impact strength may have been the stiffening of polymer chains from bonding of the wood fibers and matrix. For high-impact properties, a slightly weaker adhesion between fiber and polymer is desirable because it results in a higher degradation of impact energy and supports fiber pull-out.[47]

In composites, the effect of the reinforcement is to increase the tendency to agglomerate, generating low interfacial adhesion and leading to weakening of the interfacial regions. These agglomerates then act as sites for crack initiation. Poor interfacial bonding has been indicated in the literature as the major reason for losses in strength and elastic modulus.[48]

Adding fillers also resulted in an increment of void content that contributed to stress concentration, thus reducing strength. This behavior was consistent with that observed in the impact tests that revealed decreases in composite samples. Numerous cavities are clearly visible in Figure 7.2b (PP with wood and maleic anhydride), which shows the lowest impact strength. This indicates that the level of interfacial bonding between the fibers and matrix is weak, and the application of stress causes the fibers to be pulled out from the matrix easily, leaving gaping holes.

These two properties are indicators of the plasticity of a material and indicate that PP has a tendency to fracture with loading reinforcement. The flexural moduli of composites are influenced mainly by the adhesion of the matrix and dispersion of reinforcing fillers. The results for this mechanical property also supported the presence of a certain degree of miscibility in composite plastics.

All compositions seen in Figure 7.4 showed flexural moduli higher than that of pure PP. This increase in flexural properties was expected due to the improved adhesion of components in the blends. For some authors,[58] this result was due to the restriction of the mobility and deformability of the matrix with the introduction of mechanical restraint.

Many researchers[49–51] observed that the inclusion of wood or lignocellulosic fibers into thermoplastics such as PP and polyethylene caused a decrease in tensile strength and elongation at break but an increase in Young's modulus. This increase of flexural modulus can be attributed to the increase in volume fractions of high-modulus fibers in plastic composites.[52]

After reinforcement was increased, tensile and compression strengths constantly decreased. The presence of wood or other reinforcement in a polymeric matrix augmented the polymer's rigidity, increasing the value of the modulus in relation to the pure polymer. This phenomenon has been reported by other researchers who studied the effects of wood flour on mixtures of recycled polystyrene and polyethylene[53,54] and virgin polystyrene.[55] The flexural moduli of composites with DDGS were significantly higher than those of other composites, possibly due to removal of the oils from the DDGS materials.

The initial chemical treatments of DDGS certainly produced a positive effect on mechanical properties (flexural modulus and impact strength)—a better result than that obtained with untreated wood. These results are in agreement with those of Julson et al.[51] The improved dispersion obtained from the composite with treated DDGS was also responsible for the highest flexural modulus. For non-coupled composites (wood and PP only), the filler particles began to form aggregates. Direct physical bonds between filler particles are weak and thus easily broken during tensile loading. This explains the decrease in the flexural modulus (WPM).

Compatibilizers can change the molecular morphologies of the polymer chains near the fiber–polymer interphase. Yin et al.[56] reported that the addition of the MAPP coupling agent even at low levels (1 to 2%) increased the nucleation capacity of wood fibers for polypropylene and dramatically altered the crystal morphology of polypropylene around the fiber.

When MAPP was added, surface crystallization dominated over bulk crystallization and a transcrystalline layer could be formed around the wood fibers. Crystallites have much higher moduli as compared to the amorphous regions and can increase the modulus contribution of a polymer matrix to the composite modulus.[57] The flexural moduli of the composites can be correlated with their morphology results. Composites whose surfaces are smoother and more homogeneous exhibit the greatest flexural moduli. The resultant increase in flexural modulus properties (Figure 7.4) can be explained as improved wettability (compatibility) of the reinforcement fibers with the polymer matrix. The increased compatibility is obtained by reducing the polarity of the wood fiber surface nearer the polymer matrix.

The mechanical results of this study show that loading of PP with these natural fibers leads to a decrease in tensile and impact strengths of the pure polymer. On the other hand, the flexural modulus increased due to the higher stiffness of the fibers. The significant improvements in flexural properties of the composites made with MAPP and reinforcing fillers were further supported by SEM micrographs.

7.4 Conclusions

Wood fibers, rice husks, and DDGS, all of which originate from renewable resources, are interesting alternatives to mineral fibers. All samples with reinforcements exhibited markedly heterogeneous and very rough fracture surfaces with large voids or cavities around the filler particles due to the accumulation of stresses in the particle–matrix interface zone. This produced an adverse effect on mechanical properties such as tensile strength and impact resistance.

Scanning electron micrographs reveal that interfacial bonding between a treated filler and matrix significantly improved, suggesting that better dispersion of the filler into the matrix was achieved by treatment of rice husks and DDGS. The thermal properties revealed a strong nucleation ability of the reinforcement flour and MAPP on PP crystallization. Crystallization of all the composites with the MAPP coupling agent only or with reinforcement began earlier than the crystallization of pure PP. This suggests that MAPP and organic reinforcements act as nucleation agents and were responsible for the shift of crystallinity toward higher temperatures.

Tensile and impact strength exhibited marked downward tendencies as reinforcements were loaded. This is due to the weak interfacial adhesion and low compatibility between matrix and filler. The weak bonding of the hydrophilic lignocellulosic agent and the hydrophobic matrix polymer obstructed stress propagation and caused decreases of these properties. The properties were not affected significantly by exchanging rice husks for DDGS. However, the flexural moduli for all composites with organic reinforcements were higher than the value for neat PP as a consequence of the high modulus of the cellulosic agent.

In summary, the use of organic reinforcements as fillers to polymer matrix composites proved a viable alternative. Reductions in tensile and impact strength properties reported with the addition of fillers may be tolerable for some applications. Increments in flexural modulus were achieved in all cases. The development of alternatives for recycling rice husks and DDGS as reinforcements in polymer matrix composites is an important step to provide a good destination for these wastes and opens an opportunity for producing a new value-added product.

This development can help reduce production costs and wood use in composites based on polymer matrices, especially with the worldwide scarcity of wood. This novel application of rice husks and DDGS for bio-based composites has significantly higher economic value than the traditional use of these materials as feedstocks. The DDGS materials from the corn ethanol industry and the rice husk products show immense potential for engineering new green composites to be integrated with thermoplastics. The properties of PP composites can be adjusted by mixing various reinforcement species for filler blends.

Funding

This research received no specific grant from any funding agency in the public, commercial, or not-for-profit sectors.

References

1. J. George, S.S. Bhagawan, N. Prabhakaran et al. Short Pineapple Leaf Fiber-Reinforced Low-Density Polyethylene Composites. *Journal of Applied Polymer Science*, 57 (1995): 843–854.
2. K.C.M. Nair, S.M. Diwan, and S. Thomas. Tensile Properties of Short Sisal Fiber-Reinforced Polystyrene Composites. *Journal of Applied Polymer Science*, 60 (1996): 1483–1497.
3. L.U. Devi, S.S. Bhagawan, and S. Thomas. Mechanical Properties of Pineapple Leaf Fiber-Reinforced Polyester Composites. *Journal of Applied Polymer Science*, 64 (1997): 1739–1748.
4. K. Oksman and C. Clemons. Mechanical Properties and Morphology of Impact-Modified Polypropylene–Wood Flour Composites. *Journal of Applied Polymer Science*, 67 (1998): 1503–1513.
5. H.D. Rozman, K.W. Tan, R.N. Kumar et al. Effect of Hexamethylene Diisocyanate-Modified Alcell Lignin as a Coupling Agent on the Flexural Properties of Oil Palm Empty Fruit Bunch–Polypropylene Composites. *Polymer International*, 50 (2001): 561–567.
6. T.G. Rials, M.P. Wolcott, and J.M. Nassar. Interfacial Contributions in Ligno-cellulosic Fiber-Reinforced Polyurethane Composites. *Journal of Applied Polymer Science*, 80 (2001): 546–555.
7. O.S.B. Saleh. Enhancement of Polyolefin Compatibility with Natural Fibers through Chemical Modification. *American Journal of Polymer Science*, 2 (2012): 102–108.
8. R. Chandra and R. Rustgi. Biodegradable Polymers. *Progress in Polymer Science*, 23 (1998): 1273–1335.
9. R. Malkapuram, V. Kumar, and Y.S. Negi. Recent Development in Natural Fiber-Reinforced Polypropylene Composites. *Journal of Reinforced Plastics and Composites*, 28 (2009): 1169–1189.
10. A.K. Mohanty, M.A. Khan, S. Sahoo et al. Effect of Chemical Modification on the Performance of Biodegradable Jute Yarn–Biopol® Composites. *Journal of Materials Science*, 35 (2000): 2589–2595.
11. J.M. Pilarski and L.M. Matuana. Durability of Wood Flour–Plastic Composites Exposed to Accelerated Freeze–Thaw Cycling. I: Rigid PVC Matrix. *Journal of Vinyl and Additive Technology*, 11 (2005): 1–8.
12. C.R. Di Franco, V.P. Cyras, J.P. Busalmen et al. Degradation of Polycaprolactone–Starch Blends and Composites with Sisal Fibre. *Polymer Degradation and Stability*, 86 (2004): 95–103.
13. L. Tilstra and D. Johnsonbaugh. Biodegradation of Blends of Polycaprolactone and Polyethylene Exposed to a Defined Consortium of Fungi. *Journal of Environmental Polymer Degradation*, 1 (1993): 257–267.
14. R.T. Woodhams, G. Thomas, and D.K. Rodgers. Wood Fibers as Reinforcing Fillers for Polyolefins. *Polymer Engineering & Science*, 24 (1984): 1166–1171.
15. M. Kazayawoko, J.J. Balatinecz, and L.M. Matuana. Surface Modification and Adhesion Mechanisms in Wood Fiber–Polypropylene Composites. *Journal of Materials Science*, 34 (1999): 6189–6199.

16. D. Ndiaye, L.M. Matuana, S. Morlat-Therias et al. Thermal and Mechanical Properties of Polypropylene–Wood Flour Composites. *Journal of Applied Polymer Science*, 119 (2011): 3321–3328.

17. T.A. Bullions, D. Hoffman, R.A. Gillespie et al. Contributions of Feather Fibers and Various Cellulose Fibers to the Mechanical Properties of Polypropylene Matrix Composites. *Composites Science and Technology*, 66 (2006): 102–114.

18. S.M. Lai, F.C. Yeh, Y. Wang et al. Comparative Study of Maleated Polyolefins as Compatibilizers for Polyethylene–Wood Flour Composites. *Journal of Applied Polymer Science*, 87 (2003): 487–496.

19. A.K. Bledzki and J. Gassan. Composites Reinforced with Cellulose-Based Fibres. *Progress in Polymer Science*, 24 (1999): 221–274.

20. Q. Li and L.M. Matuana. Effectiveness of Maleated and Acrylic Acid-Functionalized Polyolefin Coupling Agents for HDPE Wood Flour Composites. *Journal of Thermoplastic Composite Materials*, 16 (2003): 551–564.

21. L. Chotirat, K. Chaochanchaikul, and N. Sombatsompop. On Adhesion Mechanisms and Interfacial Strength in Acrylonitrile–, Butadiene–, and Styrene–Wood Sawdust Composites. *International Journal of Adhesion and Adhesives*, 27 (2007): 669–678.

22. W. Qiu, F. Zhang, T. Endo et al. Preparation and Characteristics of Composites of High-Crystalline Cellulose with Polypropylene: Effects of Maleated Polypropylene and Cellulose Content. *Journal of Applied Polymer Science*, 87 (2003): 337–345.

23. D.M. Panaitescu, D. Donescu, C. Bercu et al. Polymer Composites with Cellulose Microfibrils. *Polymer Engineering & Science*, 47 (2007): 1228–1234.

24. J.A. Trejo-O'Reilly, J.Y. Cavaillé, M. Paillet et al. Interfacial Properties of Regenerated Cellulose Fiber–Polystyrene Composite Materials: Effect of Coupling Agent's Structure on the Micromechanical Behavior. *Polymer Composites*, 21 (2000): 65–71.

25. T.A. Bullions, R.A. Gillespie, J. Price-O'Brien et al. Effect of Maleic Anhydride-Modified Polypropylene on the Mechanical Properties of Feather Fiber, Kraft Pulp, and Polypropylene Composites. *Journal of Applied Polymer Science*, 92 (2004): 3771–3783.

26. K.N. Kim, H. Kimm, and J.W. Lee. Effect of Interlayer Structure, Matrix Viscosity, and Composition of a Functionalized Polymer on the Phase Structure of Polypropylene–Montmorillonite Nanocomposites. *Polymer Engineering & Science*, 41 (2001): 1963–1969.

27. T.J. Keener, R.K. Stuart, and T.K. Brown. Maleated Coupling Agents for Natural Fibre Composites. *Composites Part A*, 35 (2004): 357–362.

28. M. Bengtsson and K. Oksman. Silane Crosslinked Wood Plastic Composites: Processing and Properties. *Composites Science and Technology*, 66 (2006): 2177–2186.

29. R.G. Raj, B.V. Kokta, D. Maldas et al. Use of Wood Fibers in Thermoplastics. VII: Effect of Coupling Agents in Polyethylene–Wood Fiber Composites. *Journal of Applied Polymer Science*, 37 (1989): 1089–1103.

30. B.V. Kokta, R.G. Raj, and C. Daneault. Use of Wood Flour as Filler in Polypropylene: Studies on Mechanical Properties. *Polymer–Plastics Technology and Engineering*, 28 (1989): 247–259.

31. H.S. Yang, H.J. Kim, J. Son et al. Rice Husk Flour-Filled Polypropylene Composites: Mechanical and Morphological Study. *Composite Structures*, 63 (2004): 305–312.

32. A.M. Piva, S.H. Steudner, and H. Wiebeck. Physicomechanical Properties of Rice Husk Powder-Filled Polypropylene Composites with Coupling Agent Study. Paper presented Fifth International Symposium on Natural Polymer and Composites, São Pedro, Brazil, 2004.

33. Q. Wu and A.K. Mohanty. Renewable Resource-Based Biocomposites from Co-Product of Dry Milling Corn Ethanol Industry and Castor Oil-Based Biopolyurethanes. *Journal of Biobased Materials and Bioenergy*, 1 (2007): 257–265.

34. V. Cheesbrough, K.A. Rosentrater, and J. Visser. Properties of Distillers' Grains Composites: Preliminary Investigation. *Journal of Polymers and the Environment*, 16 (2008): 40–50.

35. R.A. Tatara, K.A. Rosentrater, and S. Suraparaju. Design Properties for Molded, Corn-Based DDGS-Filled Phenolic Resin. *Industrial Crops and Products*, 29 (2009): 9–15.

36. H. Ismail and R.M. Jaffri. Physicomechanical Properties of Oil Palm Wood Flour-Filled Natural Rubber Composites. *Polymer Testing*, 18 (1999): 381–388.

37. N. Chevanan, K.A. Rosentrater, and K. Muthukumarappan. Effect of DDGS, Moisture Content, and Screw Speed on Physical Properties of Extrudates in Single-Screw Extrusion. *Cereal Chemistry Journal*, 85 (2008): 132–139.

38. D. Ray and B.K. Sarkar. Characterization of Alkali-Treated Jute Fibers for Physical and Mechanical Properties. *Journal of Applied Polymer Science*, 80 (2001): 1013–1020.

39. L. Valentini, J. Biagiotti, J.M. Kenny et al. Morphological Characterization of Single-Walled Carbon Nanotube–PP Composites. *Composite Science and Technology*, 63 (2003): 1149–1153.

40. B. Wunderlich. *Thermal Analysis of Polymeric Materials*. New York: Academic Press, 1990.

41. A. Pramanick and M. Sain. Nonlinear Viscoelastic Creep Prediction of HDPE–Agro-Fiber Composites. *Journal of Composite Materials*, 40 (2006): 417–431.

42. X.F. Yao, H.P. Zhao, and H.Y. Yeh. Micro and Nanoscopic Characterizations of Epoxy–Silica Nanocomposites. *Journal of Reinforced Plastics and Composites*, 25 (2006): 189–196.

43. S. Kalia, B.S. Kaith, and I. Kaur. Pretreatments of Natural Fibers and Their Application as Reinforcing Material in Polymer Composites: A Review. *Polymer Engineering & Science*, 49 (2009): 1253–1272.

44. H.G.B. Premalal, H. Ismail, and A. Baharin. Effect of Processing Time on the Tensile, Morphological, and Thermal Properties of Rice Husk Powder-Filled Polypropylene Composites. *Polymer–Plastics Technology and Engineering*, 42 (2003): 827–851.

45. P. Zugenmaier. Materials of Cellulose Derivatives and Fiber-Reinforced Cellulose–Polypropylene Composites: Characterization and Application. *Pure and Applied Chemistry*, 78 (2006): 1843–1855.

46. L. Mandelkern. Crystallization of Polymers. In *Advanced Chemistry*. New York: McGraw Hil, 1964, Chap. 5.

47. L.M. Matuana and J.W. Kim. Fusion Characteristics of Rigid PVC–Wood Flour Composites by Torque Rheometry. *Journal of Vinyl and Additive Technology*, 13 (2007): 7–13.

48. M. Bengtsson, M.L. Baillif, and K. Oksman. Extrusion and Mechanical Properties of Highly Filled Cellulose Fibre–Polypropylene Composites. *Composites Part A*, 38 (2007): 1922–1931.

49. H. Ismail, M.N. Nasaruddin, and H.D. Rozman. Effect of Multifunctional Additive in White Rice Husk Ash-Filled Natural Rubber Compounds. *European Polymer Journal*, 35 (1999): 1429–1437.
50. N.M. Stark and M.J. Berger. Effect of Particle Size on Properties of Wood Flour-Reinforced Polypropylene Composites. Paper presented at Fourth International Conference on Wood Fiber–Plastic Composites. Forest Products Society, Madison, WI, May 12–14, 1997.
51. J.L. Julson, G. Subbarao, D.D. Stokke et al. Mechanical Properties of Biorenewable Fiber–Plastic Composites. *Journal of Applied Polymer Science*, 93 (2004): 2484–2493.
52. F. Febrianto, D. Setyawati, M. Karina et al. Influence of Wood Flour and Modifier Contents on the Physical and Mechanical Properties of Wood Flour–Recycled Polypropylene Composites. *Journal of Biological Sciences*, 6 (2006): 337–343.
53. A. Sanadi, D.F. Caulfield, and R.M. Rowell. *Lignocellulosics and Plastic Composites*. Washington, DC: American Chemical Society, 1998.
54. J. Simonsen and T.G. Rials. Morphology and Properties of Wood Fiber-Reinforced Blends of Recycled Polystyrene and Polyethylene. *Journal of Thermoplastic Composite Materials*, 9 (1996): 292–302.
55. D. Maldas and B.V. Kokta. Effect of Recycling on the Mechanical Properties of Wood Fiber–Polystyrene Composites. I: Chemithermomechanical Pulp as a Reinforcing Filler. *Polymer Composites*, 11 (1990): 77–83.
56. S. Yin, T.G. Rials, and M.P. Wolcott. Crystallization Behavior of Polypropylene and Its Effect on Wood Fiber Composite Properties. Paper presented at Fifth International Conference on Wood Fiber–Plastic Composites. Forest Products Society, Madison, WI 1999.
57. S.S. Ray, P. Maiti, M. Okamoto et al. New Polylactide–Layered Silicate Nanocomposites. I: Preparation, Characterization, and Properties. *Macromolecules*, 35 (2002): 3104–3110.

8

Natural Fiber Polyolefin Composites: Processing, Melt Rheology, and Properties

Haroutioun Askanian, Ya Feng, Sophie Commereuc,
Roman Cermak, Florence Delor-Jestin, Kristyna Montagova,
Valérie Massardier, and Vincent Verney

CONTENTS

8.1 Introduction

Today's world faces many environmental problems, and two of them are plastic wastes and excessive felling. The former is connected with tremendous production and use of plastics in practically all areas of everyday living. The latter arises from people's demands and desires to be surrounded by natural products, particularly wood. Wood is used in furniture manufacture and also in the construction industry.[1,2]

The combinations of these materials—synthetic polymers, plastics, and woods—can reduce the environmental problems and also present other advantages such as low specific weight and better specific strength and stiffness in comparison to glass-reinforced composites.[3] From this view, wood fibers provide safer handling and working conditions, and they are less abrasive to mixing and molding equipment.

The most interesting characteristic of about natural fillers is their potentially positive environmental impact. Wood flour originates from plants and is thus considered as a renewable resource whose production requires minimum energy. Natural fillers are carbon dioxide-neutral; their contribution to CO_2 concentration in the atmosphere after combustion equals the CO_2 consumption during wood growth.[1,4]

In the past two decades, wood–plastic composites (WPCs) gained the attention of both scientific centers and industry and interest is growing significantly. Potential applications are primarily in the automotive, building, furniture and household equipment industries.[5] However, one limitation is a low rate of interaction of filler and matrix. Wood flour (WF) is a cellulose that bears hydroxyl groups, so its surface is hydrophilic, while most commodity plastics are hydrophobic, particularly polyolefins. This problem could be handled, for example, by using coupling agents or chemical modification of the WF or matrix or by effective filling and mixing to optimize the properties of a composite.[1,6,7]

The quality of a composite material depends strongly on the origin of the wood that may have different compositions and structures of the cellulose fibers.[3] The purity of the material in relation to the amounts of accompanying substances is even more crucial.[8,9] This chapter treats pine wood flour as representative of soft woods. This material was investigated after compounding with polymeric matrices by means of rheology, differential scanning calorimetry (DSC), and tensile testing.

Isotactic polypropylene (iPP) was used in our study because it is a widely used plastic that has applications in many areas including those for which WPC composites currently designed, i.e., the construction and automotive industries. iPP is an interesting material from both the scientific and industrial views.

Several PPs with different melt flow indices (MFIs) were selected for the study. These materials were filled with various concentrations and types of WFs. Prepared composites should show a number of attractive properties that offer a wide range of applications. A fundamental presumption for the use of composites is a need to understand the interrelations of process, structure, and properties. Thus, the main goal of the study was to determine experimentally these interrelations based on PPs and WFs. This work focused on the following tasks:

- Preparing PP–WF composites with various concentrations of fillers in PPs with various MFIs
- Comparing the properties of composites with different types of woods

- Evaluating the influences of various wood concentrations and MFIs on structural, thermal, and processing properties of composites and verifying the amounts of renewable carbon

8.2 Experimental Details

8.2.1 Materials

Six PP homopolymers were selected as matrices in this study. The materials were provided by Borealis group. The descriptions of the materials used are summarized in Table 8.1. The MFI values were measured at 230°C with a load of 2.16 kg according to ISO 1133.

The PPs were filled with industrial natural French pine (*Pinus sylvestris*) wood flour (WF). Some had undergone a thermal treatment to remove polar molecules and render the internal structures more accessible. PP is mainly a non-polar hydrophobic molecule and wood is a polar hydrophilic material and they are incompatible. A third component (compatibilizer) was not added because of a compromise between environment and cost issues. By taking these factors into account, we chose to use heat-treated wood flour; the thermal treatment process is called *retification* and is carried out in an inert (argon) atmosphere.[10] The chemical structure of pine was modified strongly by the process. The treated wood exhibited particular properties: pronounced hydrophobic character, improved energetic power, and a decrease of the volatile components.[11-13] The average diameter of wood flour is 350 μm.

8.2.2 Preparation of Composites

Before mixing, the WF was dried in a drying oven for 24 hr at 80°C. Composites were prepared using a Haake MiniLab Rheomex CTW5 twin screw extruder filled with manually pre-mixed wood flour and polypropylene granulate. The extruder was heated to 180°C and screw rotation was set

TABLE 8.1

Characteristics of Polypropylene Samples

Material	M_w	M_n	PDI	MFI (g/10 min)
HB205TF	830 000	300 000	2.8	1
HE125MO	490 000	130 000	3.8	12
HF136MO	410 000	170 000	2.5	20
HG455FB	360 000	150 000	2.4	27
HK060AE	–	–	–	100
HL508FB	–	–	–	800

– Data not available.

to 80 or 100 rpm. After about 5 to 10 min of mixing, the machine was set to extrusion mode and molten composite was discharged. The samples for mechanical properties evaluation were prepared by compression molding at 180°C under a pressure of 200 bars for 1 min. Teflon films were used to avoid the adhesion of PP to the stainless surface of the mold.

8.2.3 Rheological Measurements

Shear dynamic measurements were carried out with an ARES rheometer (Rheometric Scientific) manufactured by TA Instruments. Parallel plate geometry with a diameter of 25 mm was used for frequency sweeps. The range of frequency sweeps was from 0.1 to 100 rad/s, and a strain of 5% was used to ensure that all measurements were conducted within the linear viscoelastic region. The shear dynamic measurements were conducted at a constant temperature of 180 or 200°C and a gap height of 1 mm.

8.2.4 Differential Scanning Calorimetry

Mettler Toledo (Mettler DSC-30) differential scanning calorimetry was used to measure the thermal behaviors of prepared composites. About 10 mg of each composite was used for analysis. The measurements were performed under inert atmosphere (N_2 flow) at a heating rate of 10°C/min in a temperature range of 25 to 200°C. The samples were initially heated to 200°C for the first scan, then cooled to 25°C and heated again to 200°C. The degree of crystallinity (X_c) was determined from the second heating scan. The crystalline phase can be estimated by division of measured melting enthalpy with enthalpy of theoretically fully crystallized PP:

$$X_c = \frac{\Delta H_m}{\Delta H_m^0} \cdot 100 [\%] \tag{8.1}$$

8.2.5 Tensile Testing

The tensile strength properties were evaluated according to ASTM D63, using a MTS 200/M mechanical machine at a cross-head speed of 30 mm/min. The specimen dimensions were 20 mm length, 4 mm width, and 0.15 mm thickness. Tests were performed at room temperature. At least five specimens were tested for each composite blend, and the results were averaged to obtain a mean value.

8.2.6 Water Absorption

The samples were oven dried at 80°C for 24 hr. Water absorption (WA) of the composites was measured by immersion of samples in distilled water at room temperature. Samples were removed periodically from the water, the increase in the weight was measured, then the samples were immersed in water again. The *WA* values were calculated from the following equation:

$$WA(\%) = M_t - M_0/M_0 \times 100 \tag{8.2}$$

where M_0 and M_t are the weights of the sample before and after immersion in distilled water, respectively.

8.3 Results and Discussion

8.3.1 Rheological Characterization

8.3.1.1 Effects of MFI on Viscoelastic Behavior

Viscoelastic behaviors of various PP matrices were examined. For non-cross-linked polymer melts with monomodal distributions, a Cole-Cole plot shows the arc of a circle. Its extrapolation to the real axis indicates zero shear viscosity η_0 (the classical power law $\eta_0 = K.M_w^a$ is exemplified on this figure with a power law index $a = 3.4$) proportional to M_W. All the samples exhibit the same range of values for the distribution relaxation time parameter h ($0.38 < h < 0.46$). They are classified in the range of polydispersity indexes that are in good agreement with the values expected. The comparison of Cole-Cole plots for PPs with various MFIs is depicted in Figure 8.1. The arc decrease confirms the fact that PPs with increasing MFI materials flow better.

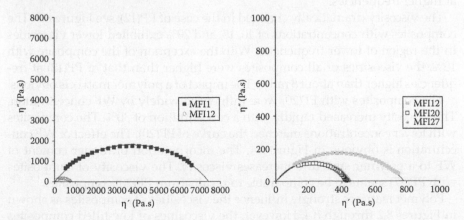

FIGURE 8.1
Cole-Cole plots for PP (1, 12, 20, 27).

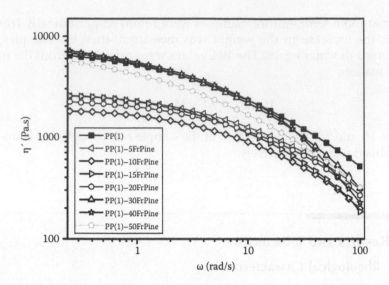

FIGURE 8.2
Dependence of viscous dissipation (η') on frequency in composites with PP(1), at 200°C.

The viscoelasticity results of composites with four MFIs and seven concentrations of WF (French pine in this case) are illustrated in the following figures. The relation of η' and frequency was used to describe viscosities of composites. WF concentration did not influence the viscosity of composites with PP(1) considerably (Figure 8.2). The composites with lower concentrations exhibited lower values of η' than PP(1). However, the high concentration of WF did not exhibit higher viscosity than PP(1) as could be expected. The viscosities of composites with 30 and 50% were slightly higher than that of PP(1). Nevertheless, decreases of viscosity in all the composites were noted at higher frequencies.

The viscosity dramatically changed in the case of PP(12); see Figure 8.3. The composites with concentrations of 10, 15, and 20% exhibited lower viscosities in the region of lower frequencies. With the exception of the composite with 10%, the viscosities of all composites were higher than that of PP(12) at frequencies higher than about 8 rad/s. The impact of a polymer matrix is obvious.

The composites with PP(20) were influenced widely by WF concentration. The viscosity increased rapidly with a concentration of 30%. The composites with lower concentrations matched the curve of PP(20). The effect of WF concentration is obvious in Figure 8.4. The incorporation of higher content of WF to a polymer structure increases viscosity. The viscosity of composites with PP(27) seems to be comparable to results from PP(20) composites.

Polymer matrices strongly influence the viscosities of composites as shown in Figures 8.2 through 8.4. However, the viscosities of low-filled composites were permanently lower than the viscosity of PP. Nevertheless, a polymer with higher MFI can influence the flow of composites and thus the viscosity

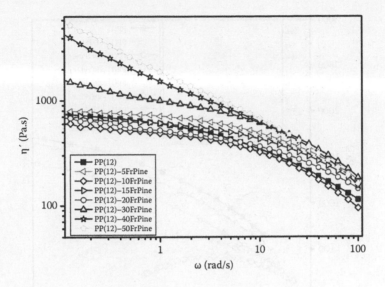

FIGURE 8.3
Dependence of viscous dissipation (η′) on frequency in composites with PP(12) at 200°C.

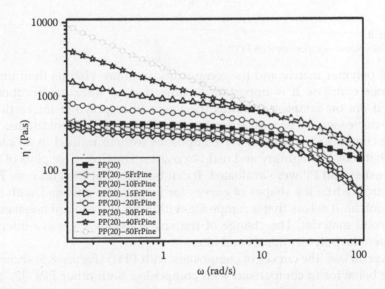

FIGURE 8.4
Dependence of viscous dissipation (η′) on frequency in composites with PP(20) at 200°C.

starts to increase and approach that of PP. In this case, the occurrence of curves of low-filled composites for PP could mean lower interactions between filler and matrix.

However, it should be considered that filled systems require certain critical amounts of filling. The low concentrations are not able to overcome the flow

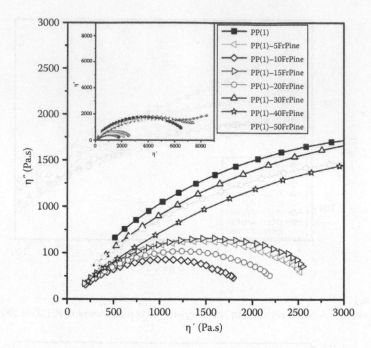

FIGURE 8.5
Cole-Cole plot of composites with PP(1).

of the polymer matrix and the composites are more viscous than un-filled polymer matrices. It is important to note that no filler modification was carried out on samples, and thus the lower filler–polymer interaction can cause decreases of viscosity in composites with lower concentrations.

The complex viscosities of WF composites were examined in a Cole-Cole plot that shows imaginary and real viscosities. The Cole-Cole plots of all the composites and PP were evaluated. It can be seen that the curve for PP is a semicircle while the shapes of curves for composites changed with higher WF content. It seems that a composite with high WF content behaves more as a solid material. The change of the shape means a greater interaction between the polymer matrix and filler.

Nevertheless, the curves of composites with PP(1) (Figure 8.5) show a different behavior in comparison with composites with other PPs (12, 20, 27). The curves of composites with PP(1) with higher concentrations of WF do not change their shapes from a semicircle. Composites with PP(1) exhibit low interactions. This can be explained by the shapes of the curves of composites that copy the curve of PP(1). In this case, the particles are more organized and create a network—the composites behave as a gel and not a solid. The change from viscous to gel behavior depends on the MFI of PP.

The composites start to show different evolutions of Cole-Cole plots with higher MFIs. In composites with PPs (12, 20, and 27), double distribution

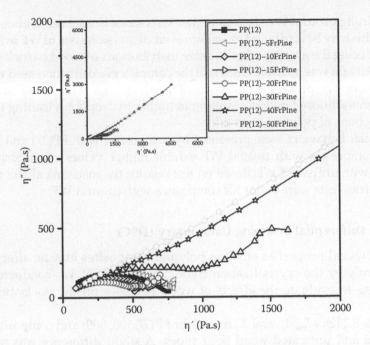

FIGURE 8.6
Cole-Cole plot of composites with PP(12).

occurs via higher restriction of molecules. However, the shapes of the curves result from the intersection of a small circle arc and a straight line. These results correspond to the strong interaction between WF and PP.

Figure 8.6 shows that PP(12) has a strong reinforcement effect in combination with WF. For PPs < PP(12), WF disturbs the chain friction movements and acts more as a lubricating agent; this confirms the decrease of viscosity. For PPs > PP(12), WF acts as a reinforcement agent and promotes chain interaction. This is described by increasing viscosity and low frequency elastic relaxation.

Finally, these results suggest that the molecular dynamic of polypropylene macromolecular chains is disturbed strongly by wood fillers, depending on their initial macromolecular mobilities. However, strong interactions (gel behavior) and lubricating processes (viscosity reduction) can be observed and more generally a mix of both phenomena can be generated.

8.3.1.2 Effects of WF Thermal Treatment and Filler Size

The influence of WF treatment on viscoelastic properties of composites was investigated with three PPs (MFI 27, 100, 800). The WF used was ground to obtain a filler particle size around 80 μm to improve dispersion in the polymer matrix.

Figure 8.6 shows the effects of WF particle size and treatment on the complex viscosities of PP–wood composites. At low frequencies, samples containing ground WF showed higher viscosities than those without grinding. This may be attributed to an improved dispersion state of WF in a matrix by reducing the average sizes and size distributions of wood particles. At the same time, it was possible to see that the complex viscosity increased with WF treatment. The process led to chemical modifications of the wood, mainly on the hemicellulose fraction, resulting in higher hydrophoby, leading to better interactions of polymers and fillers.

Similar behaviors were presented by composites with PP(27) and PP(100). The composites with treated WF exhibit higher values of viscosity than those with untreated WF. Based on test results, the mechanical and thermal properties tests were better for composites with ground WF.

8.3.2 Differential Scanning Calorimetry (DSC)

The physical properties of wood polymer composites may be affected significantly by the crystallization characteristics of PP. We conducted DSC analysis to evaluate the effects of wood particles on the nucleation rates of PPs.

Table 8.2 lists T_m, T_c and X_c results for PP(27, 100, 800) and composites with treated and untreated wood flour types. A slight difference was noted in thermal properties of the composites compared to pure PP. The T_m and T_c values of PP were slightly shifted to higher temperatures when both types of wood flours were added. An increase in the X_c of the composites can also be observed.

The improvement in the X_c values of the composites confirms the nucleation effect of the wood. No significant change was observed in the nucleation effect of the wood flour after thermal treatment, except with PP(800) in which the degree of crystallinity increased from 49 to 55% (Figure 8.7). This result may be explained by the increase in the roughness of wood flour after thermal treatment, leading to the increase of its nucleation ability.[14]

TABLE 8.2

Thermal Properties of Various Samples

Sample	T_m (°C)	T_c (°C)	X_c (%)
PP800	159.6	113.7	45
PP800 + untreated WF	162.4	116.5	49
PP800 + treated WF	161.3	118.3	55
PP27	163.3	112.5	45
PP27 = untreated WF	164.9	115.4	49
PP27 = treated WF	164.2	115.2	50
PP100	162.5	117.9	50
PP100 = untreated WF	163.5	118.1	57
PP100 = treated WF	163.5	120.2	57

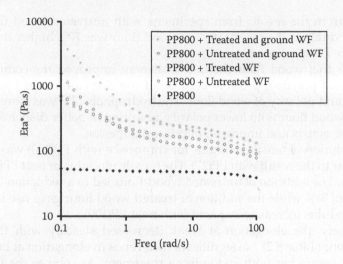

FIGURE 8.7
Variation of complex shear viscosity (η) of PP(800) and different WPCs at 180°C.

8.3.3 Mechanical Properties

The tensile tests were conducted on composites based on PP(27) and PP(100). The relationships of the tensile modulus of PP(27) and the filler loading with and without treatment are shown in Table 8.3. The tensile modulus of the composite specimens without treatment was 344 MPa. The addition of 40 wt% untreated wood flour resulted in a decrease in tensile modulus of approximately 64%, compared with pure PP(27). Because PPs are non-polar (hydrophobic) substances, their bonding strength with polar (hydrophilic) wood flour is weak.[15] The decrease of tensile modulus can be attributed to the incompatibility between wood and PP.

On the other hand, the addition of treated wood fibers led to significant improvement in the tensile moduli of the composites in comparison to results with pure PP(27). The tensile modulus of the composite specimens with treatment was 1,612 MPa at 40 wt% of wood flour. This result is totally

TABLE 8.3

Tensile Modulus and Elongation at Break of Composites Extruded with PP27 and PP100

Sample	Tensile Modulus (MPa)	Elongation at Break (%)
PP27	943	465
PP27 + untreated WF	344	1.8
PP27 + treated WF	1612	1.9
PP100	1900	5
PP100 + untreated WF	2600	1.0
PP100 + treated WF	2990	1.2

different from the results from specimens with untreated wood flour. The tensile modulus in the case of treated wood flour was 70% higher than those of the pure PP specimens.

It seems that wood flour thermal treatment improved the compatibility of wood and polymer, possibly because of the decrease in the hydrophilic character and polarity of wood flour. The hydrophobic PP was more compatible with wood flour with lower polarity, resulting in better dispersion of the fiber in the matrix and improved interfacial adhesion.

The evolution of tensile modulus for composite with PP(100) was more or less similar to the result with PP(27). The tensile modulus of neat PP(100) was 1,149 MPa. The addition of untreated wood flour led to a reduction in tensile modulus of 31% while the addition of treated wood flour gave rise to an 11% tensile modulus increase compared with neat PP(100).

Conversely, the elongation at break decreased steadily with the wood fiber content (Table 8.2). No significant difference in elongation at break was noted for composites with and without treatment. As soon as the filler was added, a decrease of elongation was observed because wood fibers have low elongation at break and restrict the polymer molecules from flowing past one another. This behavior is typical of reinforced thermoplastics in general and has been reported by many researchers.[16]

8.3.4 Water Absorption

The moisture absorption behaviors of composites with treated and untreated wood flour types and PP(27,100,800) matrices were examined.

The hydroxyl groups in the wood flour interact with water molecules by hydrogen bonding. The water uptake by hydrophobic PP is negligible,[1] so water absorption by the composites is limited to sorption unequally due to fibers. In the case of weak interface adhesion in a composite, the absorbed water can reside in the microvoids between wood flour and PP. By adding 40 wt% wood flour, the voids may become as a result of the poorer dispersion of wood flour leading to a high degree of moisture absorption for the composite.

Figure 8.8 shows the water absorption results for untreated and treated wood flour with PP(27) as a function of time. As noted previously, the treated wood flour creates a better interfacial adhesion with the PP matrix. Improving the adhesion limits the penetration of water through the interface.

The results in Figure 8.8 confirm the hypothesis that there is an important difference in water absorption ability between the two types of wood flours used in composites. Treated wood flour exhibits lower values of water absorption in comparison with untreated wood flour. In composites with untreated wood flour, the water absorption rate was ~15% after about 40 days of immersion in distilled water; the water absorption rate with treated wood flour was ~7%.

Similar evolution was observed with composites using a PP(100) matrix. The water absorption rate after ~40 days for composites with untreated wood flour was ~14% and for composites with treated wood flour was ~9%.

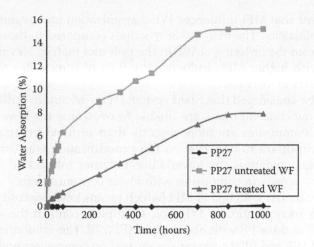

FIGURE 8.8
Water absorption kinetics for different samples (T = 20°C).

Similarly, composites based on PP(800) matrices showed water absorption rates of ~14% and ~6% for untreated and treated wood flours, respectively.

8.4 Conclusions

The goal of this chapter was to investigate on a molecular scale the primary physical and chemical interactions of a natural filler (WF) and a polypropylene matrix (PP) without other chemical molecules. The materials were studied by monitoring, through linear small oscillations, their molten viscoelastic behaviors, the changes in the macromolecular mobilities of the polymeric chains due to the inclusions of fillers, and their interactions. A second step was to analyze the consequences of these molecular dynamic changes in conditions on macromolecular reorganization in the solid state.

The first task was to obtain evidence of the dynamics of these systems. Of course, classical parameters such as filler size and concentration are yet to be described in literature and the influence of the molecular mobility of a matrix, and its impact on macromolecular recrystallization is even less known. This work demonstrated that both gel behavior (expected) and viscous reduction (quite unexpected!) can be observed and both processes compete.

The effect of MFI on viscoelastic behavior was investigated on composites with seven concentrations of French pine that were incorporated into four PPs. We found that higher concentrations of WF strongly influenced η' and η". Since the lower concentration exerted no influence, this behavior of viscosity occurs in relation to frequency in composites where the substantial effect of MFI is observed.

It is evident that MFI influences WF concentration and evolutions of viscosity and elasticity. The increase of η' values compared to those of neat PP may arise from the ordering of WF in the polymer matrix. Nevertheless, the polymers with higher MFIs influence the flow of composites and thus the viscosity starts to increase.

It should be considered that filled systems require certain critical amounts of filling. Low concentrations are unable to overcome the flow of polymer matrix and composites are more viscous than unfilled polymer matrices. Also, it is important to note that no filler modification was carried out on these samples, and thus the lower filler–polymer interaction can cause a decrease of viscosity in composites with lower concentrations.

A polymer matrix with higher MFI (which means lower viscosity) has better potential for incorporation of WF. Cole-Cole plots confirm the strong interaction of WF and the PP with highest MFI [PP(27)]. The influences of thermal treatment of WF and filler size were examined on composites with PP(27, 100, 800). WF with smaller particle size enhanced the properties of wood–polymer composites (WPCs) by improving the adhesion and dispersion of the particles. The results of thermal treatment of WF showed that composites with treated WF were able to influence viscoelastic behavior. An increase in viscosity was observed in comparison to composites with untreated WF. It can be concluded that interparticle interactions are more important when treated WF is used.

The results of DSC analysis showed an influence of the French pine WF presence on overall crystallinity. Both types of WF when present in 40 wt% concentrations in PP matrices showed increased crystallinity and slight upward shifts of the temperatures of melting and crystallization. Pine WF plays the role of nucleating agent.

Tensile tests showed that the addition of treated WF led to significant improvement in the tensile moduli of the composites in comparison to pure PP. Treated wood fibers provided rigidity and tenacity. On the other hand, the addition of untreated WF decreased tensile modulus, showing poor compatibility of matrices and fibers. Water absorption testing showed the best results with treated wood flour; the thermal treatment reduced the hygroscopicity of WF so the water uptake of composites was lower than the uptakes of composites with untreated WF.

In summary, treated wood with small particle sizes seems to be more suitable than untreated for use in WPCs based on our study.

References

1. A.K. Bledzki, M. Letman, A. Viksne et al. A Comparison of Compounding Processes and Wood Type for Wood Fibre–PP Composites. *Composites Part A,* 36 (2005): 789–797.

2. C. Clemons. Wood–Plastic Composites in the United States. *Forest Products Journal*, 52 (2002): 10–18.
3. H. Bouafif, A. Koubaa, P. Perré et al. Effects of Fiber Characteristics on the Physical and Mechanical Properties of Wood–Plastic Composites. *Composites Part A*, 40 (2009): 1975–1981.
4. M.J. John and S. Thomas. Biofibres and Biocomposites. *Carbohydrate Polymers*, 71 (2008): 343–364.
5. Z. Dominkovics, L. Dányádi, and B. Pukánszky. Surface Modification of Wood Flour and Its Effect on the Properties of PP–Wood Composites. *Composites Part A*, 38 (2007): 1893–1901.
6. R. Bouza, C. Marco, M. Naffakh et al. Effect of Particle Size and a Processing Aid on the Crystallization and Melting Behavior of IPP–Red Pine Wood Flour Composites. *Composites Part A*, 42 (2011): 935–949.
7. V. Steckel, C.M. Clemons, and H. Thoemen. Effects of Material Parameters on the Diffusion and Sorption Properties of Wood Flour–Polypropylene Composites. *Journal of Applied Polymer Science*, 103 (2007): 752–763.
8. J.S. Fabiyi and A.G. Mcdonald. Effect of Wood Species on Property and Weathering Performance of Wood–Plastic Composites. *Composites Part A*, 41 (2010): 1434–1440.
9. H. Saputra, J. Simonsen, and K. Li. Effect of Extractives on the Flexural Properties of Wood–Plastic Composites. *Composite Interfaces*, 11 (2004): 515–524.
10. R. Guyonnet and J. Bourgois. Procédé De Fabrication D'un Matériau Ligno-Cellulosique Par Traitement Thermique Et Matériau Obtenu Par Ce Procédé. 1988.
11. M. Jaziri, J.F. May, J. Bourgois et al. Matériaux Composites à Base De Bois Rétifié: Contribution à l'etude Des Propriétés Rhéologiques à l'etat Fondu En Régime Stationnaire. *Die Angewandte Makromolekulare Chemie*, 155 (1987): 67–81.
12. J. Bourgois and R. Guyonnet. Characterization and Analysis of Torrefied Wood. *Wood Science and Technology*, 22 (1988): 143–155.
13. J. Bourgois, M.C. Bartholin, and R. Guyonnet. Thermal Treatment of Wood: Analysis of the Obtained Product. *Wood Science and Technology*, 23 (1989): 303–310.
14. O. Hosseinaei, S. Wang, A.A. Enayati et al. Effects of Hemicellulose Extraction on Properties of Wood Flour and Wood–Plastic Composites. *Composites Part A*, 43 (2012): 686–694.
15. C. Klason, J. Kubát, and H.E. Strömvall. The Efficiency of Cellulosic Fillers in Common Thermoplastics. Part 1: Filling without Processing Aids or Coupling Agents. *International Journal of Polymeric Materials and Polymeric Biomaterials*, 10 (1984): 159–187.
16. J.M. Felix and P. Gatenholm. The Nature of Adhesion in Composites of Modified Cellulose Fibers and Polypropylene. *Journal of Applied Polymer Science*, 42 (1991): 609–620.

9

Polysaccharide Bio-Based Composites: Nanofiber Fabrication and Application

Jackapon Sunthonrvarabhas, Klanarong Sriroth, and Hyun-Joong Kim

CONTENTS

9.1 Introduction

Nanofibers[1] are useful for filtration systems, micro- and nano-electronic sensors, protective clothing, and many applications in the life sciences because of their large surface area per unit of mass or volume. The most promising processes, among others, include chemical vapor deposition, electrical arc discharge, and self-assembly nanostructures is an electrospinning process.[2] From an engineering view, the electrospinning process offers low temperature operation, ease of process set-up and modification, multiple process parameters, and the ability to produce continuous fibers and various fiber morphologies. Other processes exhibit only some of these advantages.

Life science applications such as tissue scaffolding and wound dressing[3–7] are common devices made from electrospinning. The process is suitable for making fibrous articles due to its abilities to connect pores in fiber sheets, control fiber sizes and characteristics, and create aligned-fiber product patterns. These qualities have attracted great interest in research and development targeted at fabricating even more advanced materials. At present, synthetic biodegradable polymers (polylactic acid, polycaprolactone, and polyglycolic acid) and natural polymers (chitosan, starch, and cellulose) are widely used in tissue scaffolding and wound dressing applications.

Starch[8,9] has great potential for use as a biomaterial in biomedical applications due to its biocompatibility and biodegradability. However, starch composite articles exhibit inferior mechanical properties, and this limits the use of starch in many applications. Two possible approaches were introduced to solve this problem. The first is the modification of starch into a higher form possessing properties suitable for specific products. The second is modifying the engineering approach. By introducing innovative methods and implementing chemical modifications into the fabrication process, new forms of materials are being developed continuously.

9.2 Polysaccharide Composites

The use of polysaccharide composites has expanded into many new areas, for example, non-food applications such as agricultural and packaging products in the forms of thermoplastic starch composites. Thermal processing allows the use of reduced size starch granules as reinforcing agents in rubber composites.[10]

Many applications have been developed to replace synthetic polymers because of environmental concerns about their biodegradability and the need to use fossil-based materials to produce them. Polysaccharides, especially starch, have recently gained interest for biodegradable composite applications. Polysaccharides can be obtained from a wide variety of plants; they are renewable and economical to produce. They are versatile enough to be used for many applications.

Due to their low strength and water absorption properties, starch composites have received much attention as fillers and degradation regulators for packaging applications. Research and development focuses on developing non-food products made of pure starch that exhibit sufficient strength for handling and suitable chemical and biological profiles.

Another area of interest targeting starch as a vital component is the biomedical field. Because of its nontoxic structure and natural occurrence, starch is safer to use than synthetic materials. Its primary disadvantages concern product fabrication and properties. Due to inferior mechanical properties, starch is not the main component in many products. Starch composites have limited working range, and the protocol for crafting articles is tedious. The next section will guide readers through a brief report on polysaccharide composite development in the biomedical field, with an emphasis on electrospinning process advances.

9.3 Medical Product Development

Specialized polymer materials and engineering structures dedicated to specific applications are vital for developing state-of-the-art materials and devices for the medical industry and redefining traditional approaches to human health care.

9.3.1 Requirements

Biomaterials are intended to interface with biological systems to evaluate, treat, augment, or replace tissues, organs, or body functions.[11] Many researchers have reported the use of biodegradable polymers as tissue scaffolds and wound dressings. Biodegradable polymers for use in medical applications should meet the following suitability requirements to be defined as materials for medical use.[12]

- A material should not trigger inflammatory or toxic responses upon implantation.
- A material should have an acceptable shelf life appropriate to the healing process.
- The degradation time should be compatible with the healing or regeneration process.
- A material should have the mechanical properties required for the specific application; variations in mechanical properties through degradation should be compatible with the healing or regeneration process.
- The degradation products should be non-toxic and capable of being metabolized and excreted from the body.
- A material should have appropriate permeability and processibility for the targeted application.

The degradation times and reactions of biodegradable materials for tissue scaffold and wound dressing applications have been reported in a review paper elsewhere.[13] Many polymers such as polylactic acid polycaprolactone and co-grafted biodegradable polymers have served as base materials.[3,14] Most of the synthetic biodegradable materials developed for these applications involved polylactic acid, polyglycolic acid, and polycaprolactone because these compounds have been approved for use by the U.S. Food and Drug Administration (FDA).[15,16] Mechanical properties, degradation profile, and biocompatibility are the key factors that must be considered when incorporating polysaccharides into products.

9.3.2 Fabrication

Many techniques have been proposed for fabricating sub-micron fibrous articles.[17] Each fabrication process generates products with unique properties, fabrication requirements, and limitations (Table 9.1). Fibrous articles produced by electrospinning are popular among researchers because of pore connectivity and the ability to control pore size distribution to suit fiber dimensions and characteristics.[17]

Another reason to use electrospinning to make fibrous articles is that most drugs and proteins are charged materials that tend to attach to the large surface areas of fibers. This makes them suitable for loading materials required for tissue scaffolding and wound dressing applications.[18] An important aspect in the production of tissue scaffolds is need for reproducibility of the micron and sub-micron structures;[19] this is another capability of the electrospinning process.

9.3.3 Electrospinning Technology

Electrostatic spinning or electrospinning utilizes electrical force as a means to create a charged liquid jet from a charged liquid surface and stretch it to produce a dry fiber.[2,20–26] It is possible to create sub-micron fibers of multiple components and different fiber shapes by modification of the basic process to include additional materials such as salts and surfactants in the final product.[26–32]

Process equipment for traditional single-nozzle electrospinning consists of four primary components (1) a high voltage power supply, (2) metal nozzle, (3) system to deliver fluid from a reservoir to a metal nozzle, and (4) a collector (Figure 9.1). The process starts by forming a pendant drop of polymer on the tip of a nozzle (common metal needle) and applying high voltage to the reservoir or metal nozzle.

Charges are delivered from the power supply through an electrode and transferred into the liquid. The charges also create differences in voltage that later generate electric field strength between electrode and collector. This field subjugates electrical force on charges on the liquid surface. When sufficient charges are produced to create an adequate electrical force to act on the liquid surface, a charged liquid jet will launch from the apex of the deformed charged polymeric pendant droplet. The droplet is formed by the electrical force acting on both the vertical and horizontal liquid surfaces.[33]

The charged liquid jet launched from the apex of the pendant droplet travels a few centimeters or more in a straight line; the traveling distance of charged jet depends on operating conditions, polymer properties, and environmental parameters. A bending instability mechanism caused by surrounding noise and perturbation frequency[34] will create an imbalance of net electrical force in a direction perpendicular to the path of the jet.

When instability occurs, the jet is stretched in a plane direction. Inertia causes a downward motion similar to a swirl or cone shape. Stretching of

TABLE 9.1

Selected Techniques for Porous and Fibrous Ariticle Fabrication

Process	Features	Advantages	Disadvantages	References
Electrospinning	Utilization of electrical force to form fibrous sheet of nanofibers	Many materials can be spun; variety of sizes and fiber morphologies	Low production rate; toxic solvent waste	2, 20, 22, 34, 63, 64
Drop-on-demand printing (SFF)	Based on thermoplastic ink jetting technology; pump molten polymer into mold to create porous product	Narrow pore distribution can be achieved	Not appropriate for multiple materials due to difficult process set-up	65
Three-dimensional printing (SFF)	Mold injection technique on complex platform can be used as polymer solution	Designed article pore size can be achieved; multiple materials are possible in solution format	Multiple sequence technique; not applicable for materials containing solids	66
Paraffin sphere dissolution	Bonded paraffin spheres formed by melting and filling template with desired material and dissolving paraffin to form porous product	Ensures interconnected pores	Broad pore size distribution due to difficulty of filling sphere-bonded paraffin void	67
Selective laser sintering (RP)	Build material layer by layer using laser to bond powder materials	Evenly dispersed pore sizes; commercial non-biocompatible modeling materials are available and may be used as guides for biocompatible materials	Requires material that does not degrade over laser beam	68, 69
Room temperature injection molding with particle leaching (IM)	Inject polymer and other materials into mold; leach particles out to form porous structure	Low temperature avoids degradation of material; can shape complex article architectures	Shrinkage possible if material is soft; caution required to avoid deformation during demolding	70

Continued

TABLE 9.1 (Continued)

Selected Techniques for Porous and Fibrous Ariticle Fabrication

Process	Features	Advantages	Disadvantages	References
GELPOR3D (IM)	Create slurry of material and mold it in a 3D structure to form porous product	Non-toxic process; low cost, uses variety of materials	Random and non-interconnected pores	71
Air pressure jet solidification (IM)	Molten material is extruded through a small nozzle by compressed air controlled by computer; porous structure is created by adding filament layer by layer	Non-toxic process; no distortion of material after solidification	Caution required during curing to prevent destruction of internal material	72
Microstereolithography (LBS)	Use physical masking to create a point where proton is in contact with light-sensitive polymer and use it to cure polymer to produce a 3D fibrous and porous structure	Can create extremely fine structures ($\sim$2 to 3 μm)	More material limitations than other techniques	73

SFF = solid free form. RP = rapid prototyping. IM = injection molding. LBS = laser-based scanning.

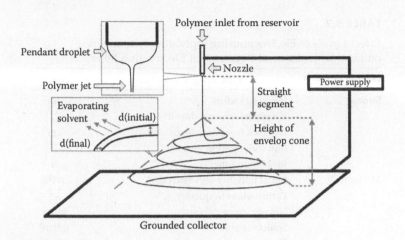

FIGURE 9.1
Electrospinning process.

the charged liquid jet decreases jet diameter or increases the surface area; these changes are also caused by solvent evaporation. Evaporating solvent evaporates as the charged and solidifying liquid jet moves downward. This produces dried fiber at the collector.[35]

9.3.3.1 Process Parameters

The details of electrospinning processing parameters have been reported in the literature and will not be discussed in great detail here.[35–39] To determine the relationship of operating parameters and fiber size, researchers collect relevant experimental results and then devise mathematical models to predict fiber size to a certain level of accuracy.

Thompson[34] studied the effect of electrospinning parameters on average fiber diameter by gathering vast amounts of experimental data and using a mathematical model based on the Navier-Stokes equations in a cylindrical coordinate system developed by Yarin[2,22,23,25] to determine effects on average fiber diameter. The effects of operating parameters are reported in Table 9.2. The effects are divided into three categories and should serve as guidelines for parameter adjustment to achieve the desired fiber diameter and morphology and ensure experiment feasibility.

9.3.3.2 Fiber Morphology

Electrospinning creates smooth fibers from many materials and can also fabricate various shapes such as core-shell fibers, beaded fibers, flat fibers, and porous fibers (Table 9.3).[17,40,41] The shape characteristics of electrospun fibers result in different properties such as high aspect ratio and low resistance to mass transport and the characteristics determine applications.[42]

TABLE 9.2

Effect Levels of Electrospinning Operating Parameters
on Final Jet Radius and Diameter of Electrospun Fiber

Degree of Effect	Operating Parameter	Unit
Strong	Initial jet radius	cm
	Volumetric charge density	C/l
	Distance from nozzle to collector	cm
	Initial elongation viscosity	P
	Relaxation time	S
Moderate	Initial polymer concentration	wt%
	Perturbation frequency	s^{-1}
	Solvent vapor pressure	mbar
	Solution density	g/cm^3
	Electrical potential	kV
Minor	Vapor diffusivity	cm^2/s
	Relative humidity of solvent vapor in air	%
	Surface tension	dyne/cm

TABLE 9.3

Principles of and Applications for Electrospun Fibers of Various Morphologies

Fiber Morphology	Principle	Effects and Uses
Porous	Use multiple solvents to create phase separation in fiber matrix; follow by solvent leaching	Affects wettability; better mechanical strength due to wet fiber bonding prior to secondary solvent leaching
Core-shell	Selection of immiscible solvent system; follow by solvent leaching	Material encapsulation; surface area increase
Beaded	Balance of capillary force using low viscosity polymer solution; beads form on electrospun fibers	Material encapsulation; superhydrophobicity from variation of surface properties of rough structure

The formation of a porous fiber[43] system normally requires mixing a minimum of three components systems with two types of solvents to create a spinnable solution.[44] If the evaporating rates of first and second are slightly different, the first evaporating material will cause the jet to form a phase-separated structure.[45] The evaporation of the second material leaves a porous structure fiber at the collector. Normally the second material can be removed by selective dissolution, thermal degradation, or photodegradation depending on material characteristics and application.[45] Before the second material is evaporated, it usually leaves a wet fiber that results in bonding of intersected fibers. The result is better mechanical strength of a porous fiber after the second material is removed.[46]

Fabrication of hollow or core-shell fiber is performed utilizes the concept of selective removal of the inner liquid through extraction or calcination.[47] A spinneret nozzle is used with appropriate materials (two immiscible or poorly miscible liquids with appropriate compositions[48]) to separate fiber material into inner and outer shells.[49,50] After the fibers are formed at the collector, they are subjected to another process to remove the inner material.

Beaded fiber or beaded-on-string fiber fabrication does not require multiple solvents. Fibers are formed by the capillary break-up of spinning jets by surface tension.[51,52] The higher viscosity of a polymer solution favors the formation of smooth fibers but in the opposite direction to polymer surface tension.[30,53] To be more specific, the coiled macromolecules of the dissolved polymer are transformed by the elongation flow of the jet into oriented entangled networks that consist of dried fibers. The contraction of the radius of the jet aroused by surface tension utilizes accounts the remaining solution to form beads.[53]

9.3.4 Electrospinning Fabrication Development

Many research groups have reported success in fabricating material combinations that are accepted by the human body, particularly nanofiber blends. Figure 9.2 summarizes material blending developments for medical nanofiber applications.[54] Natural and synthetic materials were compared and advantages noted for each combination. The figure demonstrates a great technical strategy for material development.

Traditional electrospinning is compatible with solution-based processing; several material combinations were tested for optimum properties during product development. Different techniques to create spinnable solutions were developed. A blending technology for mixed polymers was developed for various applications based on sequence and material combination as shown in Figure 9.3.

For grafted copolymerization, this technique was utilized to improve properties of electrospun fiber sheets through chemical modification. One limitation is that the chemical reaction has an optimum point, but the method provides great material cohesion—an improvement over previous techniques. An emulsion electrospinning technique was introduced and implemented in cases when both solutions A and B were immiscible. The benefit of this technique is the ease of preparation, but its main disadvantage lies in quality of output fibers and contamination of emulsifiers. Phase separation can result from poor distribution and non-optimum material mixing due to inadequate emulsifier. This separation can be viewed as both an advantage and a disadvantage because it can be used to create porous fibers and core-shell fibers after some process modification.

Finally, the conjugated solvent technique was developed to help dissolve two immiscible polymer solutions through polarity index adjustment.[55,56]

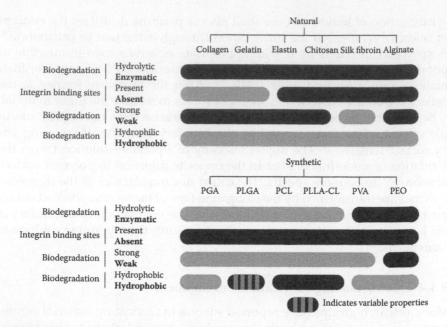

FIGURE 9.2

Blend nanofibers: materials and advantages. (*Source:* Adapted from J. Dunn and M. Zhang. *Biomaterials*, 26 [2005]: 4281–4289. With permission.)

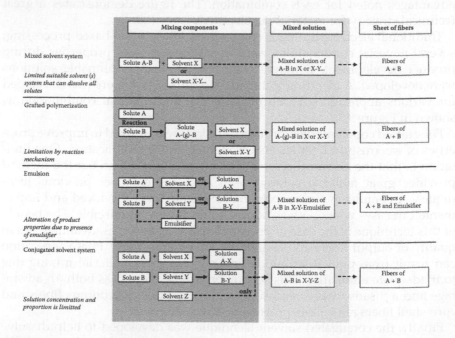

FIGURE 9.3

Mixed solution scheme.

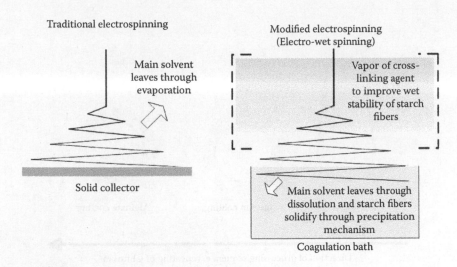

Traditional electrospinning

Modified electrospinning
(Electro-wet spinning)

Main solvent
leaves through
evaporation

Vapor of cross-
linking agent
to improve wet
stability of starch
fibers

Solid collector

Main solvent leaves through
dissolution and starch fibers
solidify through precipitation
mechanism

Coagulation bath

FIGURE 9.4
Electrowet electrospinning.

While the method is useful in some situations, it has a very limited working range and requires further optimization.

As noted previously, researchers involved in developing material combinations and mixed solution techniques are seeking other options for mixed fiber fabrication to improve polysaccharide composite properties by combinations with other biodegradable and synthetic polymers. The following ideas were reported as potential techniques for physical modification of composites.

Electro-wet electrospinning[57,58] was developed to solve pure starch fabrication difficulties; see Figure 9.4. Pure natural starch lacks sufficient physical strength to withstand electrical force by electrospinning. Its chemical structure involving varied amylose and amylopectin ratios in linear and branched chain structures makes fabrication of pure starch difficult. A commercial modified corn starch known as Hylon 7 contains amylose (linear chain structure), around 70 wt%. Because starch becomes a precipitated solid after contact with alcohol, an alcohol bath was adopted to create a precipitation mechanism by which a wetted starch-charged jet came into contact with liquid. The precipitation promotes solvent leaching and thus solidifies starch fibers as the charges transfer to grounded electrodes through the liquid surface. To improve wet stability during the process sequence, fiber was introduced into a liquid bath containing a cross-linking agent to strengthen fiber integrity. As charged jets of starch fibers were exposed to glutaraldehyde vapor, fiber without the cross-linking agent was unable to form a rigid structure and disintegrated upon collection from the liquid bath. This work was the first to report successful fabrication of pure starch fibers. Previous studies only claimed the possibility of creating starch fibers using biodegradable polymers

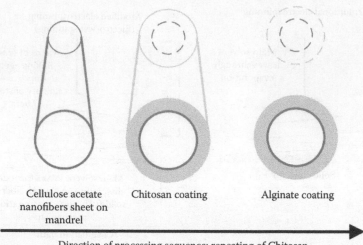

Cellulose acetate   Chitosan coating   Alginate coating
nanofibers sheet on
mandrel

Direction of processing sequence; repeating of Chitosan
and Alginate coating until desired layering is achieved

FIGURE 9.5
Electrospinning with alternate electrostatic coating.

as backbone structures.[55,59] These combinations opened new opportunities for research and development in the starch community.

Another interesting technique is the utilization of material coating or electrostatic layer-by-layer self-assembly and electrospinning.[60] Electrostatic coating was promoted in the past for thin-film preparation involving alternating adsorption of opposite charged ions, particles, and polyelectrolyte onto a substrate and used mostly for electronic application (Figure 9.5). Deng's work introduced a combination of coating and electrospinning fiber sheets and demonstrated a unique way of depositing a polysaccharide composite on biomaterial (chitosan) for tissue scaffold uses. The study showed that biocompatibility was strongly influenced by the outer layer material and the number of coating layers. Increasing number of layers through electrostatic coating increased cell binding site density accordingly and also promoted cellular biocompatibility.

Third concept showed the potential of utilizing electrospinning (Figure 9.6) and electrospraying (Figures 9.6 and 9.7) in an electrospinning set-up. The two mechanisms differ mainly by the natures and concentrations of the polymers used. When a solute is dissolved into a suitable solvent, a random coil of polymer is untwisted and the polymer solution is swollen, producing a greater polymer viscosity than that produced by a pure solvent. High polymer concentration equates to more polymer chain entanglement within the body of a polymer solution. Lower polymer concentration demonstrates lower polymer chain entanglement in proportion to concentration value. Each polymer-and-solvent pair that becomes a solution has a critical concentration at which the

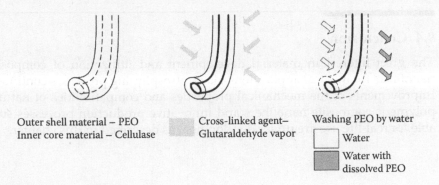

Outer shell material – PEO
Inner core material – Cellulase

Cross-linked agent–
Glutaraldehyde vapor

Washing PEO by water

Water

Water with
dissolved PEO

FIGURE 9.6
Electrospin spray mechanism scheme.

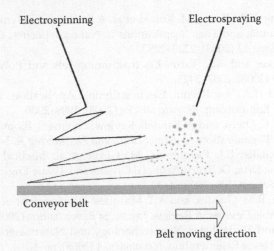

Electrospinning

Electrospraying

Conveyor belt

Belt moving direction

FIGURE 9.7
Alternative electrospin spray mechanism scheme.

mechanisms of spraying and spinning are distinguished. This concept makes it possible to optimize polymer solution properties for both electrospinning and electrospraying within a single system.

By altering surface characteristics through depositing small particles via electrospraying on an electrospun fibrous surface, surface properties can be regulated based on particle surface roughness and surface composition.[61] This ability allows an experimenter to determine optimum surface properties for products through physical modification schemes.

The combination of spin and spray techniques is not limited to surface regulation of physical structure. Spraying affects surface properties via chemical reactions. Cross-linking agents can be sprayed on top of electrospun fibrous sheets to enhance and alter their surface properties.[62]

9.4 Conclusion

The great interest in material development and utilization of composite nanofibers for medical applications via electrospinning processes continues. Improvements of the mechanical properties and compatibilities of natural polymeric composite nanofibers and innovative production processes suitable for real-life use are under development as described in this chapter.

References

1. Z.M. Huang, Y.Z. Zhang, M. Kotaki et al. A Review on Polymer Nanofibers by Electrospinning and Their Applications in Nanocomposites. *Composites Science and Technology,* 63 (2003): 2223–2253.
2. D.H. Reneker and A.L. Yarin. Electrospinning Jets and Polymer Nanofibers. *Polymer,* 49 (2008): 2387–2425.
3. T.J. Sill and H.A. Von Recum. Electrospinning: Applications in Drug Delivery and Tissue Engineering. *Biomaterials,* 29 (2008): 1989–2006.
4. B.L. Seal, T.C. Otero, and A. Panitch. Review: Polymeric Biomaterials for Tissue and Organ Regeneration. *Material Science and Engineering R,* 34 (2001): 147–230.
5. D.F. Stamatialis, B.J. Papenburg, M. Giron et al. Medical Applications of Membranes: Drug Delivery, Artificial Organs and Tissue Engineering. *Journal of Membrane Science,* 308 (2008): 1–34.
6. A.C. Vieira, R.M. Guedes, and A.T. Marques. Development of Ligament Tissue Biodegradable Devices: A Review. *Journal of Biomechanics,* (2009).
7. L. Zhang and T.J. Webster. Nanotechnology and Nanomaterials: Promises for Improved Tissue Regeneration. *Nanotoday,* 4 (2009): 66–80.
8. R.F. Tester, J. Karkalas, and X. Qi. Starch: Composition, Fine Structure, and Architecture. *Journal of Cereal Science,* 39 (2004): 151–165.
9. I. Šimkovic. What Could Be Greener Than Composites Made from Polysaccharides? *Carbohydrate Polymers,* 74 (2008): 759–762.
10. J.L. Willett. *Starch Chemistry and Technology.* 2009.
11. D.F. Williams. *The Williams Dictionary of Biomaterials.* Liverpool University Press, 1999.
12. A.W. Lloyd. Interfacial Bioengineering to Enhance Surface Biocompatibility. *Medical Device Technology,* 13 (2002): 18–21.
13. K.F. Leong, C.K. Chua, Sudarmadji, K. et al. Engineering Functionally Graded Tissue Engineering Scaffolds. *Journal of the Mechanical Behavior of Biomedical Materials,* 1 (2008): 140–152.
14. S. Agarwal, J.H. Wendoroff, and A. Greiner. Use of Electrospinning Technique for Biomedical Applications. *Polymer,* 49 (2008): 19.
15. D. Rohner, D.W. Hutmacher, T.K. Cheng et al. In Vivo Efficacy of Bone Marrow-Coated Polycaprolactone Scaffolds for the Reconstruction of Orbital Defects in the Pig. *Journal of Applied Biomaterials,* 66B (2003): 574–580.

16. L.J. Smith, J.S. Swaim, C. Yao et al. Increased Osteoblast Cell Density on Nano-structured PLGA-Coated Nanostructured Titanium for Orthopedic Applications. *International Journal of Nanomedicine*, 2 (2007): 493–499.
17. K. Tuzlakoglu and R.L. Reis. Biodegradable Polymeric Fiber Structures in Tissue Engineering. *Tissue Engineering Part B*, 15 (2009): 17–27.
18. X Zong, K. Kim, D. Fang et al. Structure and Process Relationship of Electrospun Bioabsorbable Nanofiber Membranes. *Polymer*, 43 (2002): 4403–4412.
19. E. Stratakis, A. Ranella, M. Farsari et al. Lasser-Based Micro and Nano Engineering for Biological Applications. *Progress in Quantum Electronics*, 33 (2009): 127–163.
20. A. Formhals. Method and apparatus for the production of fibers. US 2123992A, 1938.
21. A. Formhals. Method and apparatus for spinning. US 2349950A. 1944.
22. D.H. Reneker, A.L. Yarin, H. Fong et al. Bending Instability of Electrically Charged Liquid Jets of Polymer Solutions in Electrospinning. *Journal of Applied Physics*, 87 (2000): 4531–4547.
23. A. L. Yarin, S. Koombhongse, and D. H. Reneker. Taylor Cone and Jetting from Liquid Droplets in Electrospinning of Nanofibers. *Journal of Applied Physics*, 90 (2001): 4836–4846.
24. Y. Dzenis. Spinning Continuous Fibers for Nanotechnology. *Science*, 304 (2004): 1917–1918.
25. A.L. Yarin, W. Kataphinan, and D. H. Reneker. Branching in Electrospinning of Nanofibers. *Journal of Applied Physics*, 98 (2005): 064501/1–064501/12.
26. K.M. Sawicka and P. Gouma. Electrospun Composite Nanofibers for Functional Applications. *Journal of Nanoparticle Research*, 8 (2006): 769–781.
27. N. Zhao, S. Shi, G. Lu et al. Polylactide (PLA)-Layered Double Hydroxide Composite Fibers by Electrospinning Method. *Journal of Physics and Chemistry of Solids*, 69 (2008): 1564–1568.
28. S.Q. Wang, J.H. He, and L. Xu. Non-Ionic Surfactants for Enhancing Electrospin Ability and for the Preparation of Electrospun Nanofibers. *Polymer International*, 57 (2008): 1079–1082.
29. D. Fallahi, M. Rafizadeh, N. Mohammadi et al. Effect of LICL and Non-Ionic Surfactant on Morphology of Polystyrene Electrospun Nanofibers. *e–Polymer*, (2008).
30. S. Tripatanasuwan, Z. Zhong, and D.H. Reneker. Effect of Evaporation and Solidification of the Charged Jet in Electrospinning of Poly(Ethylene Oxide) Aqueous Solution. *Polymer*, 48 (2007): 5742–5746.
31. G. Eda, J. Liu, and S. Shivkumar. Solvent Effects on Jet Evolution during Electrospinning of Semi-Dilute Polystyrene Solutions. *European Polymer Journal*, 43 (2007): 1154–1167.
32. K. Lin, K.N. Chua, G.T. Christopherson et al. Reducing Electrospun Nanofiber Diameter and Variability Using Cationic Amphiphiles. *Polymer*, 48 (2007): 6384–6394.
33. T. Takamatsu, Y. Hashimoto, M. Yamaguchi et al. Theoretical and Experimental Studies of Charged Drop Formation in a Uniform Electric Field. *Journal of Chemical Engineering of Japan*, 14 (1981): 178–182.
34. C.J. Thompson, G.G. Chase, A. L. Yarin et al. Effects of Parameters on Nanofiber Diameter Determined from Electrospinning Model. *Polymer*, 48 (2007): 6913–6922.
35. J. Sunthornvarabhas, G.G. Chase. Novel Techniques for Mass Production of Electrospun Fibers: Design and Experiment. Publisher: VDM Verlag Dr. Müller. ISBN: 978-3639322989 (2011).

36. A. Kilic, F. Oruc, and A. Demir. Effects of Polarity on Electrospinning Process. *Textile Research Journal*, 78 (2008): 532–539.
37. Y. Yang, Z. Jia, J. Liu et al. Effects of Flow Rate and the Distance between the Nozzle and the Target on the Operating Conditions of Electrospinning. *Journal of Polymer Engineering*, 28 (2008): 67–86.
38. Y. Yang, Z. Jia, J. Liu et al. Effect of Solution Rate of Electrospinning. *Annual Report: Conference on Electrical Insulation and Dielectric Phenomena*, 2 (2007): 615–618.
39. D. Fallahi, M. Rafizadeh, N. Mohammadi et al. Effect of Applied Voltage on Jet Electric Current and Flow Rate in Electrospinning of Polyacrylonitrile Solutions. *Polymer International*, 57 (2008): 1363–1368.
40. D. Liang, B.S. Hsiao, and B. Chu. Functional Electrospun Nanofibrous Scaffolds for Biomedical Applications. *Advanced Drug Delivery Reviews*, 59 (2007): 1392–1412.
41. S. Soliman, S. Pagliari, A. Rinaldi et al. Multiscale Three–Dimensional Scaffolds for Soft Tissue Engineering via Multimodal Electrospinning. *Acta Biomaterialia*, (2009).
42. H. Kang, Y. Zhu, X. Yang et al. Novel Catalyst Based on Electrospun Silver-Doped Silica Fibers with Ribbon Morphology. *Journal of Colloid and Interface Science*, 341 (2010): 303–310.
43. L. Li and Y.L. Hsieh. Chitosan Bicomponent Nanofibers and Nanoporous Fibers. *Carbohydrate Research*, 341 (2006): 374–381.
44. Z. Qi, H. Yu, Y. Chen et al. Highly Porous Fibers Prepared by Electrospinning a Ternary System of Nonsolvent–Solvent Poly(L–Lactic Acid). *Materials Letters*, 63 (2009): 415–418.
45. Y. You, S.W. Lee, J.H. Youk et al. In Vitro Degradation Behaviour of Non-Porous Ultra-Fine Poly(Glycolic Acid)–Poly(L–Lactic Acid) Fibres and Porous Ultra-Fine Poly(Glycolic Acid) Fibres. *Polymer Degradation and Stability*, 90 (2005): 441–448.
46. S.O. Han, W.K. Son, J.H. Youk et al. Ultrafine Porous Fibers Electrospun from Cellulose Triacetate. *Materials Letters*, 59 (2005): 2998–3001.
47. S. Zhan, D. Chen, and X. Jiao. Co-Electrospun SiO_2 Hollow Nanostructured Fibers with Hierarchical Walls. *Journal of Colloid and Interface Science*, 318 (2008): 331–336.
48. H. Na, X. Liu, J. Li et al. Formation of Core-Shell Ultrafine Fibers of PVDF–PC by Electrospinning via Introduction of PMMA or BTEAC. *Polymer*, 50 (2009): 6340–6349.
49. L. Ge, C. Pan, H. Chen et al. Fabrication of Hollow Multilayered Polyelectrolyte Fibrous Mats and Morphology Study. *Colloids and Surfaces A*, 293 (2007): 272–277.
50. X. Li, X. Hao, H. Yu et al. Fabrication of Polyacrylonitrile–Polypyrrole (PAN-PPY) Composite Nanofibres and Nanospheres with Core-Shell Structures by Electrospinning. *Materials Letters*, 62 (2008): 1155–1158.
51. A.L. Yarin. *Free Liquid Jets and Films: Hydrodynamics and Rheology*. New York: John Wiley & Sons, 1993.
52. Y.I. Yoon, H.S. Moon, W.S. Lyoo et al. Superhydrophobicity of PHBV Fbrous Surface with Bead-on-String Structure. *Journal of Colloid and Interface Science*, 320 (2008): 91–95.
53. H. Fong, I. Chun, and D. H. Reneker. Beaded Nanofibers Formed during Electrospinning. *Polymer*, 40 (1999): 4585–4592.

54. J. Gunn and M. Zhang. Polyblend Nanofibers for Biomedical Applications: Perspectives and Challenges. *Trends in Biotechnology,* 28 (2010): 189–197.
55. J. Sunthornvarabhas, P. Chatakanonda, K. Piyachomkwan et al. Electrospun Polylactic Acid and Cassava Starch Fiber by Conjugated Solvent Technique. *Materials Letters,* 65 (2011): 985–987.
56. J. Sunthornvarabhas, P. Chatakanonda, K. Piyachomkwan et al. Electrospun Starch Composite Nanofibers: Fabrication and Characterization. Paper presented at Sixth International Conference on Starch Technology, Bangkok, 2011.
57. L. Kong and G.R. Ziegler. Fabrication of Pure Starch Fibers by Electrospinning. *Food Hydrocolloids,* 36 (2014): 20–25.
58. L. Kong and G.R. Ziegler. Quantitative Relationship between Electrospinning Parameters and Starch Fiber Diameter. *Carbohydrate Polymers,* 92 (2013): 1416–1422.
59. J. Sunthornvarabhas, P. Chatakanonda, K. Piyachomkwan et al. Method of Preparation of Electrospun Composite Fiber between Starch and Polylactic Acid from Multicomponents Solution. Thailand Patent 1101000490, 2010.
60. H. Deng, X. Zhou, X. Wang et al. Layer-by-Layer Structured Polysaccharide Film-Coated Cellulose Nanofibrous Mats for Cell Culture. *Carbohydrate Polymers,* 80 (2010): 474–479.
61. J. Sunthornvarabhas, P. Chatakanonda, K. Piyachomkwan et al. Physical Structure Behavior to Wettability of Electrospun Poly(Lactic Acid)–Polysaccharide Composite Nanofibers. *Advanced Composite Materials,* 22 (2013): 401–409.
62. L. Francis, J. Venugopal, M.P. Prabhakaran et al. Simultaneous Electrospin–Electrosprayed Biocomposite Nanofibrous Scaffolds for Bone Tissue Regeneration. *Acta Biomaterialia,* 6 (2010): 4100–4109.
63. A.L. Yarin, S. Koombhongse, and D.H. Reneker. Taylor Cone and Jetting from Liquid Droplets in Electrospinning of Nanofibers. *Journal of Applied Physics,* 90 (2001): 4836–4846.
64. P.D. Dalton, D. Grafahrend, K. Klinkhammer et al. Electrospinning of Polymer Melts: Phenomenological Observations. *Polymer,* 48 (2007): 6823–6833.
65. M.J. Mondrinos, R. Dembzynski, L. Lu et al. Porogen-Based Solid Freeform Fabrication of Polycaprolactone–Calcium Phosphate Scaffolds for Tissue Engineering. *Biomaterials,* 27 (2006): 4399–4408.
66. M. Lee, J.C.Y. Dunn, and B.M. Wu. Scaffold Fabrication by Indirect Three-Dimensional Printing. *Biomaterials,* 26 (2005): 4281–4289.
67. A.W.T. Shum, J. Li, and A.F.T. Mak. Fabrication and Structural Characterization of Porous Biodegradable Poly(Dl-Lactic-Co-Glycolic Acid) Scaffolds with Controlled Range of Pore Sizes. *Polymer Degradation and Stability,* 87 (2005): 487–493.
68. F.E. Wiria, K.F. Leong, C.K. Chua et al. Poly-E-Caprolactone–Hydroxyapatite for Tissue Engineering Scaffold Fabrication via Selective Laser Sintering. *Acta Biomaterialia,* 3 (2007): 1–12.
69. L. Liulan, H. Qingxi, H. Xianxu et al. Design and Fabrication of Bone Tissue Engineering Scaffolds via Rapid Prototyping and CAD. *Journal of Rare Earths,* 25 (2007): 379–383.
70. J. Wu, D. Jing, and J. Ding. A Room-Temperature Injection Molding–Particulate Leaching Approach for Fabrication of Biodegradable Three-Dimensional Porous Scaffolds. *Biomaterials,* 27 (2006): 185–191.

71. J. Peña, J. Román, M.V. Cabañas et al. Alternative Technique to Shape Scaffolds with Hierarchical Porosity at Physiological Temperature. *Acta Biomaterialia*, (2009).

72. Z. Chen, D. Li, B. Lu et al. Fabrication of Osteo-Structure Analogous Scaffolds via Fused Deposition Modeling. *Scripta Materialia*, 52 (2005): 157–161.

73. J.W. Choi, R. Wicker, S.H. Lee et al. Fabrication of 3D Biocompatible–Biodegradable Micro-Scaffolds Using Dynamic Mask Projection Microstereolithography. *Journal of Materials Processing Technology*, 209 (2009): 5494–5503.

10

Recent Advances in Cellulose Nanocomposites

Ali Faghihnejad and Hongbo Zeng

CONTENTS

10.1 Introduction

Composites are combinations of two or more components mixed together to produce materials with enhanced thermal, mechanical, or electrical properties different from the properties of the individual original components.[1,2] A wide range of composite materials such as metals, ceramics, and polymers have been used in construction materials, electronics components, and biomaterials.[3–7] With the rapid advance of nanotechnology, new composite materials known as nanocomposites have emerged; at least one component must be smaller than 100 nm.

Nanocomposites are made of various materials such as polymer electrolytes[8] and biological materials.[9,10] One recent approach in the development of nanocomposites is based on cellulose. Cellulose belongs to a large class of substances known as carbohydrates consisting only of carbon, hydrogen, and oxygen. Cellulose is a linear polysaccharide polymer consisting of glucose units connected via β-1,4 links.[7] Plants are the main sources of cellulose in nature. Cellulose is a major component of plant cell walls and is also produced by some bacteria such as *Acetobacter xylinum* (*A. xylinum*).

Figure 10.1 shows the chemical structure of cellulose and the corresponding hydrogen bond network among its polymer chains. The unique properties of cellulose such as high structural strength and stiffness, high degree of

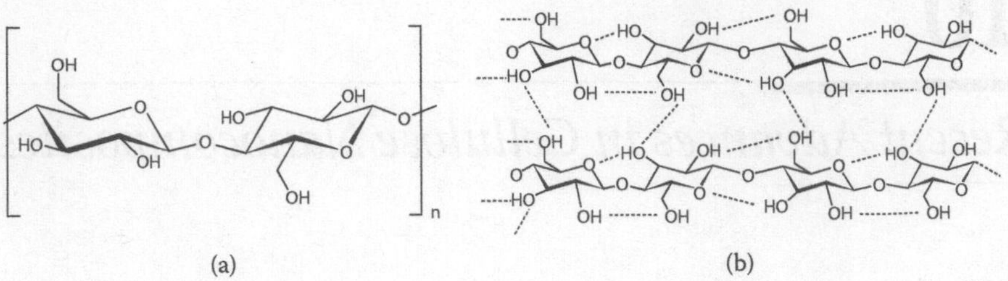

(a)   (b)

FIGURE 10.1
(a) Chemical structure of cellulose. (b) Hydrogen bond network of cellulose.

crystallinity, and insolubility in most organic solvents arise from its hydrogen bond network. The high stiffness level makes cellulose a suitable candidate for use in composite materials to improve their mechanical properties.

In this chapter, recent advances in the preparation and characterization of cellulose nanofibers, cellulose nanocomposites and their surface characterization and potential applications are reviewed.

10.2 Preparation and Characterization of Cellulose Nanofibers

Cellulose in plant cell walls forms a hierarchical structure in which nanowhiskers (also known as nanocrystals or nanofibrils) form microfibrils that in turn form cellulose fibers. Two basic methods are used to prepare cellulose nanofibers (CNFs or nanowhiskers) for inclusion in cellulose nanocomposites. The first method is based on extracting nanofibers from plant cell walls. The second method exploits biological species such as certain strains of bacteria that intrinsically produce cellulose nanofibers.

Many natural sources such as cotton, straw, sisal, soybean, and spruce have been used to prepare cellulose nanowhiskers.[7,11] Cellulose fibers are treated with strong acids such as sulfuric acid (H_2SO_4) or hydrochloric acid (HCl). The treatment leads to hydrolysis of amorphous regions of the cellulose and leaves crystalline parts intact. The solution is further mixed with deionized water and then centrifuged and sonicated to remove sulfuric acid and other residues, leaving a suspension of cellulose nanowhiskers in water. Other materials, such as N,N-dimethyl acetamide containing lithium chloride, have also been used for the extraction of cellulose nanowhiskers from plant cellulose fibrils.

Acetobacter xylinum is useful for bacterial cellulose extraction. Bacterial growth is performed in static or agitated cultures in the presence of abundant oxygen. During cultivation, the bacteria synthesize cellulose pellicles (networks of cellulose fibers). The cellulose nanofibers can then be purified and extracted in 1% NaOH solution.[11]

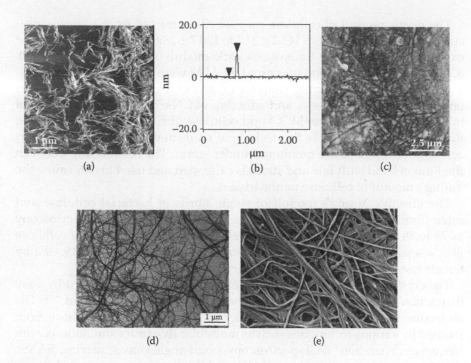

FIGURE 10.2
(a) AFM topography image of cellulose nanowhiskers after drying on mica surface. (b) Line scan across individual whisker. (c) AFM image of cellulose fibrils dispersed within polymer matrix of hydroxypropyl cellulose (HPC). (d) TEM image of individual cellulose microfibrils. (e) SEM image of bacterial cellulose. (Sources: I. Kvien, B.S. Tanem, and K. Oksman. *Biomacromolecules*, 6 [2005]: 3160–3165; T. Zimmermann, E. Pöhler, and P. Schwaller. *Advanced Engineering Materials*, 7 [2005]: 1156–1161; A. Dufresne, J. Cavaille, and M.R. Vignon. *Journal of Applied Polymer Science*, 64 [1997]: 1185–1194; S. Ifuku, M. Nogi, K. Abe et al. *Biomacromolecules*, 8 [2007]: 1973–1978. With permission.)

Cellulose nanowhiskers have been characterized by different techniques. Nanowhisker dimensions are generally between 100 and 400 nm in length and 10 nm in width as measured by transmission electron microscopy (TEM), scanning electron microscopy (SEM), and field emission scanning electron microscopy (FESEM).[7,11–15] The morphology and mechanical properties of cellulose nanowhiskers have been examined using atomic force microscopy (AFM).[12,16–18] Figure 10.2 shows typical morphologies of different cellulose fibrils revealed by AFM, SEM, and TEM.

The mechanical properties (i.e., Young's modulus) of cellulose nanocomposites can exceed the properties of traditional microfiber cellulose because of the performance of cellulose nanowhiskers. The Young's moduli of nanocomposites can be enhanced only if the length-to-diameter aspect ratio of the nanowhiskers is larger than 50.[7] The Young's moduli of bacterial nanowhiskers were measured by AFM to be 78 ± 17 GPa with diameters of nanofibers ranging from 35 to 90 nm.[16]

The elastic moduli of cellulose microfibrils extracted from tunicate measured by AFM ranged from 145.2 ± 31.1 to 150.7 ± 28.8 GPa, depending on the extraction method.[17] The transverse elastic moduli of cellulose nanocrystals (CNs) extracted from wood measured by AFM were 18 to 50 GPa.

Relative humidity ranging from 0.1 to 30% exerted no significant effects on the topography, stiffness, and adhesion of CNs.[18] The Young's moduli of microcrystalline cellulose (MCC) and cellulose fibrils were estimated using Raman spectroscopy.[19–21] In this technique, shifts in the 1,095 cm⁻¹ band characteristic of cellulose are monitored under strain. The relationship between the Raman band shift rate and strain is calibrated and used to determine the Young's moduli of cellulose nanowhiskers.

The effective Young's moduli of single fibrils of bacterial cellulose and microfibrillated cellulose (MFC) were measured using Raman spectroscopy as 79 to 88 and 29 to 36 GPa, respectively.[20] The Young's moduli of cellulose fibrils separated from sulfite pulp were measured to be of 6 ± 0.9 GPa by tensile tests and nanoindentation experiments.[22,23]

The crystalline structures of cellulose nanowhiskers were studied by x-ray diffraction (XRD) and degrees of crystallinity were determined.[19,24] The molecular and structural characteristics of cellulose nanofibers have been studied by various techniques such as molecular dynamics simulations, sum frequency generation (SFG) spectroscopy, small-angle x-ray scattering (SAXS), small-angle neutron scattering (SANS), and Raman spectroscopy.[25–28]

10.3 Preparation and Characterization of Cellulose Nanocomposites

The two basic routes for the preparation of cellulose nanocomposites are (1) adding cellulose nanofibers to matrix material to produce reinforced matrix nanocomposites in which cellulose typically constitutes less than 30 wt% of the nanocomposites and (2) modification of bare cellulose nanofibers to produce nanocomposites in which cellulose normally comprises more than 70 wt%.[2]

10.3.1 Polymer-Based Cellulose Nanocomposites

Different polymers such as poly-β-hydroxyoctanoate (PHO),[29,30] polyvinyl chloride (PVC),[31] polyaniline,[32] polylactic acid (PLA),[11,33,34] polymethyl methacrylate (PMMA),[35,36] poly-L-lactide,[37] polypropylene carbonate,[38] ethylene–vinyl alcohol copolymer,[39] polypropylene,[40] polysaccharide,[15] and polyacrylic acid[41] have been used as matrices in the preparation of cellulose nanocomposites. Among the methods for the preparation of polymer-based cellulose nanocomposites are solution casting, melt compounding, sol–gel processing, and electrospinning.[1,2]

In the solution casting method, cellulose nanofibers are dispersed in a solvent and mixed with the polymer solution to a desired ratio. The mixture is then cast on a substrate surface and solvent is allowed to evaporate to leave a nanocomposite film. A key step in this method is thorough mixing of cellulose nanofibers and polymer solution, commonly performed by magnetic stirring.[10]

The most common polymer solutions are water-based due to the hydrophilic nature of cellulose nanofibers. One of the challenges in the solution casting method is to produce uniform dispersion of cellulose nanofibers in the solvent and in the final polymer matrix. It has been reported that dried cellulose nanofibers are more difficult to disperse and tend to form aggregates that deteriorate the thermal and mechanical properties of the nanocomposites.[1,11]

In general, polar solvents such as N,N-dimethylformamide (DMF), formic acid, and dimethyl sulfoxide (DMSO) that can disrupt hydrogen bond networks of cellulose are applied to disperse cellulose nanofibers. One study[11] showed that MCC was swelled in water, re-dispersed in chloroform, and mixed with a solution of poly(lactic acid) (PLA) in chloroform with a ratio of 95 wt% PLA to 5 wt% MCC. The PLA–MCC mixture was cast in silicon-coated Petri dishes and left at room temperature for a week.

The dispersion of MCC in PLA matrix was confirmed by TEM images, although some MCC aggregates (<50 nm) remained in the matrix. It was found that the tensile strength of PLA–MCC nanocomposites was enhanced by 12% as compared to pure PLA.[11] The storage modulus of the nanocomposites was improved over that of pure PLA at a temperature range of 30 to 90°C. Optical transparency was reduced from more than 86% for pure PLA to less than 5% as measured by UV/Vis spectroscopy (determined as the amount of light transmitted at 550 nm).

In another study, MFC–amylopectin–glycerol nanocomposites were prepared by mixing a MFC suspension in glycerol with an amylopectin aqueous solution. The mixture was then cast in Teflon-coated Petri dishes and dried.[15] The micro-structure analysis of these nanocomposites was performed by FESEM. The MFCs were well dispersed in the composite structure even with high MFC content (70 wt%) with a typical width of 30 ± 10 nm and lengths of several micrometers.

Figure 10.3 shows the storage modulus versus temperature and typical stress–strain curves for MFC–amylopectin–glycerol nanocomposites with different MFC contents.[15] It is clear from Figure 10.3a that the storage modulus dramatically increases as the MFC content increases from 0 to 70 wt%. Moreover, the storage modulus becomes almost constant in a temperature range of –96 to 200°C as the MFC content increases to 70 wt%. This demonstrates the role of MFC in enhancing the thermal stability of the nanocomposite.

The results in Figure 10.3b show that the mechanical properties (tensile strength and Young's modulus) of MFC–amylopectin–glycerol nanocomposites are enhanced dramatically as the MFC content increases due to the

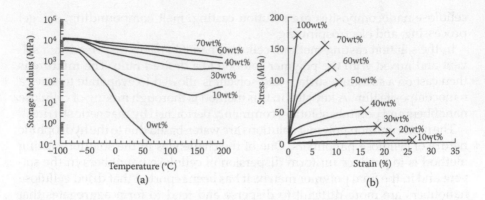

FIGURE 10.3
(a) Storage modulus versus temperature. (b) Typical stress–strain curves of MFC–amylopectin–glycerol nanocomposites with different MFC contents. (*Source:* A.J. Svagan, M.A. Azizi-Samir, and L.A. Berglund. *Biomacromolecules,* 8 (2007): 2556–2563. With permission.)

natural mechanical properties of MFC. This demonstrates the homogeneous dispersion of MFC in the nanocomposites. The tensile strength increased from 0.35 ± 0.05 to 160 ± 7.9 MPa and the Young's modulus increased from 1.6 ± 0.88 to $6,200 \pm 240$ MPa as the MFC content increased from 0 to 70 wt%.[15]

Nanocomposites of ethylene oxide–epichlorohydrin copolymer (EO-EPI)–CNF with a CNF content of 0 to 19 v% were prepared by solution casting.[42] It was found that the tensile storage moduli of dry EO-EPI/CNF nanocomposites increased from ~3.7 MPa for the neat polymer to ~800 MPa for the nanocomposite with 19 v% of CNF.

A unique feature of EO-EPI–CNF nanocomposites is their stimuli-responsiveness, for example, the tensile modulus can be changed by three orders of magnitude upon exposure to different environmental conditions. The tensile storage moduli of EO-EPI–CNF nanocomposites (with 19 v% of CNF) decreased from ~800 MPa for dried samples to 20 MPa after immersion in water for 48 hr that led to the swelling of the nanocomposites. Similarly, the tensile strength of 14.3 v% CNF nanocomposites decreased from 1.71 ± 0.23 to 0.37 ± 0.11 MPa after immersion in water. Nevertheless, the mechanical properties of the EO-EPI/CNF nanocomposites recovered to the original values after drying.

The EO-EPI–CNF nanocomposites were also immersed in isopropanol, which is capable of swelling the polymer but cannot disperse CNF. The tensile storage moduli did not change in comparison with dry samples. These observations demonstrate the solvent specificity of the switching mechanism of the EO-EPI–CNF nanocomposites: water molecules can compete with the hydrogen bonding between cellulose nanofibers and thus the mechanical properties of the nanocomposites are affected by water. Isopropanol cannot disrupt the hydrogen bonding networks of cellulose nanofibers and produces no considerable effect.[42]

An aqueous suspension of cellulose nanofibers (CNFs) was added to a poly(DL-lactide) (PDLLA) suspension with a desired ratio and stirred for 1 hr, vacuum filtered, and then air-dried for 2 to 3 days.[34] The PDLLA–CNF composites were produced with CNF contents of 8, 15, and 32 wt% by compression molding of dried pastes. The width of the CNFs was 20 to 30 nm as determined by TEM images. The storage modulus of PDLLA–CNF composites increased from 3.8 to 6.2 GPa as the CNF content increased from 0 to 32 wt%. The flexural moduli and strengths of PDLLA–CNF composites increased by 58 and 210%, respectively, as compared to the neat PDLLA polymer. This shows the role of CNF networks in enhancing the mechanical capabilities of nanocomposites.[34]

In the sol–gel process, a homogeneous aqueous solution of CNFs is gelated by solvent exchange with a water miscible solvent (acetone) to produce a CNF template gel. The CNF gel is then immersed in a polymer solution so that it is filled with the matrix polymer, then dried to remove the solvent. For successful preparation of cellulose nanocomposites via the sol–gel process, the polymer solvent should not re-disperse the CNFs and must be miscible with the gel solvent.[1,43] An important feature of the sol-gel approach is that it allows the use of hydrophobic polymers such as polypropylene that normally cannot be used in the solution casting method to be used as matrices.

Nanocomposites of polybutadiene (PBD) reinforced with CNF were prepared by the sol–gel method. The shear moduli at 25°C increased from 0.2 MPa for neat PBD to 102 MPa for nanocomposites with 15.8 v% CNF.[43] Nanocomposites of EO-EPI–CNF were prepared by the sol–gel and solution casting methods. Both preparation methods enhanced the mechanical properties of the nanocomposites. The shear moduli of EO-EPI–CNF composites at 25°C increased from 1.3 MPa for the neat polymer to 300 MPa for nanocomposites with 23 v% of CNF.[43] The polystyrene (PS)–CNF nanocomposites were prepared by the sol–gel method and the shear moduli of nanocomposites at 125°C increased from ~1 MPa for the neat PS to ~200 MPa for the nanocomposites with 10 v% of CNF.[43]

Electrospinning has been used for the preparation of polymer-based cellulose nanocomposites.[44–49] A syringe attached to a metallic needle is filled with a mixture of polymer solution and cellulose nanofibers. A high electric voltage is then applied between the needle and a target (i.e., a conductive vertical plate normally 15 to 30 cm away) and the solution is injected out toward the target. The solvent is evaporated as it travels between the needle and the target, and the polymer coagulates and forms a composite.[2]

Aqueous solutions of polyvinyl alcohol (PVA) were added to the CN suspension and stirred. The PVA–CN suspension was then electrospun to produce nanocomposite fibers that were dried under vacuum at 40°C to remove residual water.[44] The morphology of PVA–CN composites was studied using SEM and TEM. The average diameter of the nanofibers was ~235 to 275 nm. The presence of CNs in the electrospun fibers was confirmed by analyzing the cross sections of nanofibers by cryo-SEM and FESEM and further analysis with Fourier-transform infrared spectroscopy (FTIR).

Differential scanning calorimetry (DSC) results showed that the degree of crystallinity of fully hydrolyzed PVA increased from 0.50 to 0.70 upon electrospinning. However, the degree of crystallinity decreased in PVA–CN nanocomposites to 0.56 and 0.54 for CN content of 10 and 15 wt%, respectively. The drop was attributed to the decrease of PVA nucleation in the presence of CNs.[44]

Composites of polymethyl methacrylate (PMMA) with cellulose microfibrils (CMF) extracted from bacterial cellulose were prepared by electrospinning, and CMF content of as high as 20 wt% could be reached. Thermogravimetric analysis (TGA) of PMMA–CMF composites revealed the improved thermal stability of the composite. SEM imaging confirmed the uniform distribution of CMF in the composites.[46]

Nanocomposites of cellulose nanocrystals (CNs) with poly(acrylic acid) (PAA) were prepared by electrospinning of CN–PAA mixtures dissolved in ethanol. The CN–PAA nanocomposites were fabricated with different CN loadings of 5 to 20 wt%. The successful dispersion of CNs into the PAA matrix was confirmed by SEM, TEM, XRD, and FTIR. The average diameters of CN–PAA nanocomposite fibers reduced dramatically from 349 to 69 nm at 5 wt% and 20 wt% CN loading, respectively.

The Young's moduli of the nanocomposites increased dramatically with CN loading and could be enhanced 35-fold by the composite with 20 wt% CNs loading in comparison with results from bare PAA. The CN–PAA nanocomposites were heated at 140°C for 1 hr under N_2 environment, which caused chemical cross-linking of CNs with PAA and made the nanocomposite insoluble in water. The Young's moduli of cross-linked CN–PAA nanocomposites with 20 wt% CNs increased 77-fold over bare PAA.[49]

The composites of poly(vinyl alcohol) (PVA) with cellulose nanowhiskers (CNWs) were prepared by electrospinning and the morphologies of the fibers were studied by SEM. The tensile strength of the CNW–PVA nanocomposites doubled at 15 wt% loading of CNWs over bare PVA.[47]

Electrospinning was also used to fabricate nanocomposites of cellulose nanocrystals (CNs) and poly(vinylidene fluoride-co-hexafluoropropylene) (PH) with CN loading of 1 to 5 wt%. The nanocomposites demonstrated optimum mechanical properties after the addition of 2 wt% of CNs. The tensile modulus of 2 wt% CN–PH increased by 75% over that of bare PH. The thermal and mechanical stabilities of CN–PH nanocomposites in a temperature range of 30 to 150°C were confirmed by dynamic mechanical analysis tests.[48]

10.3.2 Nanocomposites Based on Bare and Modified Cellulose Nanofibers

Bare cellulose nanocrystal (CN) films are usually fabricated by solution casting, which produces films with thicknesses of 25 to 100 μm and typical densities of 0.8 to 1.5 g/cm³.[2] The bare or neat cellulose films are composed of 100% cellulose. The mechanical properties of neat CN films are affected by the orientations, morphologies, and sizes of the CNs along with their interfacial properties and moisture content.[2]

Pulp sheets of 45 μm thick were prepared by solution casting of an aqueous solution with a cellulose fiber content of 0.2 wt%. The Young's modulus and the tensile strength of the pulp sheets were reported as ~15 GPa and 230 MPa, respectively.[50] Neat MFC films were prepared by casting of an amylopectin starch solution in Teflon-coated Petri dishes. The Young's moduli and tensile strengths of the neat MFC films were measured as 13 GPa and 180 MPa, respectively.[15]

MFC films 60 to 80 μm thick were fabricated by vacuum filtration of 0.2 wt% MFC suspension in water. The Young's moduli, yield stresses, and tensile strengths of the films ranged from 10 to 14 GPa, 82 to 92 MPa, and 129 to 214 MPa, respectively. The mechanical properties of the MFC films were enhanced as the molar mass of nanofibrils increased.[51]

In another study, MFC films of ~70 μm thick were prepared by vacuum filtration of 0.5% aqueous suspension. The average Young's moduli of dried MFC films were measured as 14 GPa. The storage moduli of the films decreased from 6.5 to 3.5 GPa as temperature increased from 25 to 225°C. It was also reported that the incorporation of melamine formaldehyde into the MFC films increased the Young's moduli and thermal stability of the films.[52] CNW sheets were prepared by solution casting and the effect of the magnetic field on the mechanical properties of the composite was investigated. The magnetic field affected the orientation of nanowhiskers during the preparation steps and thus affected the mechanical properties of the composite films.[28]

The optical properties of neat and modified cellulose nanofiber films have been investigated.[53,54] Bacterial cellulose (BC) sheets 65 μm thick were prepared by incubating *A. xylinum* bacteria. Acrylic resin and epoxy resin were incorporated into the sheets. Optical appearance and corresponding light transmittance are shown in Figure 10.4.[53]

Cellulose nanofibers were extracted from wood flour to produce a 0.1 wt% aqueous suspension of nanofibers. Cellulose nanofiber sheets were then prepared by vacuum filtration of the aqueous suspension and followed by hot pressing.[54] The results of a light transmittance tests for cellulose nanofiber sheets are shown in Figure 10.4c. The figure shows that the light transmittance of neat cellulose sheets is less than 50% in a wavelength range of 200 to 1000 nm, which leads to loss of transparency.

The low transparency of neat cellulose sheets is mainly due to increased light scattering caused by the large difference between refractive indices of cellulose nanofibers and the porosities within the nanofiber network (i.e., air). The refractive indices of cellulose nanofibers were reported as 1.618 along the fibers and 1.544 in the transverse direction.[53] The transparency of BC sheets could be significantly improved to more than 80% in a wavelength range of 500 to 1000 nm after incorporating resins due to small differences in the refractive indices of cellulose fibers and resins.

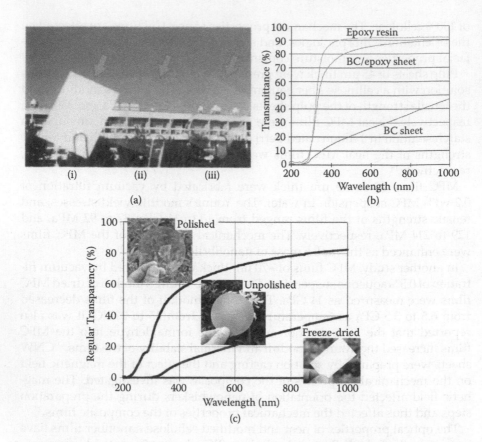

FIGURE 10.4
(a) Appearance of (i) BC sheet; (ii) BC sheet with added acrylic resin; (iii) BC sheet with added epoxy resin. (b) Light transmittance of 65 μm thick BC sheet, BC–epoxy sheet, and epoxy resin sheet. (c) Light transmittance of 60-μm-thick cellulose nanofiber sheets before and after polishing. (*Sources*: H. Yano, J. Sugiyama, A.N. Nakagaito et al. *Advanced Materials*, 17 [2005]: 153–155; M. Nogi, S. Iwamoto, A.N. Nakagaito et al. *Advanced Materials*, 21 [2009]: 1595–1598. With permission.)

It should be noted that the refractive indices of the epoxy resin and acrylic resin are 1.522 and 1.596, respectively.[53] Figure 10.4c shows that the light transmittance of neat cellulose nanofiber sheets significantly improves to more than 70% at a wavelength of 600 nm upon polishing the cellulose sheets by emery paper. Thus, it was concluded that the low transparency of neat cellulose sheets was due to surface light scattering.[54] In summary, the factors that can affect the optical properties of cellulose nanofiber sheets are the sizes of cellulose nanofibrils, film thickness, film roughness, refractive index mismatches of cellulose fibrils and matrix material, and light wavelengths.[2]

As previously noted, the cellulose content of modified cellulose nano-composites is usually more than 70%. Cellulose sheets can be prepared by solution casting of an aqueous suspension of cellulose nanofibrils extracted from sugar beet chips. The corresponding nanocomposites were fabricated by mixing the aqueous suspension of cellulose fibrils with water-soluble PVA and phenol-formaldehyde resin (PF) followed by casting the mixtures into Petri dishes.[55]

Young's moduli of neat cellulose sheets, cellulose (90%)–PVA (10%), and cellulose (90%)–PF (10%) composites were reported as 9.3, 7.7, and 9.5 GPa, respectively. Tensile strengths were measured as 104, 84, and 127 MPa, respectively.[55]

MFC sheets were prepared by vacuum filtration of aqueous suspension of cellulose fibers and then immersed in PF–methanol solution for 48 hr. The MFC–PF composite sheets were dried in ambient conditions for 3 days, then hot pressed at 160°C for 30 min.[56] The Young's modulus of the composites with cellulose contents of 77 wt% was ~16 GPa.[56]

Nanoclays (CLs) have also been used as reinforcing agents in cellulose-based nanocomposites.[57-59] MFC suspension was added to an aqueous suspension of bentonite ($H_2Al_2O_6Si$) CLs and stirred for 2 days. The mixture was then vacuum-filtered for 24 hr. The MFC–CL sheets with clay contents of 0 to 50 wt% were then dried under vacuum for 2 days, then hot pressed at 160°C and 100 MPa.

Tensile strengths and Young's moduli of MFC–CL composites with clay contents of 5 wt% increased to ~105 MPa and 6 GPa, respectively, as compared to initial values of 90 MPa and 5 GPa for neat MFC sheets. However, these parameters decreased as the clay content exceeded 5%, reportedly due to agglomeration of nanoclays.[59] Laponite (LP) clay was incorporated in bacterial cellulose (BC) sheets by immersing the sheets in aqueous suspensions of clay that were then stirred for 3 days. The BC–LP composites with clay contents of 0 to 40 wt% were dried at room temperature for 56 days.

Tensile strengths and Young's moduli of the BC–LP (30 wt%) composites were reported as 227 MPa and 21 GPa—increases of 38% and 81%, respectively, as compared to neat BC sheets.[57] XRD data confirmed the increases of crystalline structures in the nanocomposites as the clay content increased. TGA data further revealed the role of clays in enhancing the thermal stability of BC–LP composites.[57]

A recent study also investigated the application of oxidized cellulose nanofibrils in the preparation of nanocomposites and their mechanical properties.[58] Cellulose nanofibrils were oxidized by 2,2,6,6-tetramethylpiperidine-1-oxyl (TEMPO) radicals and suspended in water. The TEMPO-oxidized cellulose nanofibrils (TOCNs) were then mixed with a montmorillonite (MTM) clay suspension, cast in Petri dishes, and oven-dried at 40°C for 3 days. The TOCN–MTM composites were prepared with MTM contents of 1 to 50 wt% and their optical and mechanical properties were investigated.

Light transmittance decreased with MTM content from ~90% for neat TOCN sheets to less than 50% for TOCN–MTM composites with MTM contents of 50 wt%. Young's modulus increased from 12 to 18 GPa and tensile strength increased from 210 to 510 MPa as the MTM contents increased from 0 to 5%. As the MTM content exceeded 5%, Young's modulus almost remained constant and tensile strength slightly decreased to 450 MPa for an MTM content of 50%.[58]

10.4 Surface Properties of Cellulose and Its Nanocomposites

Understanding the basic molecular interactions between cellulose–cellulose and cellulose–matrix materials and their corresponding surface and interfacial properties can aid the development of new nanocomposites with unique characteristics. Two important properties of cellulose and its derivatives are the Hamaker constant and surface energy as reported in the literature.

In general, various types of intermolecular and surface dynamics such as van der Waals, electrostatic, steric, solvation, and hydrophobic forces can act among molecules, particles, and surfaces in complex systems as shown in Figure 10.5. Van der Waals forces exist between any two particles or molecules in a medium. They can be attractive or repulsive and originate from the interactions of electric dipole moments of the molecules.[60] The van der Waals interaction energy per unit area E_{vdw} between two parallel plates separated by distance D is given by:

$$E_{vdw} = -A/12\pi D^2 \tag{10.1}$$

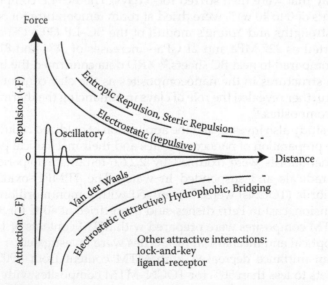

FIGURE 10.5
Basic intermolecular and surface forces between two particles.

where A is the Hamaker constant. The Hamaker constant A_{132} of two micro-scopic bodies 1 and 2 interacting across medium 3 can be given by the Lifshitz theory approximately as

$$A_{132} \approx \frac{3}{4}k_BT\left(\frac{\varepsilon_1 - \varepsilon_3}{\varepsilon_1 + \varepsilon_3}\right)\left(\frac{\varepsilon_2 - \varepsilon_3}{\varepsilon_2 + \varepsilon_3}\right)$$

$$+\frac{3h_P\nu_e}{8\sqrt{2}}\frac{\left(n_1^2 - n_3^2\right)\left(n_2^2 - n_3^2\right)}{\left(n_1^2 + n_3^2\right)^{1/2}\left(n_2^2 + n_3^2\right)^{1/2}\left\{\left(n_1^2 + n_3^2\right)^{1/2} + \left(n_2^2 + n_3^2\right)^{1/2}\right\}} \quad (10.2)$$

where k_B is the Boltzmann constant ($1.381 * 10^{-23}$ J/K), ε is the dielectric per-mittivity, n is the refractive index, h_P is Planck's constant ($6.626 * 10^{-34}$ m^2 kg/s), and ν_e is main electronic absorption frequency. Other methods for the cal-culation of a Hamaker constant include experimental force measurements and measurement of the critical zeta potential.[61,62] The reported values of the Hamaker constant of cellulose cited in the literature are within the ranges of 5.8 to $8.4 * 10^{-20}$ J in a vacuum and 8.0 to $9.9 * 10^{-21}$ J in water.[61]

One of the most widely used methods to determine the surface energies of materials is based on contact angle measurements. Different theories are proposed for determinations of surface energy based on contact angle measurements including the Zisman and van Oss methods. In the van Oss method, the surface energy comprises of two terms: the Lifshitz-van der Waals component γ^{LW} and the Lewis acid–base component γ^{AB} as

$$\gamma = \gamma^{LW} + \gamma^{AB} \quad (10.3)$$

The Lewis acid–base γ^{LW} is given by:

$$\gamma^{AB} = 2\sqrt{\gamma^+\gamma^-} \quad (10.4)$$

where γ^+ and γ^- represent the electron–acceptor and electron–donor inter-actions, respectively. The relationship between contact angle θ of liquid L on a solid surface S is then given by:

$$\gamma_L(\cos\theta + 1) = 2\left(\sqrt{\gamma_S^{LW}\gamma_L^{LW}} + \sqrt{\gamma_S^+\gamma_L^-} + \sqrt{\gamma_S^-\gamma_L^+}\right) \quad (10.5)$$

where γ_L and γ_S represent the surface energy of liquid and solid, respectively. To determine the surface energy of a solid γ_S based on Equations (10.3) to (10.5), at least three different liquids should be used for contact angle measurements. The reported value of the surface energy of cellulose in literature is within the range of 24 to 74 mJ/m^2, depending on the types of cellulose samples and

TABLE 10.1

Surface Energy and Corresponding Components of Cellulose (mJ/m^2)

Sample	Sample Code	γ_s	γ_s^{LW}	γ_s^{AB}	γ_s^+	γ_s^-	Reference
Cellulose powder	Sigmacell 101	58.98	54.49	4.49	0.11	47.83	70
α-Cellulose	Sigma C8002	57.43	55.73	1.7	0.013	56.31	71
Cellulose fiber	Kraft pulp	46.4	43.6	2.7	0.07	27.4	72
Cellulose fiber	NSSC	57.2	45.7	11.6	1.59	21.0	72
Cotton linter	DP 1356	63.3	59.7	3.64	0.13	24.21	73
Microcrystalline cellulose	Avicel	51.82	51.82	0	0	50.14	70

Source: Q. Shen. In *Model Cellulosic Surfaces*, M. Roman, Ed. Washington: American Chemical Society, 2009, pp. 259–289. With permission.

measurement techniques.[61] Some of the reported values of the surface energies of cellulose and its corresponding components are shown in Table 10.1.

10.5 Applications of Cellulose Nanocomposites

Cellulose nanocomposites have potential for use in many diverse applications. This chapter provided a brief overview of these applications. One of the main benefits of cellulose-based composites comes from their enhanced mechanical properties (Young's modulus and tensile strength) that make them good candidates for structural materials.[63]

Cellulose-reinforced polymer matrices have been reported to have enhanced mechanical and ion conduction properties that make them suitable for use in lithium batteries.[64,65] Solidified liquid crystals of cellulose are reported to have certain optical properties that can be used in security papers. [10,66] Oxidized cellulose was reported to be capable of sorption of uranium and thorium ions and thus may have utility in environmental remediation applications.[67]

Cellulose nanocomposites have also been used in biomedical applications as scaffolds for the growth of bones and other tissues.[63] Nanocomposites of BC and PVA have been fabricated to mimic the mechanical properties of cardiovascular tissues.[68] Bacterial cellulose-based composites have been used for skin tissue repair.[69] Optically transparent cellulose nanocomposites that have been reinforced mechanically can be used in electronic devices.[53,54] Some cellulose nanocomposites have been reported to be effective gas barrier films useful in the paper product and packaging industries.[58,63]

Despite the significant advances in the development of cellulose nanocomposites, certain challenges hinder the widespread use of these composite materials. Some challenges that should be addressed in future studies include easy thermal degradation of the materials, the hygroscopic nature of cellulose, high manufacturing costs, and a lack of fundamental understanding of molecular interaction mechanisms involved in fabricating nanocomposite materials.[63]

Acknowledgment

This work was supported by the Natural Sciences and Engineering Research Council of Canada (NSERC).

References

1. Y. Habibi, L.A. Lucia, and O.J. Rojas. Cellulose Nanocrystals: Chemistry, Self-Assembly, and Applications. *Chemical Reviews,* 110 (2010): 3479–3500.
2. R.J. Moon, A. Martini, J. Nairn et al. Cellulose Nanomaterials Review: Structure, Properties, and Nanocomposites. *Chemical Society Reviews,* 40 (2011): 3941–3994.
3. J.N. Coleman, U. Khan, W. J. Blau et al. Small but Strong: A Review of the Mechanical Properties of Carbon Nanotube–Polymer Composites. *Carbon,* 44 (2006): 1624–1652.
4. A.K. Bledzki and J. Gassan. Composites Reinforced with Cellulose Based Fibres. *Progress in Polymer Science,* 24 (1999): 221–274.
5. B.C. Thompson and J.M. Frechet. Polymer–Fullerene Composite Solar Cells. *Angewandte Chemie International Edition,* 47 (2008): 58–77.
6. E.T. Thostenson, Z. Ren, and T. Chou. Advances in the Science and Technology of Carbon Nanotubes and Their Composites: A Review. *Composites Science and Technology,* 61 (2001): 1899–1912.
7. S.J. Eichhorn, A. Dufresne, M. Aranguren et al. Review: Current International Research into Cellulose Nanofibres and Nanocomposites. *Journal of Materials Science,* 45 (2010): 1–33.
8. F. Croce, G.B. Appetecchi, L. Persi et al. Nanocomposite Polymer Electrolytes for Lithium Batteries. *Nature,* 394 (1998): 456–458.
9. M. Kikuchi, S. Itoh, S. Ichinose et al. Self–Organization Mechanism in a Bone-Like Hydroxyapatite–Collagen Nanocomposite Synthesized in Vitro and Its Biological Reaction in Vivo. *Biomaterials,* 22 (2001): 1705–1711.
10. M.A. Samir, F. Alloin, and A. Dufresne. Review of Recent Research into Cellulosic Whiskers, Their Properties, and Their Applications in the Nanocomposite Field. *Biomacromolecules,* 6 (2005): 612–626.
11. K. Oksman and M. Sain. *Cellulose Nanocomposites.* Washington, DC: American Chemical Society, 2006.
12. I. Kvien, B. S. Tanem, and K. Oksman. Characterization of Cellulose Whiskers and Their Nanocomposites by Atomic Force and Electron Microscopy. *Biomacromolecules,* 6 (2005): 3160–3165.

13. A. Dufresne, J. Cavaille, and M.R. Vignon. Mechanical Behavior of Sheets Prepared from Sugar Beet Cellulose Microfibrils. *Journal of Applied Polymer Science*, 64 (1997): 1185–2294.
14. S. Ifuku, M. Nogi, K. Abe et al. Surface Modification of Bacterial Cellulose Nanofibers for Property Enhancement of Optically Transparent Composites: Dependence on Acetyl Group Ds. *Biomacromolecules*, 8 (2007): 1973–1978.
15. A.J. Svagan, M.A. Azizi-Samir, and L.A. Berglund. Biomimetic Polysaccharide Nanocomposites of High Cellulose Content and High Toughness. *Biomacromolecules*, 8 (2007): 2556–2563.
16. G. Guhados, W. Wan, and J.L. Hutter. Measurement of the Elastic Modulus of Single Bacterial Cellulose Fibers Using Atomic Force Microscopy. *Langmuir*, 21 (2005): 6642–6646.
17. S. Iwamoto, W. Kai, A. Isogai et al. Elastic Modulus of Single Cellulose Microfibrils from Tunicate Measured by Atomic Force Microscopy. *Biomacromolecules*, 10 (2009): 2571–2576.
18. R.R. Lahiji, X. Xu, R. Reifenberger et al. Atomic Force Microscopy Characterization of Cellulose Nanocrystals. *Langmuir*, 26 (2010): 4480–4488.
19. S. Eichhorn and R.J. Young. The Young's Modulus of a Microcrystalline Cellulose. *Cellulose*, 8 (2001): 197–207.
20. S. Tanpichai, F. Quero, M. Nogi et al. Effective Young's Modulus of Bacterial and Microfibrillated Cellulose Fibrils in Fibrous Networks. *Biomacromolecules*, 13 (2012): 1340–1349.
21. Y.C. Hsieh, H. Yano, M. Nogi et al. Estimation of the Young's Modulus of Bacterial Cellulose Filaments. *Cellulose*, 15 (2008): 507–513.
22. T. Zimmermann, E. Pöhler, and T. Geiger. Cellulose Fibrils for Polymer Reinforcement. *Advanced Engineering Materials*, 6 (2004): 754–761.
23. T. Zimmermann, E. Pöhler, and P. Schwaller. Mechanical and Morphological Properties of Cellulose Fibril Reinforced Nanocomposites. *Advanced Engineering Materials*, 7 (2005): 1156–1161.
24. W. Kunihiko, T. Mari, M. Tasushi et al. Structural Features and Properties of Bacterial Cellulose Produced in Agitated Culture. *Cellulose*, 5 (1998): 187–200.
25. P.A. Penttila, A. Varnai, M. Fernandez et al. Small-Angle Scattering Study of Structural Changes in the Microfibril Network of Nanocellulose during Enzymatic Hydrolysis. *Cellulose*, 20 (2013): 1031–1040.
26. H. Miyamoto, C. Yamane, and K. Ueda. Structural Changes in the Molecular Sheets along (Hk0) Planes Derived from Cellulose Ib by Molecular Dynamics Simulations. *Cellulose*, 20 (2013): 1089–1098.
27. C.M. Lee, A. Mittal, A. L. Barnette et al. Cellulose Polymorphism Study with Sum Frequency Generation (SFG) Vibration Spectroscopy: Identification of Exocyclic CH_2OH Conformation and Chain Orientation. *Cellulose*, 20 (2013): 991–1000.
28. T. Pullawan, A.N. Wilkinson, and S.J. Eichhorn. Influence of Magnetic Field Alignment of Cellulose Whiskers on the Mechanics of All-Cellulose Nanocomposites. *Biomacromolecules*, 13 (2012): 2528–2536.
29. D. Dubief, E. Samain, and A. Dufresne. Polysaccharide Microcrystals Reinforced Amorphous Poly(α-Hydroxyoctanoate) Nanocomposite Materials. *Macromolecules*, 32 (1999): 5765–5771.
30. A. Dufresne, M.B. Kellerhals, and B. Witholt. Transcrystallization in MCL-PHAS– Cellulose Whisker Composites. *Macromolecules*, 32 (1999): 7396–7401.

31. L. Chazeau, J.Y. Cavaille, G. Canova et al. Viscoelastic Properties of Plasticized PVC Reinforced with Cellulose Whiskers. *Journal of Applied Polymer Science,* 71 (1999): 1797–1808.

32. V. Janaki, K. Vijayaraghavan, B. Oh et al. Synthesis, Characterization, and Application of Cellulose–Polyaniline Nanocomposite for the Treatment of Simulated Textile Effluent. *Cellulose,* 20 (2013): 1153–1166.

33. M.J. John, R. Anandjiwala, K. Oksman et al. Melt-Spun Polylactic Acid Fibers: Effect of Cellulose Nanowhiskers on Processing and Properties. *Journal of Applied Polymer Science,* 127 (2013): 274–281.

34. T. Wang and L.T. Drzal. Cellulose–Nanofiber-Reinforced Poly(Lactic Acid) Composites Prepared by a Water-Based Approach. *ACS Applied Materials & Interfaces,* 4 (2012): 5079–5085.

35. F. Fahma, N. Hori, T. Iwata et al. Morphology and Properties of Poly(Methyl Methacrylate)–Cellulose Nanocomposites Prepared by Immersion Precipitation. *Journal of Applied Polymer Science,* 128 (2013): 1563–1568.

36. S. Sain, D. Ray, A. Mukhopadhyay et al. Synthesis and Characterization of PMMA–Cellulose Nanocomposites by in Situ Polymerization Technique. *Journal of Applied Polymer Science,* 126 (2012): E127–E134.

37. S. Fujisawa, T. Saito, S. Kimura et al. Surface Engineering of Ultrafine Cellulose Nanofibrils toward Polymer Nanocomposite Materials. *Biomacromolecules,* 14 (2013): 1541–1546.

38. X. Hu, C. Xu, J. Gao et al. Toward Environment-Friendly Composites of Poly(Propylene Carbonate) Reinforced with Cellulose Nanocrystals. *Composites Science and Technology,* 78 (2013): 63–68.

39. M. Martínez–Sanz, A. Lopez–Rubio, and J. M. Lagaron. Nanocomposites of Ethylene Vinyl Alcohol Copolymer with Thermally Resistant Cellulose Nanowhiskers by Melt Compounding. I: Morphology and Thermal Properties. *Journal of Applied Polymer Science,* 128 (2013): 2666–2678.

40. E. Bahar, N. Ucar, Λ. Onen et al. Thermal and Mechanical Properties of Poly-propylene Nanocomposite Materials Reinforced with Cellulose Nanowhiskers. *Journal of Applied Polymer Science,* 125 (2012): 2882–2889.

41. J. Yang, C. Han, J. Duan et al. Studies on the Properties and Formation Mechanism of Flexible Nanocomposite Hydrogels from Cellulose Nanocrystals and Poly(Acrylic Acid). *Journal of Materials Chemistry,* 22 (2012): 22467–22480.

42. J.R. Capadona, K. Shanmuganathan, D.J. Tyler et al. Stimuli-Responsive Polymer Nanocomposites Inspired by Sea Cucumber Dermis. *Science,* 319 (2008): 1370–1374.

43. J.R. Capadona, O. Van Den Berg, L.A. Capadona et al. Versatile Approach for the Processing of Polymer Nanocomposites with Self-Assembled Nanofibre Templates. *Nature Nanotechnology,* 2 (2007): 765–769.

44. M.S. Peresin, Y. Habibi, J.O. Zoppe et al. Nanofiber Composites of Polyvinyl Alcohol and Cellulose Nanocrystals: Manufacture and Characterization. *Biomacromolecules,* 11 (2010): 674–681.

45. W.L. Magalhaes, X. Cao, and L.A. Lucia. Cellulose Nanocrystals–Cellulose Core-in-Shell Nanocomposite Assemblies. *Langmuir,* 25 (2009): 13250–13257.

46. R.T. Olsson, R. Kraemer, A. Lopez–Rubio et al. Extraction of Microfibrils from Bacterial Cellulose Networks for Electrospinning of Anisotropic Biohybrid Fiber Yarns. *Macromolecules,* 43 (2010): 4201–4209.

47. J. Lee and Y. Deng. Increased Mechanical Properties of Aligned and Isotropic Electrospun PVA Nanofiber Webs by Cellulose Nanowhisker Reinforcement. *Macromolecular Research*, 20 (2012): 76–83.
48. B.S. Lalia, Y.A. Samad, and R. Hashaikeh. Nanocrystalline–Cellulose-Reinforced Poly(Vinylidenefluoride-Co-Hexafluoropropylene) Nanocomposite Films as a Separator for Lithium Ion Batteries. *Journal of Applied Polymer Science*, 126 (2012): E441–E447.
49. P. Lu and Y. Hsieh. Cellulose Nanocrystal-Filled Poly(Acrylic Acid) Nanocomposite Fibrous Membranes. *Nanotechnology*, 20 (2009): 415604.
50. S. Iwamoto, K. Abe, and H. Yano. Effect of Hemicelluloses on Wood Pulp Nanofibrillation and Nanofiber Network Characteristics. *Biomacromolecules*, 9 (2008): 1022–1026.
51. M. Henriksson, L.A. Berglund, P. Isaksson et al. Cellulose Nanopaper Structures of High Toughness. *Biomacromolecules*, 9 (2008): 1579–1585.
52. M. Henriksson and L.A. Berglund. Structure and Properties of Cellulose Nanocomposite Films Containing Melamine Formaldehyde. *Journal of Applied Polymer Science*, 106 (2007): 2817–2824.
53. H. Yano, J. Sugiyama, A.N. Nakagaito et al. Optically Transparent Composites Reinforced with Networks of Bacterial Nanofibers. *Advanced Materials*, 17 (2005): 153–155.
54. M. Nogi, S. Iwamoto, A.N. Nakagaito et al. Optically Transparent Nanofiber Paper. *Advanced Materials*, 21 (2009): 1595–1598.
55. J. Leitner, B. Hinterstoisser, M. Wastyn et al. Sugar Beet Cellulose Nanofibril-Reinforced Composites. *Cellulose*, 14 (2007): 419–425.
56. A.N. Nakagaito and H. Yano. Effect of Fiber Content on the Mechanical and Thermal Expansion Properties of Biocomposites Based on Microfibrillated Cellulose. *Cellulose*, 15 (2008): 555–559.
57. G.F. Perotti, H.S. Barud, Y. Messaddeq et al. Bacterial Cellulose–Laponite Clay Nanocomposites. *Polymer*, 52 (2011): 157–163.
58. C. Wu, T. Saito, S. Fujisawa et al. Ultrastrong and High Gas Barrier Nanocellulose–Clay-Layered Composites. *Biomacromolecules*, 13 (2012): 1927–1932.
59. M.H. Gabr, N.T. Phong, M.A. Abdelkareem et al. Mechanical, Thermal, and Moisture Absorption Properties of Nano-Clay Reinforced Nano-Cellulose Biocomposites. *Cellulose*, 20 (2013): 819–826.
60. A. Faghihnejad and H. Zeng. Fundamentals of Surface Adhesion, Friction, and Lubrication. In *Polymer Adhesion, Friction and Lubrication*, H. Zeng, Ed. Toronto: John Wiley & Sons, 2013, pp. 1–57.
61. Q. Shen. Surface Properties of Cellulose and Cellulose Derivatives: A Review. In *Model Cellulosic Surfaces*, M. Roman, Ed. Washington, DC: American Chemical Society, 2009, pp. 259–289.
62. J.N. Israelachvili. *Intermolecular and Surface Forces*, Vol. 3. SanDiego, CA: Academic Press, 2011.
63. M.A. Hubbe, O.J. Rojas, L.A. Lucia et al. Cellulosiac Nanocomposites: A Review. *Bioresources*, 3 (2008): 929–980.
64. M.A. Samir, F. Alloin, W. Gorecki et al. Nanocomposite Polymer Electrolytes Based on Poly(Oxyethylene) and Cellulose Nanocrystals. *Journal of Physical Chemistry B*, 108 (2004): 10845–10852.

65. M.A. Samir, F. Alloin, J. Sanchez et al. Cross-Linked Nanocomposite Polymer Electrolytes Reinforced with Cellulose Whiskers. *Macromolecules,* 37 (2004): 4839–4844.
66. United States Patent no. 5629055.
67. D. Han, G.P. Halada, B. Spalding et al. Electrospun and Oxidized Cellulosic Materials for Environmental Remediation of Heavy Metals in Groundwater. In *Model Cellulosic Surfaces,* M. Roman, Ed. Washington, DC: American Chemical Society, 2009, pp. 243–257.
68. W.K. Wan, J.L. Hutter, L. Milton et al. Bacterial Cellulose and Its Nanocomposites for Biomedical Applications. In *Cellulose Nanocomposites: Processing, Characterization, and Properties,* K. Oksman and M. Sain, Eds. Washington, DC: American Chemical Society, 2006, pp. 221–241.
69. L. Fu, J. Zhang, and G. Yang. Present Status and Applications of Bacterial Cellulose-Based Materials for Skin Tissue Repair. *Carbohydrate Polymer,* 92 (2013): 1432–1442.
70. F. Dourado, F.M. Gama, E. Chibowski et al. Characterization of Cellulose Surface Free Energy. *Journal of Adhesion Science and Technology,* 12 (1998): 1081–1090.
71. Q. Shen, J.F. Hu, and Q.F. Gu. Examination of the Surface Free Energy and Acid–Base Properties of Cellulose by the Column Wicking Technique and Critical Packing Height and Density. *Chinese Journal of Polymer Science,* 22 (2004): 49–53.
72. W. Shen, Y.J. Sheng, and I.H. Papker. Comparison of the Surface Energetics Data of Eucalyptus Fibers and Some Polymer Obtained by Contact Angle and Inverse Gas Chromatography Methods. *Journal of Adhesion Science and Technology,* 13 (1999): 887–901.
73. Y. Xu, H.G. Ding, and Q. Shen. Influence of the Degree of Polymerization on surface Properties of Cellulose. *Chinese Cellulose Science and Technology,* 15 (2007): 53–56.

11

Improvement of Damage Resilience of Composites

Jiye Chen

CONTENTS

11.1 Introduction

Over the past few decades, bio-inspired design strategy (BIDS) has become a very successful approach in the design of engineering structures or and components with specific functions required by engineering applications.[1] In the area of application of BIDS in fiber composites, for example, a tree branch joint can be mimicked for designing synthetic composite T-piece joints in aerospace structures that will improve damage resilience under extreme loading.[2]

The skins of animal horns or antlers with special biotubule construction (Figure 11.1) can resist severe loading conditions by absorbing deformed energy.[3] The authors considered this feature to be an excellent model for the design of the deltoid regions in T-joint components.

BIDS is a potential approach for designing novel synthetic fiber composites that mimic the biostructural features of plants or animals with high damage resilience. However, BIDS should be applied only with a full understanding of the basic material characteristics, failure mechanisms, and damage resilience levels of biofiber composites.

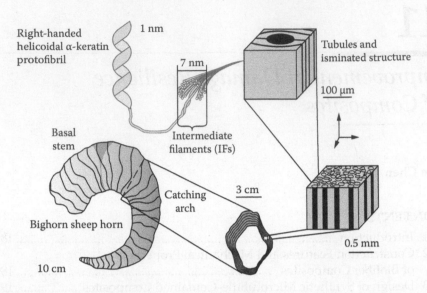

FIGURE 11.1
Microconstruction with biotubules of horns. (*Source:* L. Tombolato, E.E. Novitskaya, P.Y. Chen et al. *Acta Biomaterialia*, 6 [2010]: 319–330. With permission.)

The catalogue of natural composites includes specific animal cells and plant fiber-based composites. From a review of mechanical behavior, animal cells are usually stiffer and stronger than plant composites. However, plant fiber-based materials can be used directly in secondary engineering structures although their use as sustainable materials is questionable because their properties are not controlled easily. Animal cells are not used commonly in engineering structures although some organs, e.g., horns and tusks, teeth, and sea shells, exhibit excellent material properties.

The biofiber composites for the mimicking program in this chapter came from the horns of desert bighorn sheep because horns must be strong and durable to handle extreme loading impacts, making them superior structural materials. In 2010, L. Tombolato et al. studied the microstructural construction and mechanical properties of desert bighorn sheep horns. They explored the microconstruction of biotubules of the horns and their excellent damage resilience through compression and bending tests and comparisons with other materials.[3] The specific work of fracture of these horns ranged from 12 to 60 kJ/m². The highest work of fracture reported (32 kJ/m²) exceeded those of most other biological and synthetic materials and even mild steel (26 kJ/m²) by 23%. This was attributed to crack-stopping mechanisms such as delamination and keratin fiber pull-out.

However, the averaged Young's modulus (4 GPa) of the horn was only about 2% of the modulus of mild steel (210 GPa). Thus, such high fracture energy is determined by the microbiotubules of horns that efficiently absorb deformed energies when they are under extreme loading. Therefore, this biotubule composite is one of the best natural fiber materials for mimicking technology and is the concept used for the redesign of synthetic fiber composites in this investigation.

This investigation starts from a brief review of biotubule composite materials by Tombolato et al. Selected material properties of biotubule composites will be used as references for designing synthetic composites.[3]

11.2 Construction Features and Mechanical Properties of Biofiber Composites

Tombolato et al.[3] indicated that horns are composed of α-keratin, a fibrous structural protein found in hair, nails, claws, and hooves. Horns have lamellar structures (2 to 5 µm thick) stacked in a radial direction with tubules (40 to 100 µm in diameter) dispersed between the lamella, extending along the length in the growth direction.

Keratin molecules are held together by H-bonding and disulfide cross-linked bonds. These molecules are helically wound and assembled into microfibrils (intermediate filaments [IFs]) and form superhelical ropes 7 nm in diameter.[4] The α-helices are mainly parallel to the long axis diameter that is mainly parallel to the long axes of the ropes. The IFs are embedded in a viscoelastic protein matrix composed of two types of proteins (1) high sulfur proteins that have more cysteinyl residues and (2) high glycine–tyrosine proteins that have high contents of glycyl residues.[4]

The filaments and the matrix are organized further into circular lamellae that surround a hollow tubule (medullary cavity), similar to the configuration of osteons in compact bone. These lamellar tubules are held together by intertubular material, which is chemically the same as the tubular material but has a different orientation of IFs.[5] These can be seen in Figure 11.1. Tombolato's group found that a three-dimensional laminated composite consisted of fibrous keratin in a horn with a porosity gradient across the thickness of the horn. The presence of the tubules is surprising because a horn is a laminate.

Bending produces shear stresses in the horn that promote delamination and eventual failure. The tubules present natural regions of delamination, areas of weakness from which delamination cracks can propagate. Tombolato et al.[3] measured the toughness values for the ambient dried samples of horn and found they ranged from 56 MJ/m³ for the longitudinal and transverse directions to 74 MJ/m³ for the radial direction. The higher toughness found in samples tested in the radial direction is most likely associated with the energy absorption by compression of the tubules. This is the orientation in which horns clash during combat.

The mechanical role of the hollow tubules serves mechanical functions such as toughening by crack deflection. The tubules also prevent cooperative buckling of the structure by increasing the resistance to compressive failure of the tubule walls. Extensive tubule collapse was found for samples tested in compression in the radial direction, which enhanced the toughness. The main findings from Tombolato's work are:

1. A horn is a composite material consisting of stacked lamellae in the radial direction with a thickness of 2 to 5 mm with tubules ~4 × 100 μm in diameter interspersed between the lamellae, producing an overall cross-sectional porosity of 7%.

2. A gradient in porosity exists across the thickness of the horn and the porosity decreases from the external surface (8 to 12%) to the interior surface (~0%).

3. Rehydration has a significant effect on reducing maximum bending strength and elastic modulus of the horn, more so than the effect of sample orientation.

4. There is no significant difference between the elastic modulus and the maximum bending strength in the longitudinal or transverse directions for rehydrated specimens, suggesting that the weakened matrix dominates mechanical performance.

5. Fracture micrographs from bend tests show delamination, ligament bridging, and intermediate filament fracture as the main failure modes.

6. Compression tests revealed that the elastic moduli and yield strengths of the ambient dried samples in the longitudinal and transverse directions are the same; elastic moduli and yield strengths of the radial samples were smaller due to the ease of compressing the tubules. Hydration degraded elastic modulus and strength.

7. Due to the laminated structure of the horn, compression tests for the ambient dried condition revealed that the toughness was the same in the longitudinal and transverse directions but higher in the radial direction where compression of the tubules aids in energy absorption. The toughness was lower for the rehydrated condition, but the orientation trend was the same.

8. Delamination and microbuckling of lamellae in the longitudinal and transverse directions are the main sources of failure.

The high value of toughness and work of fracture of the horn with biotubules makes it capable of high energy absorption before breaking. The conclusion was that the biotubules serve only a mechanical function (to increase crack defection, thereby increasing the toughness), making the equine horn a highly fracture-resistant biological material.

A horn must support large compressive and impact loads and provide some shock absorption from impacts such as the 3,400 N fighting force of a bighorn sheep estimated by the ram's mass and velocity.[6] The horn keratin is described as a nanoscale composite comprised of IFs as fiber-oriented reinforcements of a hydrated keratin matrix, as shown in Figure 11.1.

For these reasons, it is of interest to investigate the microstructural features and mechanical properties that underpin the excellent fracture resistance of these horns. In other words, there must be relationships between excellent fracture resistance and the microstructure at different orientations that also must relate to tensile, compressive properties and failure mechanisms. Horns are lightweight, made of tough material, and follow a biomimetic design that ensures superior impact-resistant materials. This investigation will focus on the biomimetic design to improve damage resiliences of synthetic composites.

11.3 Design of Synthetic Microtubule-Contained Composites

L. Burns et al. investigated a way to improve fracture behavior of T-joints using the structural features of a tree branch in 2012. One structural feature of a tree branch is fiber embedded from web to flange. Burns' investigation found that fracture toughness is increased but initial damage resilience is decreased.[2] The latter part of the result is not a critical aspect of components such as primary structures in airplanes and certain types of engines. The studies described in this chapter tried to improve both fracture toughness and initial damage resilience using mimicked micro-constructional features of biotubule composites.

Figure 11.2 shows the deltoid region of a T-joint. The size of this T-joint is 100 mm for length, height, and thickness and 4.4 mm and 6.8 mm for web

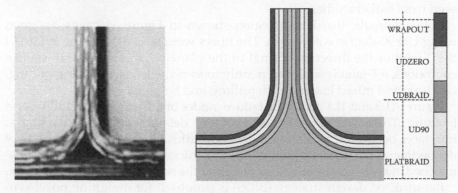

FIGURE 11.2
Composite T-piece specimen showing construction of layout.

TABLE 11.1

Comparison of Mechanical Properties of Keratinized Materials and Other Biological and Synthetic Materials

Material	Elastic Modulus (GPa)	Tensile Strength (MPa)	Bending Strength (MPa)	Toughness (MJ/m^3)	Work of Fracture (KJm^{-2})
Oryx horn	4.3		212		19
	6.1	137			
	4.3	122			
	1.8	56			
Waterbuck horn	3.3		245		20
Sheep horn	4.1		228		22
	9			56 to 74	
	1.5		127.1	12 to 18	
	2.2		39.1		
	0.81				
Bovine hoof	0.4	16.2			
Equine hoof	2.6	38.9			
	0.2				
	0.3 to 0.6	6.5 to –9.5			
Bovine femur	13.5	148	246.7		1.7
Antler bone					6.2
Red abalone shell	70	170	197	1.24	
Fiberglass	5.9	110			90
Polycarbonate	2.4	67		33	33

Source: L. Tombolato, E.E. Novitskaya, P.Y. Chen et al. *Acta Biomaterialia*, 6 (2010): 319–330. With permission.

and flange, respectively. The deltoid region is within a rough triangle area with an 8 mm base and 5 mm height. Figure 11.2 also shows the construction layers. The material of each layer is detailed in Table 11.1. More details can be seen from earlier studies.[7-11]

Conventionally, the deltoid region shown in Figure 11.2 was designed using UD 90-degree composites. The fibers were placed along the length of the T-joint or the direction normal to the plane of cross section. In service conditions, a T-joint is subjected mainly to in-plane loads, including pulling, bending, and mixed loading with pulling and bending.

Figures 11.3 and 11.4 depict the failure modes of a T-joint under pulling and bending. These failure modes are typically delamination or debonding at interfaces between the deltoid region and radius laminates.[7-11] The driver of debonding in the deltoid region is local via thickness stresses caused by mismatched materials at the interfaces that limit the achievable damage resilience.

Bio-inspired design strategy (BIDS) is proposed for designing novel synthetic composites with mimicked microtubules to replace conventional UD 90-degree composites in the deltoid regions of T-joints. The aim is to use

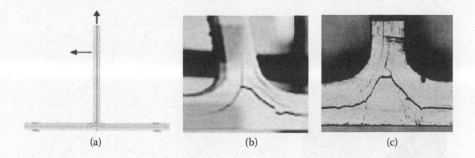

FIGURE 11.3
(a) Composite T-joint under mixed loading. (b) Bending case failure pattern. (c) Pulling case failure pattern.

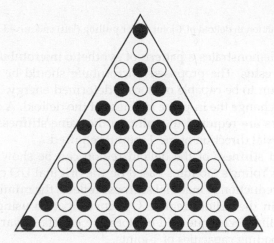

FIGURE 11.4
Mimicked deltoid containing microtubules.

mimicked microtubules containing UD composites to fill the deltoid region. This study explores how the damage resilience of T-joint relates to the stiffness of mimicked microtubule composites. It is a conceptual investigation intended to demonstrate innovations at critical areas to improve the damage resilience of T-joints in a cost-effective manner.

Based on the biotubule composites of the horn shown in Figure 11.1, a composite with microtubules as mimicked synthetic adhesives was proposed to replace traditional UD 90-degree composites in the deltoid regions of T-joints. The novel deltoid with microtubules is shown in Figure 11.5 and consists of synthetic microtubules, carbon fibers, and resins. The white circles in the figure are the proposed microtubules 0.1 to 0.5 mm in diameter. Solid black circles represent the carbon fibers that play the same role in traditional UD 90-degree composites—resisting possible loads along the lengths of T-joints. Both microtubules and carbon fibers could be mixed with resins.

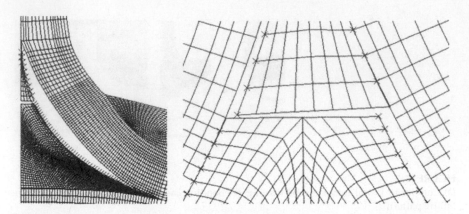

FIGURE 11.5
Predicted delamination in deltoid of T-joint under pulling (left) and mixed loading (right).

Figure 11.5 demonstrates a pattern of synthetic microtubules and carbon fibers within resins. The proposed microtubule should be made of a soft polymer or foam to be capable of storing deformed energy. The soft tubule is designed to change the in-plane stiffness of the deltoid. A certain number of carbon fibers are required to maintain the same stiffness of the deltoid region in the axial direction of the T-joint if required.

The softened stiffness of the deltoid region can be shown by reduction of the in-plane Young's modulus based on the original UD composites. The percentage of reduction of material stiffness reflects the mimicking activities of biotubules in the deltoid region. Our investigation using a progressive damage modeling technique will explore the effects of various mimicked levels on the loading capacities of T-joints.

11.4 Progressive Damage Model for Predicting Failure Loads

The damage model used in this investigation was utilized from the author's previous work.[7-11] A bilinear relationship between traction and relative displacement was employed and is expressed by Equation (11.1):

$$\sigma_j(\varepsilon) = \begin{cases} K_{j0}\varepsilon_j & \varepsilon_j \leq \varepsilon_{j0} \\ (1-d)K_{j0}\varepsilon_j & \varepsilon_{j0} \leq \varepsilon_j < \varepsilon_{jc} \\ 0 & \varepsilon_j \geq \varepsilon_{jc} \end{cases} \quad (11.1)$$

$$j = I, II, III$$

where, ε_j, ε_{j0}, and ε_{jc} (j = I, II, III) are the interface relative displacement, initial damage relative displacement, and maximum relative displacement corresponding to mode I, II, and III fractures, respectively. The equation actually expresses an individual material softening damage law. The first coupled effect in a mixed mode case was stated in a quadratic formula for predicting damage initiation:

$$F(\varepsilon_0) = \left(\frac{\varepsilon_I}{\varepsilon_{I0}}\right)^2 + \left(\frac{\varepsilon_{II}}{\varepsilon_{II0}}\right)^2 + \left(\frac{\varepsilon_{III}}{\varepsilon_{III0}}\right)^2 \tag{11.2}$$

Equation (11.2) is a mixed initial damage model that considers the coupling effects and determines the point of damage initiation when $F(\varepsilon_0) = 1$. In a single fracture case, $\varepsilon_j = \varepsilon_{j0}$ (j = I, II, III) when $F(\varepsilon_0) = 1$. The corresponding interface state $\sigma_j = \sigma_{jt}$, where σ_{jt} (j = I, II, III) is the interface normal or shear strength or yield traction. Thus, $F(\varepsilon_0) = 1$ in Equation (11.2) generally presents the interface strength state in which interface materials begin softening in the mixed mode fracture case. When a crack occurs at the interface, the traction becomes zero. This is predicted by $F(\varepsilon_c) = 1$ shown in Equation (11.3), another quadratic formula with coupled effects in mixed damage cases:

$$F(\varepsilon_c) = \left(\frac{\varepsilon_I}{\varepsilon_{Ic}}\right)^2 + \left(\frac{\varepsilon_{II}}{\varepsilon_{IIc}}\right)^2 + \left(\frac{\varepsilon_{III}}{\varepsilon_{IIIc}}\right)^2 \tag{11.3}$$

Actually, the change of material state from Equation (11.2) to Equation (11.3) is material softening progress with damage accumulation, represented by a coupled damage scale d in Equation (11.1). This includes three damage components corresponding to three fracture modes, and may be expressed by a quadratic relationship as shown in Equation (11.4):

$$d = \sqrt{(\gamma_I d_I)^2 + (\gamma_{II} d_{II})^2 + (\gamma_{III} d_{III})^2} \tag{11.4}$$

Each individual damage scale d_j (j = I, II, III) in Equation (11.4) is used to measure the reduction of stiffness by Equation (11.1) in the material softening stage ($\varepsilon_j > \varepsilon_{j0}$) and is determined by the reduction of stiffness from the elastic to the softening stage. This can be expressed by Equation (11.5):

$$K_j = (1 - d_j)K_{j0} \tag{11.5}$$

Analytically, the damage scale d_j can be calculated by Equation (11.6):

$$d_j = Q\left(1 - \frac{\varepsilon_{j0}}{\varepsilon_j}\right) \tag{11.6}$$

where Q is a material coefficient expressed as $\varepsilon_{jc}/(\varepsilon_{jc} - \varepsilon_{j0})$. The damage coupling factor γ_j in Equation (11.4) is determined as $0 \le \gamma_j \le 1.0$, $j = I, II, III$.

For the single fracture mode, for example, in opening mode, $j = I$, $\gamma_I = 1.0$, and all others are zero. Therefore, the damage scale $d = d_I$. For the mixed fracture mode, the value of γj is derived as follows. First, assume the amount of total damage scale in the mixed facture mode case equals the value of damage scale in the pure single mode I case. This can be expressed as:

$$d = \sqrt{(\gamma_I d_I)^2 + (\gamma_{II} d_{II})^2 + (\gamma_{III} d_{III})^2} = d_I \tag{11.7}$$

where the three damage coupling factors γ_I, γ_{II}, and γ_{III} would not be the same in the general mixed damage case in reflecting the contribution from each damage mode on the total damage scale. However, γ_I, γ_{II}, and γ_{III} are simply given the same values as $\gamma_I = \gamma_{II} = \gamma_{III} = \gamma$ to simplify the problem and assume the mixed damage ratio is taken as $\beta = d_{II}/d_I$ and $\eta = d_{III}/d_I$. Using Equation (11.7), γ can be worked out by Equation (11.8):

$$\gamma = \sqrt{\frac{1}{1 + \beta^2 + \eta^2}} \tag{11.8}$$

11.5 Coupled Effects in Numerical Integration

The coupled effects in a mixed damage mode case can be explored in a numerical integration process. Taking the first-order differential of Equation (11.4) yields total damage rate expressed by Equation (11.9):

$$d' = \frac{\gamma_I^2 d_I A \varepsilon_I^{-2} \varepsilon_I'(t) + \gamma_{II}^2 d_{II} B \varepsilon_{II}^{-2} \varepsilon_{II}'(t)}{\sqrt{\gamma_I^2 d_I^2 + \gamma_{II}^2 d_{II}^2}} \tag{11.9}$$

where, A and B are two material constants relating to modes I and II, respectively, given by Equation (11.10):

$$A = \frac{\varepsilon_{Ic}\varepsilon_{I0}}{\varepsilon_{Ic} - \varepsilon_{I0}}, \quad B = \frac{\varepsilon_{IIc}\varepsilon_{II0}}{\varepsilon_{IIc} - \varepsilon_{II0}} \tag{11.10}$$

The total mixed mode damage rate in current incremental steps can be expressed as:

$$d'_{i+1} = \frac{\gamma_I^2 d_{Ii} A \varepsilon_{Ii}^{-2} \varepsilon_{I(i+1)}'(t) + \gamma_{II}^2 d_{IIi} B \varepsilon_{IIi}^{-2} \varepsilon_{II(i+1)}'(t)}{\sqrt{\gamma_I^2 d_{Ii}^2 + \gamma_{II}^2 d_{IIi}^2}} \tag{11.11}$$

where the relative displacement rate in a current incremental step can be obtained by Equation (11.12):

$$\varepsilon'_{j(i+1)}(t) = \frac{\varepsilon_{j(i+1)}}{t}, \quad j = I, II \tag{11.12}$$

The current relative displacement can be calculated by Equation (11.13):

$$\varepsilon_{j(i+1)} = \varepsilon_{ji} + \int_{t_i}^{t_{i+1}} \varepsilon'_{j(i+1)} \, dt, \quad j = I, II \tag{11.13}$$

Thus, Equation (11.6) for accounting current individual damage can be rewritten as Equation (11.14):

$$d_{j(i+1)} = Q \left(1 - \frac{\varepsilon_{j0}}{\varepsilon_{j(i+1)}} \right) \tag{11.14}$$

Equation (11.11) accounts for the current total damage rate in a mixed mode damage case. It indicates that the current damage rate of a mixed damage case relates to two individual relative displacement rates in the current step and two individual relative displacements and two individual damage scales in the previous step. Material constants, ε_{I0}, ε_{Ic}, and γ yield a proportional value as a coefficient to mixed total damage rate. In single damage mode I, $d_{IIi} = \varepsilon_{IIi} = 0$, and $\gamma_I = 1.0$, and Equation (11.11) can be changed as

$$d'_{i+1} = A \frac{\varepsilon'_{I(i+1)}(t)}{\varepsilon_{Ii}^2} \tag{11.15}$$

We can see from Equation (11.15) that the current damage rate in single damage mode is relevant only to the current relative displacement rate and previous relative displacement. It should be noticed that $d_{IIi} = \varepsilon_{IIi} = 0$ for the pure mode I damage case in Equation (11.11), but the mode II-related relative displacement rate $\varepsilon'_{II}(t)$ may not be zero. Thus, the non-zero mode II-related relative displacement rate would change the state of pure mode I damage to a mixed damage case in the next step of numerical integration. Similarly, the current damage rate in a single damage mode II case can be given by Equation (11.16).

$$d'_{i+1} = B \frac{\varepsilon'_{II(i+1)}(t)}{\varepsilon_{IIi}^2} \tag{11.16}$$

In numerical integration, the total damage can be determined by Equation (11.17) using the damage rate.

$$d_{j(i+1)} = d_{ji} + \int\limits_{t_i}^{t_{i+1}} d'_{j(i+1)}dt, \quad j = I, II \tag{11.17}$$

11.6 Failure Mechanisms of Synthetic T-Joints with Mimicked Biotubules

Equations (11.1) to (11.17) were programmed into a user element in ABAQUS [17]. As an example, this mixed damage model was used to predict damage resilience of composite T-joint components. These components play a very important role in many aerospace, aviation, wind turbine, and other structures. One of the major design objectives for the composite T-joint was to improve its damage resilience. Considering cost-effectiveness, the uniformly distributed composites and laminates are still the preferred options for T-joint components and were also studied in this investigation.

One major service loading conditions, T-pulling, was investigated. Previous studies indicated that the main failure mode is a typical mixed damage-dominated delamination or debonding in the deltoid region of the T-joint.[7-11] Predicting such delamination plays a key role in the study of damage resilience and helps determine the approach to enhance damage resilience.

The basic geometric data were taken from a previous work.[7-11] The same materials used in previous work[7-11] were used in this investigation and listed in Tables 11.1 and 11.2. The basic FE model of a T-joint and the T-pulling loading conditions were similar to those used earlier (Figure 11.3). Predicted delamination in the deltoid region of a T-joint under pulling and mixed loading is shown in Figures 11.6 and 11.7, respectively, as predicted by the mixed damage model.

Load displacement curves of a T-joint with a biomimicked deltoid region in pulling and mixed loading cases together with results from the original designs are shown in Figures 11.6 and 11.7. The averaged tested failure load worked out from six originally designed specimens at various manufacturing quality levels is ~50 N/mm along the length direction.

The driver leading the delamination in the deltoid region was the strain energy release rate at the interface between the deltoid region and the radius laminates (Table 11.3). The significant strain energy release rate was caused by the mismatched materials in this region that limited the achievable damage resilience.

A deltoid with a novel material was investigated to enhance the damage resilience of a T-joint. The proposed deltoid was redesigned using a novel material with mimicked biotubules.[3] It was recognized that the high damage

TABLE 11.2

Synthetic Material Properties

Description	E11 (GPa)	E22 (GPa)	E33 (GPa)	G12 (GPa)	G13 (GPa)	G23 (GPa)	v12	v13	v23	a_{11}	a_{22}	a_{33}
Outer braided wrap	59.7	60.1	9.7	21.95	4.7	4.7	0.279	0.28	0.28	1.94e-6	2.22e-6	2.8e-5
Braided UD layer	160	9.7	9.7	5.9	5.9	4.7	0.33	0.33	0.28	-1.1e-7	2.8e-5	2.8e-5
0-degree layer	152	9.7	9.7	5.9	5.9	4.7	0.33	0.33	0.28	-1.1e-7	2.8e-5	2.8e-5
90-degree layer and deltoid	9.7	152.0	9.7	5.9	4.7	5.9	0.021	0.28	0.33	2.8e-5	-1.0e-7	2.8e-5
Platform braids	65.8	46.1	9.7	25.8	4.7	4.7	0.421	0.28	0.28	2.9e-6	1.2e-6	2.8e-5

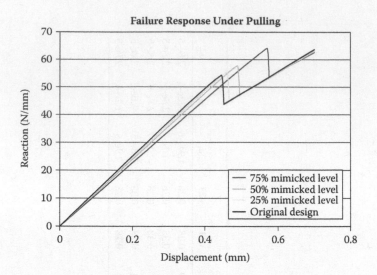

FIGURE 11.6
Failure response of T-joint under pulling.

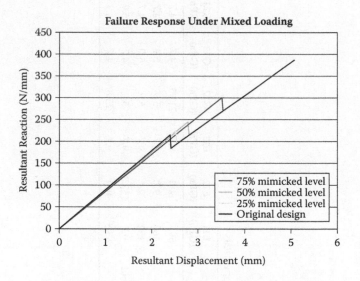

FIGURE 11.7
Failure response of T-joint under mixed load.

TABLE 11.3

Interlaminar Material Strength and Fracture Energy

σ_{33c}	σ_{13c}	σ_{23c}	G_{Ic}	G_{IIc}	G_{IIIc}
45 MPa	35 MPa	35 MPa	300 J/m²	1,000 J/m²	1,000 J/m²

resilience of natural biofiber components from horn comes from biotubules that allow internal deformation, thus reducing the transferred load. The novel material proposed for the deltoid was simply treated as a material showing a reduction of transverse Young's moduli from the original UD 90-degree material. This let the deltoid region become a soft body with material properties similar to those of the biotubule composites shown in Table 11.1.

Softening the transverse material properties in the deltoid region reflects one of the main characteristics of novel materials with microtubules required in the deltoids of the T-joints. The stiffness of the novel materials with microtubules should be less than the stiffness of current materials used in the deltoid region. In conventional designs, UD 90-degree composites are used for the deltoid region. Reduced Young's moduli of 25, 50, and 75% were used for UD 90-degree composites to investigate the effects of reduced transverse material properties on damage resilience and the stiffness of T-joints.

Tables 11.4 and 11.5 present failure responses of the T-joint with a proposed novel deltoid under pulling and mixed loading. We can see that the failure loads increased significantly in all three mimicked cases with reduced Young's moduli. The stiffness of the T-joint was reduced slightly in all three cases with reduced transverse materials as compared to the original case.

In the pulling situation, the increased percentages of failure load based on the original case were 2, 6, and 19% in three material reduction cases, 25, 50, and 75%, respectively. The corresponding deducted stiffness percentages in the three cases were 1.5, 3.6, and 7.4%, respectively. For mixed loading, the increased percentages of failure load based on the original case were 5, 14, and

TABLE 11.4

Predicted Failure Loads and Stiffness under Pulling

Material	Failure Load	Increased Load (%)	Maximum Deflection	Stiffness	Stiffness Reduction (%)
90 UD	54.1077	0	0.444725	121.6655	0
25% reduction	55.2352	2	0.461251	119.7509	1.57
50% reduction	57.4265	6	0.489296	117.3656	3.53
75% reduction	64.2069	19	0.56996	112.6516	7.40

TABLE 11.5

Predicted Failure Loads and Stiffness under Mixed Loading

Material	Failure Load	Increased Load (%)	Maximum Deflection	Stiffness	Stiffness Reduction (%)
90 UD	215.41	0	2.39483	89.94793	0
25% reduction	225.388	5	2.51957	89.45495	0.55
50% reduction	246.016	14	2.78092	88.46569	1.65
75% reduction	300.796	40	3.51346	85.61247	4.82

40% in the same three material reduction cases. Corresponding deducted percentages of stiffness in the three cases were 0.55, 1.65, and 4.82%, respectively.

The investigation revealed that the greater the reduction of transverse Young's modulus, the greater the damage resilience increase in the three cases. This investigation implied that the relatively soft deltoids, like biotubules, play a role in increasing the damage resilience of T-joints by storing deformational energy. In the case with 75% reduction, the whole deltoid behaved like biofiber tubules between radius laminates to release load transferring at the interface of deltoid and radius laminate. The target for the redesigned T-joint would be a composite for the deltoid with material properties similar to UD 90-degree composites with ~75% reduced transverse Young's modulus; this is at the expense of a slight reduction in stiffness.

11.7 Discussion and Future Work

A biomimetic approach for designing the deltoid region of a composite T-joint to improve damage resilience was introduced in this chapter. In 2010, Tombolato et al. found that some natural fiber composites such as biotubules of horn showed significant damage resilience to external loads.[3] The micromechanics of a horn allow the biotubules to resist extreme loading by internal deformation, thus absorbing the applied energy.

This concept was mimicked in the design of a deltoid of a T-joint and a novel UD composite with artificial microtubules was investigated as a replacement of conventional UD composites in the deltoids of T-joints. The mimicked level was measured as a percentage of microtubules over the area of the deltoid region. The percentage of mimicked level was based on an equivalent reduction of material stiffness of an expected novel material in the deltoid due to the flexible effects of microtubules. Three mimicked levels (25, 50, and 75%) of microtubules in the deltoid region were investigated. The effect of mimicked percentage on damage resilience was analyzed using a mixed damage model.[7-11]

Initial investigation indicated that a composite T-joint with mimicked microtubules significantly improved the damage resilience of the joint. Compared to the original design, the T-joint with mimicked microtubules increased damage resilience by 40% in mixed loading cases and by 19% in pulling cases when 75% of the deltoid region of the T-joint was mimicked with microstructural features. The more microtubules mimicked into the deltoid region, the more damage resilience increased.

This was proved by the prediction of failure response depicted in Figures 11.6 and 11.7. Significantly improved damage resilience was achieved by mimicking microtubules in a crucial region. This initial investigation shows that the biomimetic approach has great potential to produce novel composite T-joints to achieve an acceptable level of cost-effectiveness in terms of damage resilience and production costs.

Future work will include comprehensive studies of the microconstruction of biotubule composites, exploring the relationships of damage resilience and biotubule microconstruction, establishing a database for mimicking science technology, and developing multiscale modeling technologies for designing and manufacturing synthetic novel composites. The concept of mimicked biotubule composites would be used in the development of novel heterogeneous adhesive composites.

References

1. G. Mayer and M. Sarikaya. Rigid Biological Composite Materials: Structural Examples for Biomimetic Design. *Experimental Mechanics*, 42 (2002): 395–403.
2. L.A. Burns, A.P. Mouritz, D. Pook et al. Bio-Inspired Design of Aerospace Composite Joints for Improved Damage Tolerance. *Composite Structures,* 94, (2012): 995–1004.
3. L. Tombolato, E.E. Novitskaya, P.Y. Chen et al. Microstructure, Elastic Properties, and Deformation Mechanisms of Horn Keratin. *Acta Biomaterialia,* 6 (2010): 319–330.
4. M. Feughelman. Mechanical Properties and Structure of α–Keratin Fbres: Wool, Human Hair, and Related Fbres. Sydney: University of New South Wales Press, 1977.
5. M.A. Kasapi and J.M. Gosline. Design Complexity and Fracture Control in the Equine Hoof Wall. Journal of Experimental Biology, 200 (1997): 1639–1659.
6. A.C. Kitchenerm. Fighting and the Mechanical Design of Horns and Antlers. In *Biomechanics in Animal Behaviour,* B.R. Domenici, Ed., Oxford: BIOS Scientific Publishers, 2000.
7. J. Chen, E. Ravey, S. Hallett et al. Prediction of Delamination in Braided Composite T-Piece Specimens. *Composites Science and Technology,* 69 (2009): 2363–2367.
8. J. Chen and D. Fox. Numerical Investigation into Multi-Delamination Failure of Composite T-Piece Specimens under Mixed Mode Loading Using a Modified Cohesive Model. *Composite Structures,* 94 (2012): 2010–2016.
9. J. Chen. A Numerical Investigation of Thermal-Related Matrix Shrinkage Crack and Delamination in Composite T-Piece Specimens Using a Modified Interface Cohesive Model. *Journal of Thermoplastic Composite Materials,* 25 (2012): 267–282.
10. J. Chen. Simulation of Multi-Directional Crack Growth in Braided Composite T-Piece Specimens Using Cohesive Models. *Fatigue & Fracture of Engineering Materials & Structures,* 34 (2011): 123–130.
11. J. Chen. A Mixed Damage Model for Simulating Delamination of Composite T-Joint Components. Paper presented at ASME International Mechanical Engineering Congress and Exposition, 2012.

12

Lifetime Durability of Bio-Based Composites

Vincent Verney, Solène Gaudin, Sophie Commereuc,
Haroutioun Askanian, Florence Delor-Jestin, Alexandre Govin,
and René Guyonnet

CONTENTS

12.1 Introduction

The markets for applications of biocomposites are expanding internationally.[1] A biocomposite is a material made from a mixture of natural fibers and a thermoplastic polymer to obtain a product having some characteristics of both resources: (a) filler reinforcement (plus wood-like appearance of wood plastics) and (b) plastic performance in wet conditions. Wood–polymer composites (WPCs) are used mainly in four different sectors of materials applications: building, construction, automotive,[2] and marine infrastructures.

Current major scientific and technological efforts are focused on increasing the bio-based carbon content and minimizing environmental impacts associated with the use of polymeric materials. However, bio-based composites can create problems at the end of their lives. The current recycling streams are not suited to such materials. That is why the focus is increasingly on "eco-friendly" materials such as composites based on natural or synthetic biodegradable polymers and fillers such as starch, vegetable fiber, or wood flour. The most studied WPC-based biodegradable matrix systems[3] are composites based on polylactic acid (PLA).[4-6] Many studies deal with the mechanical properties and the inclusions of additives to improve the effects of reinforcing fillers.[7] However, very few studies are devoted to the durability of these new biocomposites during their service life.[8]

The challenge is to extend the use of these materials in specific applications with longer service times. Even if the biodegradable properties of bio-based composites are well studied, their photo-aging properties remain to be considered.[9]

Studies of natural filler composites—polymer—indicate several drawbacks, for example, the incompatibility of filler and polymer due to their hydrophilic character and the hydrophobic character of the matrix. The hydrophilic nature of wood largely increases the degradation kinetics of a composite. This intrinsic character of wood can be enhanced by various techniques including thermal treatments.

The purpose of this study is to implement various natural fillers (wood and starch) in biodegradable polymers and then to characterize their original properties compared to photochemical durability. After the characterization of the composites, a method of estimating their ultraviolet (UV) durability was tested. Both accelerated and natural outdoor conditions of UV exposition were investigated and melt viscoelasticity was used as a physicochemical probe of molecular evolution under UV radiation.

12.2 Experimental

12.2.1 Polymer Matrix

Three biodegradable polymer matrices were selected: poly (lactic acid) (PLA, Cargill Nature Works 4042D); Ecoflex, an aliphatic aromatic copolyester (PBAT); and Ecovio a 45:55 mixture of PLA and Ecoflex (both from BASF).

12.2.2 Fillers

12.2.2.1 Wood Flours

Various natural fillers have been used: pine or poplar flour (Robusta and I215 species). They were obtained after mechanical grinding of wood pellets with an average size of about 150 μm. These flours can be used as natural

TABLE 12.1

Compositions of M2, M3, and PT4 Composites

Wt%	M2	M3	PT4
Glycerol	15	5	4
Starch	35	22	26
PLA	0	12	0
Ecoflex	50	61	70

materials or after mild pyrolysis treatment under inert atmosphere for a short time (< 30 min) at moderate temperature (< 240°C). This treatment called *retification* leads to the cracking of hemicellulose[10,11] and gives heat-treated wood with a higher hydrophobic characteristic and better resistance to bio-degradation. In all cases the wood–polymer concentration was 50:50.

12.2.2.2 Starch Flours

The studied biocomposite samples were designated M2, M3, and PT4 and contained either starch from maize (M2 and M3) or potato (PT4). The incorporated starch in the thermoplastic matrix can be destructured (M3, PT4) or not (M2). Table 12.1 shows the exact composition of each sample.

12.2.3 Techniques

Bio-based composites were mixed in a Thermo Haake Microcompounder at various temperatures with a rotation speed of 100 rpm. After processing, samples were compression molded with a hot plate compression molding device to produce films ~100 μm thick.

Melt viscoelasticity was examined with an ARES mechanical spectrometer (with parallel plate configuration: 8 mm diameter, gap = 1 mm) and a constant stress StressTech Rheologica rheometer (with parallel plate configuration at 10 mm diameter, gap = 1 mm). In all cases, strain (or stress) amplitude were checked to ensure they remained within the linear viscoelastic region.

UV photodurability was assessed through accelerated photo-irradiation in an Atlas SEPAP 1224 at 60°C or by outdoor exposure in Clermont-Ferrand, France; samples faced the south direction.

12.2.4 Melt Viscoelasticity: Background

Rheology is the science that studies the deformation and flow of materials in liquid state (melted) or in solid state to determine the elasticity and viscosity of a material.[12] Materials respond to applied external forces (stresses) or displacements (strains) by an elastic (Hooke solid) or viscous (Newton fluid) behavior—or more usually, a combination of both (viscoelastic behavior).

The study of viscoelasticity thus establishes the laws of behavior to predict the strain of a sample under the action of a set of forces in certain conditions of time and temperature.

Elasticity is the ability of a material to store the energy of deformation, expressed more simply as its ability to regain its original shape after being deformed. Hooke's law describes the behavior of an ideal purely elastic solid: the stress required to deform such a body is proportional to the instantaneous deformation. Furthermore, when the stress is canceled, the body returns to its original form and the strain energy is fully restored. Such a body is characterized by a mechanical modulus of rigidity G, which is the ratio between the stress σ and the strain γ:

$$G = \frac{\sigma}{\gamma} \tag{12.1}$$

Viscosity is the measure of the resistance of a material to flow rate and reflects the energy dissipation of the material deformation in the flow. Newton's law is similar to Hooke's law for ideal viscous fluids: the stress required to deform such a body is proportional to the strain rate. Such a body is characterized in that its viscosity η is the ratio of the stress σ and strain rate $\dot{\gamma}$:

$$\eta = \frac{\sigma}{\dot{\gamma}} \tag{12.2}$$

A dynamic rheological measurement consists of applying to the material a sinusoidal deformation γ^* with an amplitude γ_0 and a pulsation ω, and measuring the resulting sinusoidal stress σ^* with an amplitude σ_0 that will be out of phase with a phase angle δ (when the material is not purely elastic). It is convenient to use complex quantities in the case of viscoelastic polymers. Thus, the strains and stresses are given, respectively, by:

$$\gamma^* = \gamma_0 \cdot e^{i\omega t} \tag{12.3}$$

$$\sigma^* = \sigma_0 \cdot e^{i\omega t + \delta} \tag{12.4}$$

Thus, the behavior of a viscoelastic polymer is characterized by its dynamic complex modulus G^*, which is a measure of the overall strength of the material to the deformation expressed by the equation:

$$G^*(\omega) = \frac{\sigma^*}{\gamma^*} = G'(\omega) + iG''(\omega) \tag{12.5}$$

where G' is the real component called the elastic modulus (or storage modulus) and G'' the imaginary component called viscous modulus (or loss modulus).

The ratio of these two components is the loss angle tangent (tan δ = G''/G'), and it describes the damping capacity of the material. The complex dynamic viscosity is a measure of the overall strength of a material to flow as a function of shear rate. It is expressed by the equation:

$$\eta^* = \frac{G^*(\omega)}{i\omega} = \eta'(\omega) - i\eta''(\omega) \tag{12.6}$$

and then:

$$\eta' = \frac{G''(\omega)}{\omega} \tag{12.7}$$

$$\eta'' = \frac{G'(\omega)}{\omega} \tag{12.8}$$

$$\tan \delta = \frac{\eta'(\omega)}{\eta''(\omega)} = \frac{G''(\omega)}{G'(\omega)} \tag{12.9}$$

12.2.4.1 Cole-Cole Plots

Cole-Cole plots (Figure 12.1) are complex plane representations ($\eta''(\omega)$ $\eta'(\omega)$). The original Cole-Cole distribution model (13) in exponent predicts an arc of a circle. Two parameters can be obtained easily from this representation. The first is the Newtonian viscosity $\eta 0$: intersection between the circle and the real axis. It is proportional to the molecular weight of the power as shown in the equation:

$$\eta_0 = K \cdot (M_w)^a \tag{12.10}$$

Thus, small variations in molecular weight will lead to large variations in viscosity.[14] The second parameter is the h parameter that represents the width

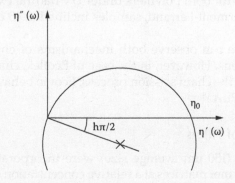

FIGURE 12.1
Cole-Cole representation.

of the relaxation time distribution instead of molecular weight. It characterizes the eccentricity with respect to the real axis and can be determined by the angle between the horizontal axis and the radius of the arc of a circle passing through the origin.[15] This angle is actually $h\pi/2$. The parameter h is involved in the analytical expression of the dynamic viscosity proposed by K.S. Cole and R.H. Cole:[13]

$$\eta^*(\omega) = \frac{\eta_0}{1 + (j\omega\lambda_m)^{1-h}} \tag{12.11}$$

where λ_m is the average relaxation time; it corresponds to the inverse of a critical frequency for which:

$$\eta' = \frac{\eta_0}{2} \tag{12.12}$$

12.3 Photo-Aging of Wood-Based Biocomposites

12.3.1 Photo-Aging of Pure Polymer Matrices

As we have to deal with complex systems (polymer matrix + filler), it is first necessary to study the behaviors of polymer matrices alone to be able to discriminate the positive and negative impacts of fillers on the photodurabilities of raw materials. Figure 12.2 illustrates the molecular evolution of PLA through the evolution of its viscoelastic behavior.

In the case of polylactide polymers, the chemical mechanism of photo-aging is a chain scission process.[16,17] The same trends are observed in natural weathering expositions. The cases of Ecoflex (aliphatic aromatic copolyester) and Ecovio (blend of Ecoflex + PLA) are very complex. Figure 12.3 shows the Cole-Cole plots for both polymers under UV natural exposition (April to August 2008 in Clermont-Ferrand, samples inclined at 45 degrees facing the south direction).

In both cases, we can observe both mechanisms of chain scissions and chain recombinations. However, in the case of Ecoflex, chain recombination predominates over the chain scission process; Ecovio behaved contrarily due to the presence of PLA.[18]

12.3.2 Influence of Fillers

Polar wood flours (150 μm average size) were incorporated by melt compounding into polymer matrices at a relative concentration of 50:50 in weight. The effect of fillers is very important if we look at the increase of G′ moduli (Figure 12.4).

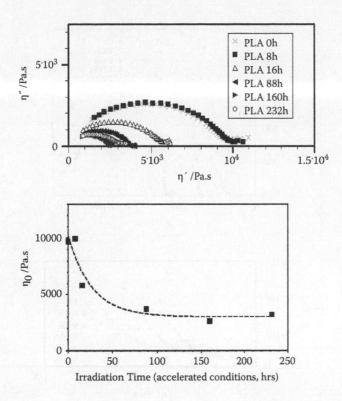

FIGURE 12.2
Cole-Cole (T = 180°C) plots for PLA photo-aging at various times of accelerated exposition in SEPAP 1224. Up: Cole-Cole plots. Down: evolution of Newtonian viscosity with time of irradiation.

12.3.3 UV Aging of Wood Polymer Composites

PLA, Ecoflex, and Ecovio wood composites (50% in weight poplar flour: I214 and Robusta, thermally treated or not) were exposed both to accelerated UV (dry exposition, Sepap 1224 at 60°C) and natural weathering (climatic exposition from April to August 2008). At regular periods, samples were taken from the exposition carousel and analyzed in melt viscoelasticity. We chose to define a criterion of photostability by the ratio of the storage modulus of the material measured at a frequency N = 10 Hz at time of exposition (*t*) versus the modulus of the initial non-exposed material:

$$R_{10\ Hz} = \frac{G'(t)}{G'(t0)} \tag{12.13}$$

Of course, a constancy of 1 means photostability; a value lower than 1 means a decrease of mechanical rigidity. Figure 12.5 demonstrates the method in the cases of Ecoflex and Ecovio poplar WPCs.

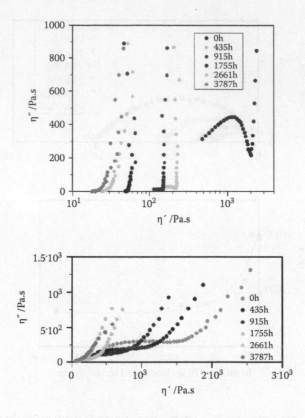

FIGURE 12.3

Cole-Cole plots of Ecoflex at different times of natural weathering exposition (spring and summer 2008). Up: T = 140°C. Down: Ecovio, T = 180°C.

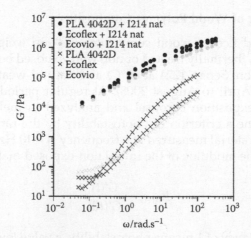

FIGURE 12.4

Variations of storage modulus G' versus frequency for different polymers and 50:50 WPC.

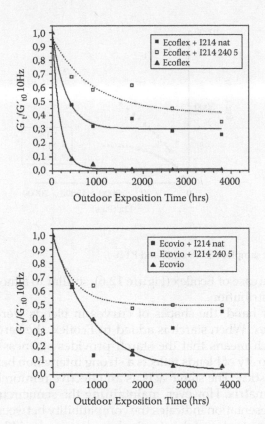

FIGURE 12.5
Storage modulus ratio at N = 10 Hz of Ecoflex and Ecovio wood poplar composites.

The main result indicated that the presence of a filler leads to a photostabilization effect of the viscoelastic behaviors of both Ecoflex- and Ecovio-based WPCs. PLA demonstrated no evidence of any effect. Moreover, thermal treatment also favors this stabilizing effect. We found the same result with pine flour PLA–Ecoflex and –Ecovio WPCs. However, there are optimum experimental conditions for the thermal treatment. The combination T = 240°C, time = 5 min seemed to be the most effective for the retification process.

12.4 Photo-Aging of Starch-Based Biocomposites

The same methodology was applied to starch-based biocomposites. The viscoelastic behavior of samples was characterized by melt rheology at 160°C. The Cole-Cole plots (η'' versus η') of all the samples were evaluated and compared to the Ecoflex matrix. M3 and PT4 show semicircle Cole-Cole

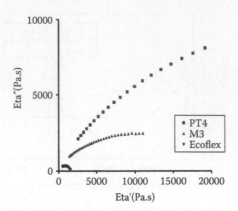

FIGURE12.6
Cole-Cole plots for samples (Ecoflex, M3 and PT4).

curves as in the case of Ecoflex (Figure 12.6), similar to a monomodal molecular weight distribution.

On the other hand, the shapes of curves in blends were different from those for Ecoflex. When starch is added to Ecoflex, the zero shear viscosity increases, which means that the starch provides stiffness to blends.[19] The zero shear viscosity of blends reflects a strong interaction between the starch and thermoplastic.[20] The starch acts as an effective reinforcing agent for the thermoplastic matrix. However, maintaining the semicircular shape of the Cole-Cole representation indicates the compatibility between the two phases due to the destructuring of the starch phase in samples M3 and PT4.

Moreover, the viscosity of PT4 is vastly superior to that of M3 despite the similarities of their starch contents Several hypotheses can be proposed with regard to the composition of PLA in M3. However, we can assume that the difference between M3 and PT4 samples could be that the starch used in PT4 was derived from potatoes (~20% amylase, ~80% amylopectin), while the starch in M3 was derived from corn (~25% amylase, ~75% amylopectin).[21] Amylopectin is a branched polymer while amylose is a linear one polymer. Therefore, more amylopectin in the samples resulted in a higher η_0 value.

Increasing the starch content in a matrix increases the structuring of the medium. It seems that a blend with a high starch content behaves more as a solid material. M2 contained a 35% starchy phase composed of non-destructured corn flour. M2 shows a linear Cole-Cole characteristic of a highly filled polymer system.

12.4.1 Accelerated Photo-Aging

Figure 12.7 displays the evolution of complex viscosity components through the accelerated photo-aging of PT4. The zero shear viscosity η_0 increases from the first hour of UV exposure (0 to 30 hr), providing evidence of a molecular weight increase due to recombination reactions (especially in the

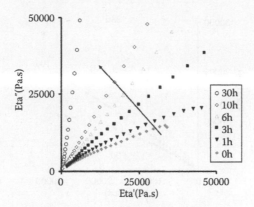

FIGURE 12.7
Cole-Cole plots for PT4 upon accelerated photo-aging.

Ecoflex phase). The gradual distortion of the Cole-Cole diagram suggests that gelation occurs through photo-oxidation. After 30 hr of UV exposure, the Cole-Cole diagram looks like a straight line, which is the characteristic feature of a fully cross-linked material.

Similar results were obtained with samples M2 and M3. Cole-Cole plots show evidence that cross-linking is the predominant process occurring through photo-aging of blends M2 and M3. It seems that the nature of the starch phase (corn in M2 and potato in PT4) does not impact behavior during photo-aging—the same behavior shown by the Ecoflex matrix. In the same way, the presence of a small portion of PLA in M3 (12%) does not change the overall behavior of the biocomposite during photo-aging; the predominant process remains cross-linking. The behavior of biocomposites under UV irradiation is the result of neat Ecoflex.

12.4.2 Natural Weathering

Due to accelerated photo-aging, the changes of viscoelastic properties through natural weathering of biocomposites M2, M3, and PT4 were monitored. Natural weathering was performed at Clermont-Ferrand in central France (latitude 45° 450′ N, longitude 3° 100′ E, altitude 329 m) in May 2008 for 4, 6, 10, 17, and 28 days.

Dynamic oscillatory measurements highlight a modification of the molecular structure of the material from the beginning of UV exposure. Photo-aging of biocomposites under natural conditions seems to be more complex. For example, in the case of M3 (Figure 12.8), likewise for accelerated photo-aging, chain recombination reactions take place first (4 days of outdoor exposure), and at 6 days a Cole-Cole curve shows a decay of molecular weight due to the chain scissions process that prevails. Then recombinations become predominant for longer exposure times (10, 17, and 28 days) possibly involving cross-linking phenomena.

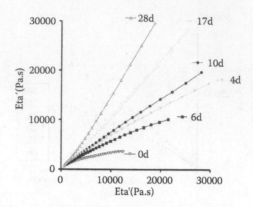

FIGURE 12.8
Cole-Cole plots for M3 upon natural weathering (0 to 28 days).

PT4 also exhibited a dual process due to the competition of chain scissions and chain recombinations. First recombinations (increase of η_0 until 6 days of exposure under natural conditions) are observed; then at 10 days, chain scissions are observed. An inversion is observed again in the second stage of natural weathering and the material is cross-linked.

M2 acts as a filled polymer in the initial state—Cole-Cole representation reflected strong structuring of the environment. It is more difficult to assess evolutions during photo-aging. The slope value $\eta'(\omega) = f(\eta''(\omega))$ changes slowly with exposure time up to 17 days. After 28 days, it evolves dramatically toward higher values; recombinations are observed over the course of the photo-aging and cross-linking takes place.

While overall good agreement between behaviors in accelerated and natural conditions was observed for the longest exposure time (recombination and cross-linking), the photo-aging of biocomposites under natural conditions is more complex and shows strong competition among various processes. Moisture can have an important impact on the starchy phase. Chain scissions can occur by breaking glycosidic bonds; the amylose molecules are shortened and amylopectin molecules tend toward a more linear form. Oxygen promotes the chain scissions of amylopectin, increasing the linearity of the polysaccharide chains. It is difficult to exploit these results and assign them only to the Ecoflex thermoplastic phase.

12.5 Conclusions

We have built an original methodology for the assessment of the mechanical stability under UV aging of biodegradable wood polymer composites based on molecular evolution. These materials are complex as they combine a natural filler phase with a biodegradable polymer matrix, itself very sensitive to any source of degradative condition.

The resulting effect is a dual molecular process involving a competition between chain scissions and chain recombinations. Melt viscoelasticity is a powerful tool for discriminating one mechanism from another along the UV degradative pathway. This approach based upon the molecular evolution at a microscale must be extended to macromechanical behavior.

References

1. R.M. Rowell. Challenges in Biomass–Thermoplastic Composites. *Journal of Polymers and the Environment,* 15 (2007): 229–235.
2. J. Holbery and D. Houston. Natural Fiber-Reinforced Polymer Composites in Automotive Applications. *JOM,* 58 (2006): 80–86.
3. B.A. Acha, N.E. Marcovich, and J. Karger-Kocsis. Biodegradable Jute Cloth-Reinforced Thermoplastic Copolyester Composites: Fracture and Failure Behaviour. *Plastics, Rubber, and Composites,* 35 (2006): 73–82.
4. R. Csizmadia, G. Faludi, K. Renner et al. PLA–Wood Biocomposites: Improving Composite Strength by Chemical Treatment of Fibers. *Composites Part A,* 53 (2013): 46–53.
5. S. Pilla, S. Gong, E. O'Neill et al. Polylactide–Pine Wood Flour Composites. *Polymer Engineering & Science,* 48 (2008): 578–587.
6. T. Mukherjee and N. Kao. PLA-Based Biopolymer Reinforced with Natural Fibre: A Review. *Journal of Polymers and the Environment,* 19 (2011): 714–725.
7. D.D. Stokke and D.J. Gardner. Fundamental Aspects of Wood as a Component of Thermoplastic Composites. *Journal of Vinyl and Additive Technology,* 9 (2003): 96–104.
8. F.P. La Mantia and M. Morreale. Accelerated Weathering of Polypropylene–Wood Flour Composites. *Polymer Degradation and Stability,* 93 (2008): 1252–1258.
9. K.K. Pandey and T. Vuorinen. Comparative Study of Photodegradation of Wood by a UV Laser and a Xenon Light Source. *Polymer Degradation and Stability,* 93 (2008): 2138–2146.
10. R. Guyonnet and J. Bourgois. Characterization and Analysis of Torrefied Wood. *Wood Science and Technology,* 22 (1988): 143–155.
11. M. C. Bartholin, R. Guyonnet, and J. Bourgois. Thermal Treatment of Wood: Analysis of the Obtained Product. *Wood Science and Technology,* 23 (1989): 303–310.
12. J.D. Ferry. *Viscoelastic Properties of Polymers.* New York: John Wiley & Sons, 1961.
13. K.S. Cole and R.H. Cole. Dispersion and Absorption in Dielectrics. I: Alternating Current Characteristics. *Journal of Chemical Physics,* 9 (1941): 341–351.
14. H. Azizi and I. Ghasemi. Investigation on the Dynamic Melt Rheological Properties of Polypropylene–Wood Flour Composites. *Polymer Composites,* 30 (2009): 429–435.

15. C. Friedrich and H. Braun. Generalized Cole-Cole Behavior and Its Rheological Relevance. *Rheologica Acta*, 31 (1992): 309–322.
16. P. Stloukal, V. Verney, S. Commereuc et al. Assessment of the Interrelation between Photooxidation and Biodegradation of Selected Polyesters after Artificial Weathering. *Chemosphere*, 88 (2012): 1214–1219.
17. S. Bocchini, K. Fukushima, A.D. Blasio et al. Polylactic Acid and Polylactic Acid-Based Nanocomposite Photooxidation. *Biomacromolecules*, 11 (2010): 2919–2926.
18. J. Rychlý, L. Rychlá, P. Stloukal et al. UV-Initiated Oxidation and Chemiluminescence from Aromatic–Aliphatic Co-Polyesters and Polylactic Acid. *Polymer Degradation and Stability*, 98 (2013): 2556–2563.
19. L. Avérous and C. Fringant. Association between Plasticized Starch and Polyesters: Processing and Performances of Injected Biodegradable Systems. *Polymer Engineering & Science*, 41 (2001): 727–734.
20. R. Mani and M. Bhattacharya. Properties of Injection Moulded Starch–Synthetic Polymer Blends. IV: Thermal and Morphological Properties. *European Polymer Journal*, 34 (1998): 1477–1487.
21. A. Buléon, P. Colonna, V. Planchot et al. Starch Granules: Structure and Biosynthesis. *International Journal of Biological Macromolecules*, 23 (1998): 85–112.

13

Mechanical Properties of Natural Fiber-Reinforced Composites

Rattana Tantatherdtam, Rungsima Chollakup, and Wirasak Smitthipong

CONTENTS

13.1 Introduction

The use of natural fibers for the reinforcement of composites has received increasing attention by the academic sector and industry.[1,2] Generally, fibers are stronger than matrices. When a composite is forced by external action, the force is transferred from the matrix to fibers, which increases its strength compared to the matrix without fiber. The sizes and shapes of fibers have important roles in the efficiency of reinforcement.[3,4] For example, pineapple fiber-reinforced plastic can be produced by various processes based on fiber size:[5]

1. Short fiber lengths (less than 1 cm) are suitable for fiber reinforcement in thermoplastics that can be prepared by extrusion and injection molding; a compatibilizer should be added between the fiber and polymer resin.
2. Fiber lengths between 1 and 10 cm are used for compression molding. The fibers are sprayed onto a resin as a template polymer or the fiber and polymer are sprayed together into a template.
3. Long fibers are aligned in a smooth mat, which is like a piece of non-woven fiber carding, then the non-woven material is laminated by layers of polymer using a compression heat press.

To achieve effective fiber reinforcement in a polymer matrix, the fiber must be placed in the direction parallel to the force action. This means that the mechanical properties of natural fiber-reinforced composites depend on the types of fiber fillers and polymer matrices.[6]

Normally in product design of bio-based composites, one must consider the structure of a material that determines its properties. The structure–property relationship of a bio-based material directly impacts its applications. Bio-based composite mechanical properties are important; a bio-based composite must biodegradable but should be strong enough to handle the function for which it was designed.[7]

When considering unidirectional fiber alignment in a matrix, the fiber direction is called the longitudinal axis or fiber axis. The longitudinal axis typically has the highest stiffness and strength direction. Any direction perpendicular to the fibers is transverse. At times, to simplify analysis and test requirements, composite properties are assumed to be the same in any transverse direction. This is the transverse isotropy assumption and is satisfied approximately for most unidirectional composites (Figure 13.1).

A multidirectional composite can exhibit many stiffness constants. Strength predictions are equally complicated because of directional differences, i.e., compression is not always equal to tension and failure theories are complex. As the complexity of matrix calculations increases, it becomes evident that errorless mathematical manipulations are impossible without the aid of computer simulations.

To test mechanical properties, we used a tensile testing machine to mimic the vertical stretching of a dumbbell sample.[8] The measurement indicates shape changes at a constant rate. The tensile testing machine can measure tension force as a function of the elongation or deformation of a material until it is broken. Generally, stress (N/m^2 or Pa) is the force acting per unit of cross-sectional area perpendicular to an applied force or the ability to

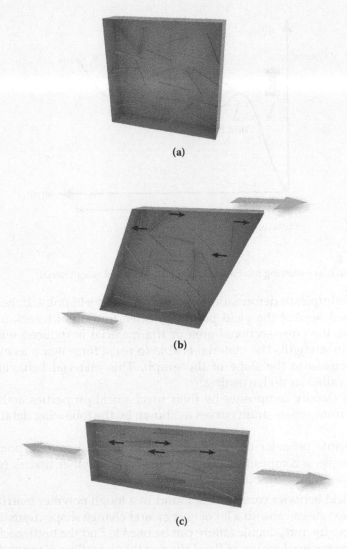

(a)

(b)

(c)

FIGURE 13.1
Fiber orientation in simple composite from impact of process flow. (a) Random distribution. (b) Rotation of fiber reinforcement during shear flow. (c) Arrangement of fiber reinforcement during traction flow.

resist an external force (load). Strain (%) is defined as the ratio of stretching (elongation) or size change per unit original length. The slope at the beginning of the stress–strain graph is a straight line called Young's modulus.[9]

The stress–strain curve of a viscoelastic material is a straight line in the beginning (linear elastic behavior at low strain). This indicates that the material can return to its original shape when we stop applying force (Figure 13.2). When a specimen is stretched until it begins to change shape

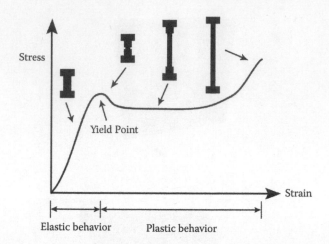

FIGURE 13.2
Stress–strain curve showing mechanical properties of viscoelastic material.

permanently (plastic deformation), we can find the yield point. If the material is stretched beyond the yield point, we often find a bottleneck (necking). After that, the cross-sectional area of the material is reduced without an increase in strength. The material is able to resist force again as evidenced by an increase in the slope of the graph. This material behavior before failure is called strain hardening.[10]

We can classify composites by their mechanical properties and fracture behavior from stress–strain curves as shown by the following details.[9]

1. A brittle behavior composite is broken without a yield point, for example, a composite with an amorphous polymer matrix (glassy polymer) is brittle but stiff.

2. A yield behavior composite is found in a tough polymer matrix. The composite can absorb a lot of energy and change shape dramatically. Before fracture, ductile failure can be used to find the bottleneck phenomenon (necking) from the sliding of the crystalline plane and elongation of the chain molecules. Examples are semicrystalline polymers.

3. Rubbery behavior is found in a rubber matrix. This type of matrix can remain flexible under high stress caused by stretching. The yield point does not appear; the material can return back to the initial state without losing its original shape. This is called a non-linear elastic behavior. Examples are rubber and elastomer matrices.

However, a single type of polymer matrix can present brittle or tough behavior, depending on test conditions such as temperature and speed. Therefore, test temperature and speed must be specified when examining the mechanical properties of bio-based composites.[11]

13.2 Viscoelastic Properties of Bio-Based Composites

Drastic changes in the physical properties of polymers caused by reinforcement also lead to pronounced changes in their viscoelastic behaviors.
Viscoelastic properties and relaxation behaviors of composites change as a
result of the formation of surface layers at the polymer–solid interface. The
molecular mobility of polymeric chains is restricted in these layers and the
restriction affects mechanical properties.

Dependence of elasticity modulus of a particulate-filled polymer has been
shown in experimental studies. To establish the dependence of modulus on
filler loading, various proposed models allow us to calculate the elasticity
modulus from corresponding values of the constituent components.[12]
A large number of equations can calculate the moduli of heterogeneous
polymeric compositions.

The influence of fillers that weakly interact with polymers on viscoelastic
properties is determined by simple filling of the polymer volume by rigid
inclusions. The filler particles are comparably large and the distance between
them even at high loading is also large compared with the usual end-to-end
length of a polymer chain. Filler particles are separated from one another
and cannot be bridged by a macromolecular chain. For such systems, the
contributions of interphase layers can be neglected.

The dynamic mechanical properties of a filled system in the absence of
interactions between components can be described by a mechanical model for
non-interacting polymer mixtures. This model is very useful for describing
properties of filled systems with interfacial layers. Based on hydrodynamic
considerations, an equation (analogous to the Einstein equation for viscosity
of suspension) was proposed to calculate the modulus of composite, E_c:[13–15]

$$E_c = E_u(1 + 2.5\phi + 1.4\phi^2) \tag{13.1}$$

where E_u is the modulus of unfilled polymer and ϕ the volume fraction of
filler. Various models have been proposed to calculate the modulus of composites. All are based on assumed morphological structures.

To calculate the bulk moduli of particulate-filled systems that incorporate the
effects of filler–matrix interactions, a molecular theory was proposed.[4,11] The
theory treats the composite as a molecular mixture, specifically as a binary polymer blend of supermacromolecules (fillers) in a polymer matrix. The model is a
lattice with vacant sites or sites occupied by segments of the two components.

The vacant sites introduce excess free volume into the system. A melt and
glassy matrix may be distinguished by the fact that the free volume fraction in
the melt at thermodynamic equilibrium is determined uniquely by minimization of the free energy. The molecular parameters of such a system are the
ratios of attraction and repulsion potentials between the filler and polymer.
This approach has enabled us to evaluate the bulk modulus of a composite and

compare it with that obtained from an interlayer model.[16,17] The experimental results of these two very different approaches were in good agreement.

When using these theories, one should keep in mind that they do not take into account the distribution of particles by size and shape. Mechanical models are very useful for calculating viscoelastic modulus and mechanical loss in particulate-filled polymers. The best-known model was proposed by Takayanagi.[18] Its application to the calculation of dynamic mechanical properties is described elsewhere.[19]

Mechanical models help us calculate the viscoelastic properties of composites with various morphologies. The dynamic modulus depends on both filler loading and morphology. A comparison of theoretical results with experimental data shows that the calculations create agreement between the dynamic properties and composition of a system. At the same time, the phase morphology of the system and structures of particles of the disperse phase should also be taken into account.[18]

One reason for deviation of theoretical equations connecting elasticity modulus with filler amount from experimental data is the formation of surface layers at the polymer–filler interface (interphase layer).[20] The properties of these layers differ from bulk properties. It is very important to estimate the contribution of the interphase to the viscoelastic properties of composites.

To estimate theoretically the contribution of the interphase, the phenomenological model of Takayanagi may be used.[21,22] Particulate-filled polymer is presented as a cube of matrix with edge length. In the center of this cube, another smaller cube made of the filler material is inserted. The smaller cube is covered by a uniform layer of polymeric material whose properties represent the interphase. This model may be considered as consisting of a two-component model (filler with interphase layer) and a three-component model in which one component is the two-component model.

In a more generalized form, the influence of the interphase layers on the viscoelastic properties may be seen from the concentration dependencies at various temperatures. If the interphase is present, the characteristic of the curve $G'' = f(\phi)$ depends on the temperature and properties of the interphase (G'' = loss shear modulus). At temperatures below the glass transition temperature of the interphase and matrix, dependence $G' = f(\phi)$ is close to linear (G' = storage shear modulus) because the properties of the interphase in the glassy state are almost the same as for the matrix.

At the same time, at transition temperatures, the same dependences are non-linear. The deviation from linearity depends on the difference (ΔT) between glass transition temperatures of the interphase T_{gi} and matrix T_{gm}. If $\Delta T > 0$, increasing ϕ will lead to more rapid growth of G' as compared with linear growth and vice versa. At a high concentration of low modulus interphase, a negative slope of concentration dependence may be observed, i.e., an increase in ϕ leads to diminishing G'. Further increase in filler loading leads again to increasing G'.[23] Clearly, the properties and the thickness of an interphase are important aspects of composite performance.

The temperature dependencies of the mechanical loss (ΔT) also change when interphase is present, especially when $T_{gi} \ll T_{gm}$. At a great difference between the properties of the interphase and matrix, the maximum of (ΔT) may be shifted along the temperature axis and a second maximum may appear, whose height and position strongly depend on the properties of the interphase.

Limited molecular mobility in the boundary or interphase layer is equivalent to increasing chain stiffness or formation of additional bonds in the structural network of a polymer. The addition of filler has the same effect on a polymer as a temperature decrease or increase of frequency of deformation. It follows from this that time–temperature relationship must apply to filled polymers along with the well-known principle of time–temperature superposition.[24]

In time–temperature superposition, the curves of frequency dependencies of modulus and losses at various temperatures in the range of transition temperatures exhibit similar characteristics and may be displaced by parallel shift along the frequency axis. The value of the shift depends on the temperature and is described by the Williams-Landell-Ferry (WLF) equation.[25] Using this analogy, one can calculate the same dependencies in a broad range of frequencies at any given temperature from the frequency dependence of viscoelastic properties in a rather narrow frequency range (master curve with shift factor a_T); see Figure 13.3.

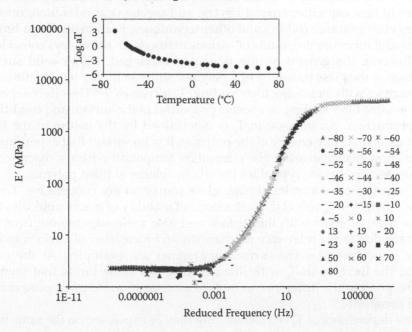

FIGURE 13.3

Storage modulus (E′) versus reduced frequency master curves for palm fiber-reinforced NR composites at different temperatures (–80°C to 80°C). Inset: Example of variation of logarithm of shift factor a_T versus temperature for this system.

The principles of time–temperature superposition are detailed below. An increase in filler loading leads to the same increase in the real part of the complex elasticity modulus as a rise in frequency of deformation or reduction in temperature. The most general result includes the dependence of stress on deformation standardized with respect to temperature and filler loading. In particular, the influence of the filler in the initial section of the stress–strain curve is greater than the prediction by the general method of reduction. The magnitude of the elongation at break may in turn be presented in a concentration-invariant form by horizontal or vertical shift of the experimental curve.

13.3 Relaxation Times of Bio-Based Composites

Glass transition temperature (T_g) corresponds to the temperature at which the mobility of segments of the polymer begins. Consequently, adsorption and adhesion are reflected in T_g. The T_g of a polymer can be changed under the influence of a solid surface.[26] The data on the change in T_g of filled polymers have been obtained by various methods, including dilatometry, measurement of heat capacities from dielectric and mechanical relaxation, nuclear magnetic resonance (NMR), and other techniques. Each method has limitations and therefore the results of various methods do not always coincide.

However, the general picture remains unchanged. Under solid surface influence, the glass transition of a polymer shifts to higher temperatures. In most cases with increasing filler loading, T_g increases.[27,28] The effect depends at the same filler loading on specific properties of the surface and conditions of production. An increase in T_g is determined by the nature of the filler surface and surface energy of the polymer. It is important that experimental data show that changes in glass transition temperature have a macroscopic characteristic, i.e., are typical for the whole volume of filled polymer.

The processes occurring during glass transition are cooperative. Therefore, changes in T_g reflect the restrictions of mobility of macromolecules that have direct contact with the surface and also molecules remote from the surface due to the relay-race mechanism and formation of various molecular structures near the surface; aggregates are examples. At the same time, the increase in T_g with filler loading has some limits that seem to correspond to the difference in molecular relaxation between pure matrix and composite.[29]

The dependence of T_g on filler loading may be expressed on the same basis as changes in the properties of the polymer phase in the boundary layer but in a different way.[30] For a theoretical description of the behavior of filled polymer, it is important to analyze the processes of structural relaxations near T_g. We know that the average relaxation time increases with increasing filler loading at constant temperature. This is very significant for selecting conditions of processing of filled polymers since optimum properties of a material depend on processing conditions (temperature, time, and pressure). However, to determine the mechanisms of the processes at the polymer–filler interface, it is desirable to compare relaxation times, not at identical temperatures, but at temperatures equidistant from the glass transition temperature (considering that it increases with filling).

The influence of the fillers on relaxation during transition from a glassy to a rubber-like state may be estimated by analyzing mechanical relaxation.[31-33] The theoretical basis is that during cooling of the melt at equilibrium from $T_1 \gg T_g$ at a constant rate, there is a certain point at which the rate of structural rearrangements in the melt determined by the heat mobility of the chain segments is lower than the cooling rate. Further cooling leads to larger deviations of the "instant" melt structure from the equilibrium state. In other words, at a rather low temperature $T_2 < T_g$, the structure or set of structures is frozen in a specimen.

Experimental data on viscoelastic properties enable the construction of the fundamental functions of viscoelastic relaxation spectra. An abundance of data indicates an increase in the average relaxation time in polymers on the surfaces of solid particles.[34-36] This is linked to changes in the structures of surface or interphase layers, adsorption interactions that limit molecular mobility, and the influence of interactions on the packing of macromolecules. From this view, we may expect that surface area will increase when filler loading increases. Thus, the thickness of the interlayer between the particles is reduced. Then molecular mobility is reduced and the average relaxation times are increased.

Accordingly, we may suppose that a change in filler loading will cause a predictable change in the spectra of mechanical relaxation times. In fact, the literature confirms this supposition.[37,38] Viscoelastic properties of filled polymers under dynamic loading conditions were investigated to verify these concepts.

A very interesting picture is revealed by analysis of the concentration dependence of average relaxation time. A non-monotonic relationship, with a broad minimum after initial increase is obtained. Such characteristics of dependence may be connected with an increase in filler loading, contributing to a decrease in the average relaxation time of the whole system since the average relaxation time of filler is lower than that of binder cured in its presence.

With the concentration of interphase increasing, its effect on relaxation times becomes obvious. The resultant contribution of these two factors is that the average relaxation time of a polymer filled with a polymeric filler of the same nature is a non-monotonous function of filler loading.[31,39] The non-monotonic changes of relaxation time may be also connected with non-uniformity of the interphase structure that on both its volume fraction and thickness.

13.4 Mechanical Properties of Various Types of Natural Fiber-Reinforced Rubber Composites

Natural and synthetic rubbers constitute interesting matrices that should be reinforced before use. Their applications range from rigid to soft composites. For example, short fibers are used in rubber compounding because of processing advantages, improvements in certain mechanical properties, and ecology considerations. Short fiber-reinforced rubber composites are used to fabricate reinforced products like V-belts, hoses, tire treads, and other complex-shaped articles. This section focuses on rubber composites with bio-based reinforcements.

13.4.1 Processing of Natural Fiber–Natural Rubber Composites

Typical processing techniques for preparation of natural rubber (NR) and various plant fiber composites include a two-roll mixing (or internal mixing) and subsequent compression molding (vulcanization of rubber compounds). Generally, standard block rubber is first mixed with fiber and other compounding ingredients in a conventional sulfur vulcanization system in a two-roll mill. The nip gap, mill:roll speed ratio, and number of passes should be kept uniform for all mixes to fiber alignment and uniform dispersion of fiber to maximize fiber orientation in the mill direction, the mill opening (nip gap) of pineapple leaf fiber–NR composites should be 1.5 mm, 1.3 mm for isora fiber NR composites, and 1.25 mm for sisal–oil palm hybrid fiber–NR composites.[40–43] The compounded rubber composites are sheeted off the mill for convenient preparation of test specimens and subsequently vulcanized typically with a mold press according to their respective optimum cure times measured with a rheometer.

Some typical problems related to the processing of natural fiber composite materials arise from the hydrophilic and hygroscopic natures of the fibers that significantly influence fiber dispersion in the matrix and interfacial adhesion. In general, the formation of water vapor during processing leads to the formation of voids in the material and poor mechanical properties. Therefore, it is a widely accepted technique to dry fibers before processing to reduce the humidity to 2 to 3%.[44]

In general, incorporating fibers with rubber is intended to improve mechanical performance. In addition, the mechanical properties of composites provide clear indications of the strength of the interface and the fiber–matrix interaction. Tensile, hardness, and tear properties are the most commonly investigated mechanical properties of natural fiber-reinforced NR composites.

Besides long-term performance, dynamic mechanical behaviors and rheological properties were also investigated for natural fiber composites. The extent of interfacial adhesion between the fiber and rubber matrix is assessed qualitatively from a study of restricted equilibrium swelling technique and by examining the fracture surface of the composite using scanning electron microscopy (SEM). Moreover, water sorption characteristics of fiber-reinforced NR composites are performed to analyze service performance when the composites are exposed to moisture or water and determine the strength of the rubber–fiber interface. Rubber degradation behavior on exposure to heat (thermal aging) after incorporating fibers has also been studied. Table 13.1 summarizes the studies of natural fiber–NR composites.

13.4.2 Effects of Fiber Loading

The fiber loading level plays a crucial role in determining the mechanical behaviors of fiber-reinforced polymer composites. To improve mechanical properties of short fibers, the matrix must be loaded beyond the optimum fiber concentration whereby the properties of the composite improve above the original matrix.

The dilution of the matrix may be a factor at low fiber loadings when the matrix is not restrained by enough fibers and the fibers are not capable of transferring loads to each other. At high levels of fiber loading, the increased number of fibers leads to agglomerations that break easily when stress is applied and then behave as stress concentrators. At optimum levels of loading, the population of the fibers provides maximum orientation and the fibers actively participate in stress transfer.

As concentration increases beyond the optimum fiber loading, the strength again decreases, matrix material is insufficient to hold the fibers together. Natural rubber has inherent high strength due to strain-induced crystallization. When fibers are incorporated, the regular arrangement of rubber molecules is disrupted and crystallization ability is lost. This may explain why fiber-reinforced natural rubber composites exhibit lower tensile strengths than gum compounds.[41,42]

The reduction of tensile strength and tear resistance after reaching maximum fiber loading was reported for white rice husk ash-filled natural rubber compounds.[45] As the fiber loading increased beyond 10 phr (part per hundred of rubber), the reduction in strength was suggested to arise from the agglomeration of filler particles that are no longer separated adequately or wetted by the rubber phase.

Scanning electron microscopy (SEM) studies revealed undispersed agglomerates of filler particles at high filler loading that acted as failure-initiating flaws and led to poor strength properties. The tensile moduli at 100 and 300% elongation and hardness were found to increase with increasing rice husk fiber loading while elongations at break decreased. This indicated the effect of increased filler particle incorporation into the rubber matrix whereby the elasticity or flexibility of the rubber chain decreased, resulting in more rigid vulcanizates.

TABLE 13.1

Fiber Surface Modification and Bonding Systems and Mechanical Property
Measurements for Natural Fiber-Reinforced NR Composites

Fiber Type	Fiber Surface Modification and Bonding System	Ref.	Property Measures
Coir	Fiber chemical treatment (NaOH)–bonding system (resorcinol and hexamethylene tetramine)	53	Cure characteristics, mechanical properties (tensile and tear strength), swelling
	Fiber chemical treatment (NaOH)–tri-component bonding system (hexaresorcinol-hydrated silica)	57	Mechanical properties (tensile, tear strength, hardness), extent and degree of fiber orientation, interfacial strength
	–	76	Tensile stress relaxation
	Fiber chemical treatment (NaOH, NR solution, toluene diisocyanate, acetylation)–bonding system (resorcinol and hexamethylene tetramine)	77	Water diffusion
	Fiber chemical treatment (NaOH, resorcinol–formaldehyde-latex, toluene diisocyanate, bleaching)–bonding system (resorcinol and hexamethylene tetramine)	78	Dynamic mechanical properties
Oil palm (short fiber)	Fiber chemical treatment (NaOH + heat)	50	Mechanical properties (tensile, tear strength, and hardness)
Oil palm wood flour	–	79	Mechanical properties (tensile, tear strength, and hardness), fatigue life
	–	80	Fatigue and hysteresis
	Bonding system (phenol formaldehyde, hexamethylene-tetramine, resorcinol formaldehyde, and silica)	81	Cure characteristics, mechanical properties (tensile and hardness), fatigue life
	–	47	Vulcanization system, cure characteristics, mechanical properties
White rice husk ash	Multifunctional additive (diamine salt of fatty acid)	79	Cure characteristics, mechanical properties (tensile, tear strength, and hardness)
	Multifunctional additive (diamine salt of fatty acid–silane coupling agent Si69	79	Cure characteristics, mechanical properties (tensile, tear strength, and hardness)
Bamboo	Silane coupling agent Si69	46	Cure characteristics, mechanical properties (tensile, tear strength, and hardness)
	Bonding agent (phenol, formaldehyde, silica, hexamethylenetetramine)	59	Cure characteristics, mechanical properties (tensile, tear strength, and hardness)
Paper sludge	Coupling agent (maleic anhydride grafted natural rubber)	28	Cure characteristics, dynamic and tensile properties

TABLE 13.1 (*Continued*)

Fiber Surface Modification and Bonding Systems and Mechanical Property
Measurements for Natural Fiber-Reinforced NR Composites

Fiber Type	Fiber Surface Modification and Bonding System	Ref.	Property Measures
Palm ash	Coupling agent (maleic anhydride grafted natural rubber)	48	Cure characteristics, tensile properties, and fatigue life
Sisal	Compatibilizer (natural rubber grafted maleic anhydride and epoxidized natural rubber)	62	Cure characteristics, mechanical properties (tensile, tear strength, and hardness)
Sisal (woven)	Fiber chemical treatment (mercerization, silanization, and heat treatment); bonding system (resorcinol and hexamethylene tetramine)	82	Mechanical properties (tensile, tear strength, and hardness), cross-link density(swelling)
Sisal–oil palm hybrid	Fiber chemical treatment (NaOH)–bonding system (resorcinol and hexamethylene tetramine	41	Cure characteristics, mechanical properties (tensile and tear strength)
	Fiber chemical treatment (mercerization and silanation)	54	Water sorption
Maize stalk	Fiber chemical treatment (acetic anhydride acetylation)	83	Cure characteristics, mechanical properties (tensile and hardness), water absorption
Banana	–	84	Mechanical properties (tensile and tear strength)
Pineapple leaf fiber	Fiber chemical treatment (sodium hydroxide, benzoyl peroxide)	40	Cure characteristics, mechanical properties (tensile), and thermal aging
Bagasse cellulose whiskers	–	85	Tensile and thermal properties, moisture sorption, water vapor permeation, and soil biodegradation
Cassava bagasse cellulose whiskers	–	86	Dynamic mechanical properties
Grass	Fiber chemical treatment (NaOH)–silane coupling agent (bis[3-(triethoxysilyl) propyl], tetrasulfide Si69)	87	Cure characteristics, mechanical properties (tensile and hardness)
	Fiber chemical treatment (NaOH)–bonding agent (resorcinol formaldehyde latex)	60	Cure characteristics, mechanical properties (tensile and hardness), swelling
Isora	Fiber chemical treatment (NaOH)– bonding agent (resorcinol- formaldehyde, resin, hexamethylene tetramine, precipitated silica)	88	Cure characteristics, mechanical properties (tensile and tear strength), morphology
Rice husk ash	Fiber chemical treatment (NaOH + HCl = hydrated silica)–silane coupling agent, bis(3 triethoxysilylpropyl)- tetrasulfane (Si-69)	89	Cure characteristics, mechanical properties (tensile, tear strength, hardness, and abrasion resistance), morphology

An optimum fiber concentration of 30 phr was reported in sisal–oil palm hybrid fibers with natural rubber compounds to achieve maximum tensile strength.[41,42] The effects of filler loading on tensile and tear strengths of the composites with and without bonding agents were studied in bamboo-filled NR composites.[46] Both properties decreased with increasing filler loading. The irregular shapes of bamboo fibers together with poor adhesion (due to hydrophilic nature) to the natural rubber matrix (hydrophobic nature) are the main factors in the deterioration of tensile and tear strengths with increasing filler loading. However, at similar filler loading, both properties improved with the addition of bonding agents (phenol formaldehyde–hexa-methylenetetramine–silica) due to an increase in adhesion of the rubber and bamboo fiber.

Moreover, the tensile moduli and hardnesses of composites increased with increasing filler loading. These results indicate that the incorporation of bamboo fiber into rubber matrices enhanced the stiffness of the composites. However, with the addition of bonding agents these properties increased further as a result of better interaction between fiber and rubber matrix.

Other studies also reported that the tensile modulus and hardness values increased when fiber loading was increased for natural rubber composites filled with various types of natural fibers. The use of oil palm wood flour (OPWF) as a filler in natural rubber compounds was also studied.[47] Increasing OPWF loading in rubber matrices reduced tensile strength, tear strength, and elongation at break, but increased moduli and hardness levels of the composites. The drop in tensile and tear strength resulted from the dilution effect of the filler in the composites and the irregular shapes of OPWF.

If OPWF incorporation is insufficient for uniform stress transmission in a matrix, the result is localized stress at the filler–matrix interface, leading to poor reinforcement. Ismail[28] found that tensile strengths of paper sludge-filled NR composites decreased with the addition of filler loading, possibly attributable to the geometry of the paper sludge filler. For irregularly shaped fillers, the strength of the composites decreases due to the inability of the fillers to support stress transferred from polymer matrices. This factor along with the poor adhesion of paper sludge fibers to natural rubber matrices accounts for the deterioration of tensile strength with increasing filler loading.

Many undispersed agglomerates arising from filler–filler interactions at high levels of filler loading appeared in SEM studies; this explains the reduction in tensile strength with increasing filler loading in paper sludge-filled NR composites. The strong influence of filler loading on the fatigue life of rubber composites is evidenced in palm ash-filled NR composites.[48] It is believed that the presence of a filler in a rubber matrix causes friction of filler particles, creating cracks in the in the rubber surface and subsequent catastrophic failure.

As more filler is added to a rubber matrix, we can expect that the filler particles and aggregates will not be dispersed and wetted efficiently by the matrix. These inherent defects can act as stress concentration points and shorten the fatigue lives of natural rubber composites. Similar observations were reported by Ishak et al.[49] and Ismail et al.[50] for rubber composites filled with white rice husk ash and oil palm wood flour, respectively.

13.4.3 Fiber–Hybrid Composites

Hybrid composites are composite materials based on reinforcement from two or more fibers in a single matrix. The properties of a hybrid composite depend primarily on fiber content, lengths of individual fibers, orientation, fiber:hybrid ratio, extent of intermingling of fibers, and fiber–matrix bonding.[51]

Hybrid composites of sisal–oil palm[41,42,51] and sisal–coir fiber[52] with natural rubber were prepared to investigate hybrid effects on composite properties. Hybrid fiber was found to improve the mechanical properties of the natural rubber composite significantly in comparison to a composite containing individual fibers. The maximum tensile strength, elongation at break, and modulus at 100% elongation were found in the hybrid composite when the ratio of sisal and oil palm was 70:30 with optimum fiber loading of 30 phr. It was also noted that the tensile properties of the NR composites were more dependent on sisal fiber than on oil palm fiber.

The reinforcing ability of sisal exceeds that of oil palm in any polymeric matrix due to the high strength and low microfibrillar angle of sisal fiber. Oil palm fibers are tough and hard, similar to coir fibers in cellular structure. Therefore, the strength properties of any composite composed of these two lignocellulosic fibers will be more dependent upon the tensile strength of sisal fiber while the toughness of the composite will be more influenced by that of oil palm fiber.

The optimum fiber lengths of 10 and 6 mm for sisal and oil palm fiber, respectively, were found effective for reinforcement in NR matrices to achieve maximum tensile properties. In addition, longitudinally oriented hybrid composites showed better mechanical properties than transversely oriented hybrids. It was also interesting that the extent of fiber breakage found during mixing and compounding was low. This is because natural lignocellulosic fibers can undergo bending and curling during milling instead of breaking as synthetic fibers do.

13.4.4 Chemical Treatment of Natural Fiber

Because the low interfacial properties of fiber and polymer matrix often reduce their potential as reinforcing agents due to the hydrophilic nature of natural fibers, chemical modifications may help optimize the interfaces of fibers. Chemical modifications of natural fibers designed to improve adhesion with rubber matrices were investigated by a number of researchers.

13.4.4.1 Effect of Alkalization

Alkaline treatment or mercerization is a common chemical treatment of natural plant fibers used in preparing fiber-reinforced polymer composites. Modification of a fiber surface with NaOH solution improved adhesion characteristics due to increased surface roughness and surface tension of the fiber. The development of a rough surface topography offered better fiber–rubber interface adhesion and increased composite mechanical properties. In addition, the removal of lignin and other surface waxy substances by alkali solution increased the chance of mechanical interlocking of matrix and fiber.

The influence of alkaline treatment on the tensile properties of an NR matrix composite reinforced with short coir fiber (10 wt%) has been reported.[53] Alkali treatment with NaOH at 5% concentrations for soaking periods of 0, 4, 24, 48, and 72 hr was conducted. The tensile strength and tensile modulus at 200% elongation improved with soaking time and reached a maximum at 48 hr of treatment period. The longer treatment time of coir fiber (72 hr with 5% NaOH) decreased tensile strength and modulus due to excessive delignification of fiber and caused fiber damage that led to weakening of strength.

Jacob et al. [41,42] examined the effect of NaOH concentrations (0.5, 1, 2, 4, and 10% for 1 hr) for treating sisal- and oil palm fiber-reinforced composites and concluded that maximum tensile strength resulted from 4% NaOH at room temperature due to the increased roughness of fiber surface, hence increasing the surface area available for contact with the matrix.

Modification of the fiber surface with aqueous alkali at elevated temperature was studied in oil palm fiber-reinforced natural rubber composites.[50] The tensile properties (tensile strength and tensile modulus, 100% and 300%), tear strength, and hardness of rubber composites filled with alkali-treated fiber exceeded those for untreated fiber at similar loadings. Strong adhesion between treated fiber and the rubber matrix due to treatment of fiber led to higher shear strength at the fiber–matrix interface.

Better strength and stiffness achieved from strong adhesion of fiber and rubber matrix consequently reduced the toughness of the composites as shown by lower elongation at break. Higher toughness was obtained from weak interfacial adhesion, as shown by higher elongation at break results for compounds filled with untreated fibers. The fracture surfaces examined by SEM revealed many holes left after fibers were pulled from the matrix when stress was applied in untreated fiber, whereas the presence of short broken fibers was seen due to stronger adhesion in treated composites.

Mathew et al.[43] found that the compression set was lower in isora fiber-reinforced natural rubber composites with alkali-treated fibers. They attributed the behavior to the buckling of the fiber that invariably takes place when closely packed fibers are compressed in the direction of their alignment. Because of the strong adhesion of treated fibers and rubber, the extent of buckling was reduced in treated fiber composites, which resulted in a low value for the set. Furthermore, the treated fiber composites showed better resistance to abrasion compared to the untreated materials. The better abrasion resistance of the treated fiber composites may have resulted from a combination of higher tear strength, tensile strength, and modulus achieved through better bonding with the rubber matrix.

Moreover, alkali modification led to lowering of moisture uptake due to increased fiber–matrix adhesion.[54] The moisture resistance of composites comprising fibers treated with alkali and different types of silane coupling agents was studied in sisal- and oil palm hybrid fiber-reinforced natural rubber biocomposites. Because of strong interfacial adhesion, water entry was hindered. The better adhesion between matrix and fibers decreased the velocity of the diffusing molecules, since there were fewer gaps in the interfacial region.

Another factor was that stronger adhesion resulted in tighter packing within the fiber–rubber network. This indicates that the distance (mean free path) traveled by the diffusing water molecules between two consecutive collisions decreased and lowered water uptake. The better adhesion of composites containing alkali-treated fiber was also revealed by SEM.

Holes were clearly visible in untreated composites, indicating that the level of adhesion of fibers and matrix was poor. The fibers were pulled out of the rubber matrix easily when stress was applied, leaving behind many holes. While a number of short broken fibers projecting from the rubber matrix were observed in treated composites indicating improved adhesion of fibers and rubber matrix. The fibers broke and did not totally leave the matrix when stress was applied.

13.4.4.2 Effect of Silanization

Ismail et al.[46] treated bamboo fibers with a bis(triethoxysilyl propyl) tetrasufide (Si69) silane coupling agent. Shorter scorch and cure times were noted for silane-treated composites. The enhanced interfacial adhesion of fiber and rubber matrix resulted in increased strength and stiffness of the composites. Scanning electron micrographs of the composites without Si69 showed many holes remaining after the fibers were pulled from the rubber matrix. Composites with Si69 showed better adhesion between bamboo fibers and rubber matrix. The fibers were well wetted by the rubber matrix and the fiber pull-out was minimized with silane treatment.

In addition to the experiments with Si69, the effects of a diamine salt of fatty acid $[RNH_2+(CH_2)_3NH_3+].[R'COO-]_2$, also known as a multifunctional additive (MFA), on the mechanical properties of white rice husk ash (WRHA)-filled natural rubber compounds were studied.[45] The additions of MFA, Si69, and a combination of both were shown to enhance tensile modulus (at 300% elongation), tensile strength, tear strength, and hardness of filled vulcanizates.

The incorporation of MFA and Si69 in silica-filled rubber compounds was reported to contribute to better dispersion and improved silica–elastomer adherence.[56] The result was expected because WRHA contains about 96% silica. The incorporation of MFA, Si69, or combination would improve WRHA dispersion in the rubber matrix, increase the cross-link density, and consequently improve mechanical properties of the WRHA-filled rubber vulcanizates.

In addition, chemical surface modification was found to decrease water uptake in the composites.[54] The moisture resistance of sisal and oil palm hybrid composites containing fibers treated with various types of silane coupling agents was investigated. The fiber–matrix adhesion of fluorosilane-treated composites was stronger compared to results from aminosilane-treated and vinyl silane-treated composites since the hydrogen bonds formed by the fluorine and rubber matrix were stronger. Therefore the water uptake was less in the fluorosilane-treated composite.

13.4.4.3 Effect of Bonding Agent

The dry bonding system commonly used in rubbers is a di- and tri-component mixture consisting of hydrated silica, resorcinol formaldehyde, and hexamethylene tetramine. Various researchers[41,42,50,57] reported that a dry bonding system is essential for promoting strong adhesion of fibers and natural rubber matrices. If the fibers do not bond properly with a matrix, they will slide past one another under tension. When the fiber–matrix interface is sufficiently strong, the load will be transferred effectively to the fibers, resulting in a high performance composite. Hence, the mechanism of load transfer may take place through the shear at the interface.

A tri-component bonding system consisting of hydrated silica, resorcinol formaldehyde, and hexamethylene tetramine (HRH) was reported to be the most effective bonding agent for OPWF (oil palm wood flour)-filled natural rubber composites.[58] The incorporation of HRH enhanced the mechanical properties (tensile strength, tensile modulus, and hardness) compared to a control composite and composites with phenol formaldehyde and resorcinol formaldehyde–hydrated silica bonding systems. Lower elongation at break came from the improved fiber–rubber interactions with bonding agents.

In addition, bonding agents in composites prolong curing times and slow curing rates for composites with bonding agents due to better bonding between fiber and rubber matrix. The combination with HRH also produced the highest maximum torque leading to strong fiber–rubber matrix interfacial adhesion. As a result, the composites became stronger, harder, and stiffer.

The effect of a bonding agent consisting of phenol formaldehyde, hexamethylenetetramine, and silica in bamboo fiber-reinforced NR composites was studied.[59] Tensile moduli and hardness increased with the addition of bonding agents as a result of better interaction between rubber matrix and bamboo fibers. In addition, tensile and tear strengths improved with the addition of bonding agents. The failure surfaces of composites showed very low levels of fiber pull-out. Failures occurred at the fibers due to strong adhesion of fiber and rubber matrix.

While holes were observed in composites without bonding agents, many holes were left after the fibers pulled away from the matrix when stress was applied; failures occurred at the weak fiber and rubber matrix interfaces.

Resorcinol formaldehyde latex (RFL) was used as a bonding agent in grass fiber-filled NR composites.[60] The presence of the RFL in the rubber vulcanizate prolonged the cure time because of better bonding between fibers and matrices. The composites with alkali-treated fibers showed higher torque than the composites without bonding agents because of better adhesion at the fiber–matrix interfaces. The composites were consequently stronger, stiffer, and harder.

Moreover, alkali-treated grass fiber-based composites showed a greater extents of cure than those of untreated composites (water-leached grass fibers). This suggests that the interaction of the RFL- and alkali-treated grass fibers improved via the generation of additional functionality on fiber surfaces after alkali treatment.

Bonding agent-treated fibers showed higher tensile strengths, increased tensile moduli, and hardness levels in addition to alkali treatment effects. An SEM micrograph of the composites with bonding agents showed that the fibers were well wetted by the rubber matrix and fiber breakage due to strong adhesion of fiber and matrix.

The use of a tri-component bonding agent in short isora-fiber-reinforced natural rubber composites[43] was reported to improve mechanical properties. Addition of bonding agents further improved moduli, tensile strength, tear strength, and hardness. Moreover, the alkali-treated isora fiber composites with bonding agents exhibited enhanced properties compared to the untreated ones.

However, a previous study[57] indicated that silica is not needed in bonding systems for coir fiber–NR composites when coir fiber is used as a reinforcement. It was shown that the untreated and treated coir fiber composites in a bonding system without silica exhibited higher tensile strengths than those with unnecessary silica components in their bonding systems. The same trend was found in tensile modulus (50, 100, and 200%) and tear strength. Therefore, silica is not essential for good interfacial adhesion in HRH bonding systems.

13.4.4.4 *Effect of Compatibilizer*

Maleic anhydride (MAH) is one of the most common vinyl monomers for graft modification of polymers. MAH-modified polymers are widely used as chemical coupling agents, impact modifiers, and compatibilizers for polymer blends and filler-reinforced composites. Natural rubber grafted with MAH (NR-g-MA) has been found to be an effective compatibilizer for natural fiber and NR composites.

The addition of maleated natural rubber (MNR) as a compatibilizer for paper sludge–natural rubber composites was examined by Ismail et al.[28] MAH was grafted onto natural rubber (NR-g-MA) before compounding. Tensile properties of paper sludge–NR composites with MNR were superior to NR composites without MNR due to improved interfacial adhesion between paper sludge fibers and NR.

Although tensile strengths of the composites showed a decrease with filler loading, the addition of an MNR compatibilizer was found to enhance the strengths. It was proposed that the natural rubber segments of MNR form miscible blends with bulk natural rubber and the polar component (MAH) of MNR forms hydrogen bonds with the hydroxyl groups of fibers in the interfacial region.

Stronger adhesion at the fiber and matrix interface caused better stress transfer from the matrix into the fibers, leading to higher tensile strength. Further enhancement of the stiffness of the composites was observed with the incorporation of MNR, as shown by increase in tensile moduli. The elongation at break for a compound with MNR was slightly higher than the natural rubber compound without MNR. An SEM study showed that composites compatibilized with MNR exhibited fewer aggregates, reductions in fiber pull-out, and less fiber–matrix debonding.

Longer scorch and cure times were observed in composites with MNR due to better binding of fibers and matrices. Longer cure times were also observed when various bonding agents were used in NR-filled composites[41,42,50] as noted earlier. The influence of MNR on the fatigue life of palm ash-filled NR composites was also investigated.[48] The addition of MNR enhanced the tensile properties and fatigue life of the composites resulting from enhanced interfacial adhesion that helped reduce the heat build-up during deformation, reducing chain scission.

Zeng et al.[61] found that cotton fiber–NR composites containing MNR exhibited better mechanical properties than composites without MNR. Increments of tensile strengths and moduli of the compatibilized NR composites with increasing grafted MA content were observed. This was due to the increased possibility of the reaction of MAH with fibers, leading to enhanced tensile strength and modulus.

In addition to NR-g-MA, epoxidized natural rubber (ENR50) was used to enhance the compatibility of sisal fibers and natural rubber.[62] Interfacial adhesion between sisal fiber and natural rubber was enhanced with the additions of NR-g-MA and ENR50. Moreover, NR-g-MA provided more improvements in mechanical properties of natural rubber composites than ENR50.

For an NR-g-MA-compatibilized composite at fiber loading of 10 phr, tensile strength, modulus at 100% strain, modulus at 300% strain, tear strength, and hardness increased 43, 44, 53, 42, and 13%, respectively, compared with values from an uncompatibilized natural rubber composite. Micrographs of ENR-compatibilized NR composites and NR-g-MA-compatibilized NR composites showed reduction in fiber pull-out from NR matrices and smaller gaps between fibers and NR matrices when compared with uncompatibilized NR composites.

13.4.4.5 Effect of Matrix Modification

Epoxidized natural rubber (ENR) was introduced as a modified form of NR. As the NR is epoxidized, its chemical and physical properties change according to the mole% of modification introduced. When the degree of epoxidation is increased, the rubber becomes more polar. This made ENR more compatible to natural fibers than NR. Interestingly, the tensile strength was reported to increase with increasing filler loading in rice husk ash-filled epoxidized natural rubber.[55] This could be due to the strong interaction between the hydroxyl group of cellulose and epoxy group of ENR rubber chain leading to higher tensile properties, tear strength, and hardness of ENR vulcanizates.

However, the increase in strength reached maximum level at which the filler aggregates are no longer adequately separated or wetted by the rubber phase and a subsequent reduction in strength. However, a decrease of tensile strength with filler loading was observed in oil palm wood flour (OPWF)-reinforced ENR composites.[50] The irregularly shaped OPWF fillers were believed to account for poor strength properties of the composites.

It has also been reported that tensile strengths and moduli of paper sludge-filled ENR composites were higher than those of paper sludge-filled NR composites at the same filler loading due to better rubber–filler interactions of both polar surfaces.[63]

Moreover, the elongation at break levels of ENR50 composites were lower than those of NR compounds. This is again due to the interfacial interaction between polar filler and matrix that increases the stiffness of the composites and consequently reduces elongation at break.

13.4.4.6 Surface Morphology

Investigation of the tensile fracture surface of a specimen by scanning electron microscopy (SEM) is performed to examine interfacial adhesion between fibers and polymer matrices. In short fiber-reinforced composites, the failure mechanisms are dominated by fiber debonding and pull-out.[50] Qualitative evidences of the failure mechanisms may be obtained from fractography studies.

As we know, when a load is applied to the fiber/polymer composites, it spreads smoothly through the matrix until reaching the fiber–rubber interface. If the interface is weak, the fibers may debond from the matrix and cause fractures. However, if the interface is well bonded, the stress can be

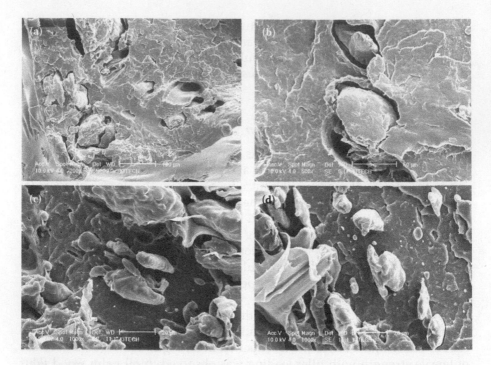

FIGURE 13.4
Scanning electron micrographs of tensile fracture surfaces. (a) and (b) Polypropylene–cassava fiber composites without MAPP compatibilizer showing cavities and large gap between fiber and matrix. (c) and (d) Composites containing MAPP compatibilizer showing better interfacial adhesion with short broken fiber ends.

transferred from the matrix to the fibers and then spread throughout the fibers. The fibers then act as main stress carriers and the composites show improved mechanical properties.

Figure 13.4 is a set of micrographs of cassava root fiber–polypropylene composites to which a maleic anhydride–polypropylene (MAPP) compatibilizing agent was added.[64] We can see clear evidence of fiber debonding with large gaps between fiber and matrix and many holes in matrices in composites without MAPP compatibilizers. The holes are attributed to fiber pull-out from the matrix during tensile testing; this indicates poor interfacial bonding of the cassava fibers and PP matrix.

Conversely, the tensile fracture surfaces of composites with added MAPP showed no clear gaps in the interfacial region between the polymer matrix and fiber and the presence of short broken fiber ends protruding from the rubber matrix (Figure 13.4c and d). When stress was applied, the fibers broke and did not entirely pull out of the matrix. This indicates that compatibilizers promote efficient fiber and matrix interfacial bonds, resulting in enhanced mechanical properties.

13.5 Effects of Thermal Annealing on Mechanical Properties of Natural Fiber-Reinforced Composites

Natural fiber-reinforced composites are used in construction and automobile applications because temperature changes are frequent in both industries. Hence, it is important to evaluate these properties when designing construction and automotive materials.

The mechanical properties of natural fiber-reinforced composites change during thermal aging. The thermal stabilities of natural fiber and polymer matrix components of composites are affected by changes in the mechanical properties of composites through thermal aging. Natural fibers are complex mixtures of organic materials and thermal treatment generates a variety of physical and chemical changes. The cellulose, hemicelluloses, and lignin components of natural fibers have their own characteristic thermal stability properties.

Based on the TGA curve, thermal degradation of natural fibers is a two-stage process—one in the 80 to 180°C temperature range caused by degradation of lignin and the other in the 280 to 380°C range due to cellulose.[65] For this degradation, significant decreases in tenacity and polymerization were observed only at temperatures above 170°C.[66] Hence, the thermal degradation of natural fibers is a crucial aspect of combining natural fibers with polymer composites.

Thermal degradation imposes a limitation on the curing temperatures of elastomers and thermosets and on extrusion temperatures of thermoplastic composites.[67,68] The inclusion of biodegradable cellulose fibers in polymer matrices results in degradation with formation of microcracks on the surface of the polymer.[69] Due to the increase of surface area, the cellulose oxidation processes leading to polymer fragmentation occur more intensively and finally cause decomposition of the composite. Thermal degradation can impact the applications of cellulose–polymer composites by affecting their mechanical properties.

13.5.1 Natural Fiber–Polyethylene (PE) Composites

Many researchers' investigations have shown that the polyethylene structure is very sensitive to thermal treatment. Woen[70] studied the effects of thermal aging on the mechanical properties of a linear low density polyethlene (LLDPE) pipe. Woen found that the prolongation of thermal exposure led to a progressive increase in tensile strength and a slight increase in hardness with a proportional decrease in elongation at break.

These results can be explained by the increase of crystallinity followed by the increase of crosslinking density and a decrease in chain mobility due to thermal oxidation as exposure time increased. In the case of cellulose materials

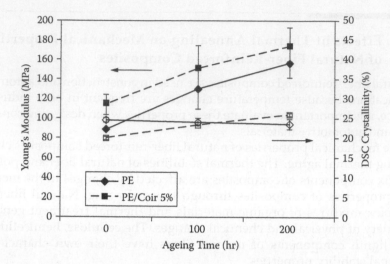

FIGURE 13.5
Young's modulus and crystallinity of pure polyethylene compared to 5 wt% coir fiber-reinforced–polyethylene composite as function of aging time. The Young's modulus increases with aging time for both types of materials, but the coir fiber-reinforced composite improved more than the pure polyethylene due to effects of crystallinity of natural fiber.

such as palm fibers with low density PE composites incubated under thermal aging at 100°C, the Young's moduli of the composites tended to increase when aging time was less than 200 hr. This can be explained by crosslinking or new crystallization of polymer molecules due to chain scission.

The hypothesis of this explanation was confirmed by increased crystallinity as shown by the DSC technique.[71] Longer aging of polyethylene composites results in higher degrees of crystallinity (Figure 13.5). However, it also interrupted the crystallization of pure PE as described elsewhere.[72] Based on this study, the crystallinity change was not found in the pure PE due to this thermoaging condition, contrary to the results for the natural fiber–polymer composites clearly due to the effects of short molecules in cellulose fibers.

13.5.2 Natural Fiber–Polypropylene (PP) Composites

The thermostabilities of PP composites with coir fibers under 100°C were investigated. The stability of pure PP was good for 1000 hr. After this induction period, IR absorbance levels for the peroxide region were 3000 to 3400 cm⁻¹ and for the carbonyl region were 1700 to 1800 cm⁻¹. The inclusion of coir fibers in the PP matrix resulted in a shortened induction period of 600 to 800 hr as shown only for carbonyl absorbance. Note that no change in the aromatic C = C absorption band at 1506 cm⁻¹ was observed.

Changes in IR spectra at the highest fiber contents led to a higher oxidation rate than PP as shown for carbonyl formation. The mechanism of oxidation

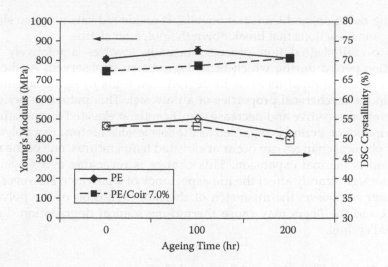

FIGURE 13.6
Young's modulus and crystallinity of pure polypropylene compared to 7 wt% coir fiber-reinforced polypropylene composite as function of aging time. The Young's modulus and crystallinity levels of both materials remain constant with aging due to main effects of polypropylene.

involving formation of hydroxide or carbonyl groups for PP also included chain scissions.[73] For the PP composites, the increased oxidation rate was explained by free radical appearance from cellulose and lignocellulose depolymerization inducing increases in the carboxyl and hydroperoxide groups.[74]

Young's moduli of the pure PP and PP composites (5% coir fiber content after thermodegradation) did not increase significantly when aging time was increased. The degrees of crystallization of the pure PP and composites increased after chemicrystallization as shown in the DSC result, but the spontaneous cracking of the surface caused by contraction of surface layers after chemicrystallization was the main reason for disappointment.[75] See Figure 13.6.

13.5.3 Natural Fiber–Polyester (PL) Composites

Thermal degradation involves the chemical and physical processes occurring at elevated temperatures in polyesters. Increased temperature accelerates most of the polyester degradation processes such as oxidation, chemical attack, and mechanical creep. Oxidation is generally considered the most serious problem arising from using polyesters at elevated temperatures. The influence of temperature on oxidation depends on the chemical structure of the polyester.

Thermal oxidation is initiated by the reaction of P° free radicals with oxygen to form peroxide radicals. All polyesters contain these free radicals due to their polymerization and processing history. However, the concentration of free radicals can be increased significantly by interactions with light, ionizing radiation, and transition metals. Peroxide radicals undergo slower propagation reactions that break down the polyester chains.

The overall degradation process normally involves a relatively long induction period during which little degradation is observed. At the end of this period, a rapid increase in degradation leads to significant reductions in the mechanical properties of a polyester. This induction period is temperature-sensitive and decrease significantly at elevated temperatures.

The induction period is regarded as the serviceable lifetime of a polyester. Other physical changes can occur at elevated temperatures; one of the most common is thermal expansion. This change is reversible and in general does not significantly affect the life expectancy of a polymer. However, in a polyester composite, the mismatch of thermal expansion of the polyester matrix and the fibers may cause thermo-mechanical degradation during thermal cycling.

13.5.4 Natural Fiber–Natural Rubber (NR) Composites

Normally, natural rubber has poor thermal resistance. Incorporation of fibers in a rubber matrix prevents crack initiation and hinders crack propagation. The fibers should be oriented perpendicular to the direction of crack propagation and good bonding should exist between the fiber and rubber.

Short isora fibers and a natural rubber composite were incubated in an air oven at 100°C for 3 to 5 days.[43] The retention of tensile strength increased continuously with fiber loading for composites with and without bonding agents during 3 days of aging. The bonding action of the resin dominated polymer chain degradation and led to a higher retention of tensile strength at shorter thermal aging times. By increasing the aging from 3 to 5 days, composites that do not contain bonding agents show lower retention of tensile properties. This indicates that polymer degradation begins after 3 days in these composites, whereas the bonding action of the resin continues in the composites containing bonding agents, but the retention is lower than that for the shorter aging period (3 days).

During thermal aging, crosslink formation and crosslink breakage can take place or an existing crosslink may break and a stable linkage may form. In composites, bonding of the resin also takes place during aging. All these reactions greatly influence the performances of composites. Tear strength and tensile moduli also show similar trends on aging.

13.6 Conclusions

Natural fibers are renewable resources that can be grown and processed within a short time. Supplies are virtually limited in comparison with supplies of traditional glass and carbon fibers needed for advanced composites. However, for some recyclable polymers, overall energy consumption during collecting, recycling, refining, and remolding processes must be considered to ensure that damage to the natural cycle is kept to a minimum.

Natural fibers are recyclable, have low density, and are eco-friendly. Their tensile properties are very good fibers and the fibers can replace conventional materials such as glass and carbon in reinforcing plastic materials. A major drawback of using natural fibers as reinforcements in plastics is incompatibility resulting in poor adhesion of fibers and matrix resins leading to low tensile properties. To improve fiber–matrix interfacial bonding and enhance tensile properties of composites, novel processing techniques involving chemical and physical modifications need to be developed.

Clearly, the strengths and stiffness values of natural fiber polymer composites depend strongly on fiber loading. Tensile strength and modulus properties increase with increasing fiber weight ratios up to a certain limit. If fiber weight ratio is below optimum value, loads are distributed to more fibers that are well bonded with a matrix, producing better tensile properties. Further increment in fiber weight ratio resulted in decreased tensile strength as described in this chapter. Mechanical properties are the main characteristics for determining the advantages of bio-based composites over conventional ones.

References

1. R. Chollakup, W. Smitthipong, W. Kongtud et al. Polyethylene Green Composites Reinforced with Cellulose Fibers (Coir and Palm Fibers): Effect of Fiber Surface Treatment and Fiber Content. *Journal of Adhesion Science and Technology,* 27 (2012): 1290–1300.
2. M. Mahdavi, P.L. Clouston, and S.R. Arwade. Development of Laminated Bamboo Lumber: Review of Processing, Performance, and Economical Considerations. *Journal of Materials in Civil Engineering,* 23 (2010): 1036–1042.
3. P.J. Herrera-Franco and A. Valadez-Gonzalez. Study of the Mechanical Properties of Short Natural Fiber-Reinforced Composites. *Composites Part B,* 36 (2005): 597–608.
4. E. Papazoglou, R. Simha, and F.H.J. Maurer. Thermal Expansivity of Particulate Composites: Interlayer versus Molecular Model. *Rheology Acta,* 28 (1989): 302–310.
5. A.B. Strong and B. Strong. Polymeric Composite Materials and Process. *Plastics: Materials and Processing* (2000): 643–680.
6. J. Ganster and H.P. Fink. Novel Cellulose Fibre-Reinforced Thermoplastic Materials. *Cellulose,* 13 (2006): 271–280.
7. K. Okubo, T. Fujii, and Y. Yamamoto. Development of Bamboo-Based Polymer Composites and Their Mechanical Properties. *Composites Part A,* 35 (2004): 377–383.

8. E.A. Elbadry, M.S. Aly-Hassan, and H. Hamada. Mechanical Properties of Natural Jute Fabric–Jute Mat Fiber-Reinforced Polymer Matrix Hybrid Composites. *Advances in Mechanical Engineering*, 20 (2012): 12.

9. G.P. Simon. *Polymer Characterization Techniques and Their Application to Blends*. Washington, American Chemical Society, 2003.

10. I.M. Ward and J. Sweeney. *Mechanical Properties of Solid Polymers*. New York, John Wiley & Sons, 2012.

11. R. Simha, E. Papazoglou, and F.H.J. Maurer. Thermal Expansivity and Bulk Modulus of Polymer Composites: Experiment versus Theory. *Polymer Composites*, 10 (1989): 409–413.

12. K. Masouras, N. Silikas, and D.C. Watts. Correlation of Filler Content and Elastic Properties of Resin Composites. *Dental Materials*, 24 (2008): 932–939.

13. C.H. Hsueh and P.F. Becher. Effective Viscosity of Suspensions of Spheres. *Journal of the American Ceramic Society*, 88 (2005): 1046–1049.

14. S.Y. Fu, X.Q. Feng, B. Lauke et al. Effects of Particle Size, Particle–Matrix Interface Adhesion, and Particle Loading on Mechanical Properties of Particulate–Polymer Composites. *Composites: Part B*, 39 (2008): 933–961.

15. E. Guth. Theory of Filler Reinforcement. *Journal of Applied Physics*, 16 (1945): 20–25.

16. P.A.M. Steeman and F.H.J. Maurer. Interlayer Model for the Complex Dielectric Constant of Composites. *Colloid and Polymer Science*, 268 (1990): 315–325.

17. M. Nardin and E. Papirer. *Powders and Fibers: Interfacial Science and Applications*. Boca Raton, FL: CRC Press, 2007.

18. S. Uemura and M. Takayanagi. Application of the Theory of Elasticity and Viscosity of Two-Phase Systems to Polymer Blends. *Journal of Applied Polymer Science*, 10 (1966): 113–125.

19. Y.S. Lipatov. *Physical Chemistry of Filled Polymers*. Shrewsbury, British Library, 1979.

20. J. Schultz and M. Nardin. Determination of the Surface Energy of Solids by the Two-Liquid Phase Method. In *Modern Approaches to Wettability: Theory and Applications*, M.E. Schrader and G. Loeb, Eds. New York, Plenum Press, 1992, pp.73–100.

21. Y.S. Lipatov. *Phase Interaction in Composite Materials*. S.A. Paipetis, Ed. England, Omega Scientific, 1992.

22. Y.S. Lipatov, V.F. Rosovizky, and V.V. Shifrin. Viscoelastic Properties of the Model System: Epoxy Resin–Glass Beads–Poly(Buthylmethacrylate). *Journal of Applied Polymer Science*, 27(1982): 455–460.

23. Y.S. Lipatov, V.F. Rosovitsky, B.V. Babich et al. On Shift and Resolution of Relaxation Maxima in Two-Phase Polymeric Systems. *Journal of Applied Polymer Science*, 25 (1980): 1029–1037.

24. J. Ferry. *Viscoelastic Properties of Polymers*, 2nd ed. New York: John Wiley & Sons, 1970.

25. M.L. Williams, R.F. Landel, and J.D. Ferry. Temperature Dependence of Relaxation Mechanisms in Amorphous Polymers and Other Glass-Forming Liquids. *Journal of the American Chemical Society*, 77 (1955): 3701–3707.

26. R.A.L. Jones and R.W. Richards. *Polymers at Surfaces and Interfaces*. Cambridge University Press, 1999.

27. M. Tajvidi, R.H. Falk, and J.C. Hermanson. Effect of Natural Fibers on Thermal and Mechanical Properties of Natural Fiber Polypropylene Composites Studied by Dynamic Mechanical Analysis. *Journal of Applied Polymer Science*, 101 (2006): 4341–4349.

28. H. Ismail, A. Rusli, and A.A. Rashid. Maleated Natural Rubber as a Coupling Agent for Paper Sludge-Filled Natural Rubber Composites. *Polymer Testing*, 24 (2005): 856–862.

29. M. Pracella, M.D. Minhaz–Ul Haque, and V. Alvarez. Functionalization, Compatibilization, and Properties of Polyolefin Composites with Natural Fibers. *Polymers*, 2 (2010): 554–574.

30. W.S. De Polo. *Dimensional Stability and Properties of Thermoplastics Reinforced with Particulate and Fiber Fillers*. Blacksburg: Virginia Polytechnic Institute and State University, 2005.

31. B. Mirzaei, M. Tajvidi, R.H. Falk et al. Stress Relaxation Behavior of Lignocellulosic High-Density Polyethylene Composites. *Journal of Reinforced Plastics and Composites*, 30 (2011): 875–881.

32. J.P. Dhal and S.C. Mishra. Processing and Properties of Natural Fiber-Reinforced Polymer Composite. *Journal of Materials*, 2013 (2012).

33. S. Luo, J. Cao, and X. Wang. Investigation of the Interfacial Compatibility of PEG and Thermally Modified Wood Flour–Polypropylene Composites Using the Stress Relaxation Approach. *BioResources*, 8 (2013): 2064–2073.

34. M. Baumgaertel and H.H. Winter. Determination of Discrete Relaxation and Retardation Time Spectra from Dynamic Mechanical Data. *Rheologica Acta*, 28, (1989): 511–519.

35. A.Y. Malkin. Continuous Relaxation Spectrum: Its Advantages and Methods of Calculation. *Applied Mechanics and Engineering*, 11 (2006): 235.

36. C.M. Roland, L.A. Archer, P.H. Mott et al. Determining Rouse Relaxation Times from the Dynamic Modulus of Entangled Polymers. *Journal of Rheology*, 48 (2004): 395.

37. J. Aneli, G. Zaikov, and O. Mukbaniani. Electric Conductivity of Polymer Composites at Mechanical Relaxation. *Journal of Characterization and Development of Novel Materials*, 3 (2011).

38. B.E. Clements and E.M. Mas. Dynamic Mechanical Behavior of Filled Polymers. I: Theoretical Developments. *Journal of Applied Physics*, 90 (2001): 5522–5534.

39. P. Shrotriya and N.R. Sottos. Viscoelastic Response of Woven Composite Substrates. *Composites Science and Technology*, 65 (2005): 621–634.

40. N. Lopattananon, K. Panawarangkul, K. Sahakaro et al. Performance of Pineapple Leaf Fiber–Natural Rubber Composites: Effect of Fiber Surface Treatments. *Journal of Applied Polymer Science*, 102 (2006): 1974–1984.

41. M. Jacob, S. Thomas, and K. Varughese. Mechanical Properties of Sisal–Oil Palm Hybrid Fiber-Reinforced Natural Rubber Composites. *Composites Science and Technology*, 64 (2004): 955–965.

42. M. Jacob, S. Thomas, and K.T. Varughese. Natural Rubber Composites Reinforced with Sisal–Oil Palm Hybrid Fibers: Tensile and Cure Characteristics. *Journal of Applied Polymer Science*, 93 (2004): 2305–2312.

43. L. Mathew and R. Joseph. Mechanical Properties of Short Isora Fiber-Reinforced Natural Rubber Composites: Effects of Fiber Length, Orientation, Loading, Alkali Treatment, and Bonding Agent. *Journal of Applied Polymer Science*, 103 (2007): 1640–1650.

44. F.P. La Mantia and M. Morreale. Green Composites: A Brief Review. *Composites Part A*, 42 (2011): 579–588.
45. H. Ismail, M.N. Nasaruddin, and U.S. Ishiaku. White Rice Husk Ash-Filled Natural Rubber Compounds: Effect of Multifunctional Additive and Silane Coupling Agents. *Polymer Testing*, 18 (1999): 287–298.
46. H. Ismail, S. Shuhelmy, and M.R. Edyham. Effects of a Silane Coupling Agent on Curing Characteristics and Mechanical Properties of Bamboo Fibre-Filled Natural Rubber Composites. *European Polymer Journal*, 38 (2002): 39–47.
47. H. Ismail, R.M. Jaffri, and H.D. Rozman. Effects of Filler Loading and Vulcanisation System on Properties of Oil Palm Wood Flour–Natural Rubber Composites. *Journal of Elastomers and Plastics*, 35 (2003): 181–192.
48. H. Ismail and F.S. Haw. Effects of Palm Ash Loading and Maleated Natural Rubber as a Coupling Agent on the Properties of Palm Ash-Filled Natural Rubber Composites. *Journal of Applied Polymer Science*, 110 (2008): 2867–2876.
49. Z.A.M. Ishak, A.A. Bakar, U.S. Ishiaku et al. Investigation of the Potential of Rice Husk Ash as a Filler for Epoxidized Natural Rubber. II: Fatigue Behaviour. *European Polymer Journal*, 33 (1997): 73–79.
50. H. Ismail, H.D. Rozman, R.M. Jaffri et al. Oil Palm Wood Flour-Reinforced Epoxidized Natural Rubber Composites: Effect of Filler Content and Size. *European Polymer Journal*, 33 (1997): 1627–1632.
51. M.J. John, K.T. Varughese, and S. Thomas. Green Composites from Natural Fibers and Natural Rubber: Effect of Fiber Ratio on Mechanical and Swelling Characteristics. *Journal of Natural Fibers*, 5 (2008): 47–60.
52. A.P. Haseena, K.P. Dasan, R. Namitha et al. Investigation on Interfacial Adhesion of Short Sisal–Coir Hybrid Fibre-Reinforced Natural Rubber Composites by Restricted Equilibrium Swelling Technique. *Composite Interfaces*, 11 (2004): 489–513.
53. V.G. Geethamma, R. Joseph, and S. Thomas. Short Coir Fiber-Reinforced Natural Rubber Composites: Effects of Fiber Length, Orientation, and Alkali Treatment. *Journal of Applied Polymer Science*, 55 (1995): 583–594.
54. M. Jacob, K.T. Varughese, and S. Thomas. Water Sorption Studies of Hybrid Biofiber-Reinforced Natural Rubber Biocomposites. *Biomacromolecules*, 6 (2005): 2969–2979.
55. Z.A.M. Ishak and A.A. Bakar. Investigation of the Potential of Rice Husk Ash as Filler for Epoxidized Natural Rubber (ENR). *European Polymer Journal*, 31 (1995): 259–269.
56. H. Ismail, P.K. Freakley, I. Sutherland et al. Effects of Multifunctional Additive on Mechanical Properties of Silica-Filled Natural Rubber Compound. *European Polymer Journal*, 31 (1995): 1109–1117.
57. V.G. Geethamma, K. Thomas-Mathew, R. Lakshminarayanan et al. Composite of Short Coir Fibres and Natural Rubber: Effect of Chemical Modification, Loading, and Orientation of Fibre. *Polymer*, 39 (1998): 1483–1491.
58. H. Ismail, M.R. Edyham, and B. Wirjosentono. Dynamic Properties and Swelling Behaviour of Bamboo-Filled Natural Rubber Composites: Effect of Bonding Agent. *Iranian Polymer Journal*, 10 (2001): 377–383.
59. H. Ismail, M.R. Edyham, and B. Wirjosentono. Bamboo Fibre-Filled Natural Rubber Composites: Effects of Filler Loading and Bonding Agent. *Polymer Testing*, 21 2 (2002): 139–144.

60. D. De, B. Adhikari, and D. De. Grass Fiber-Reinforced Phenol Formaldehyde Resin Composite: Preparation, Characterization, and Evaluation of Properties. *Polymers for Advanced Technologies,* 18 (2006): 72–81.
61. Z. Zeng, W. Ren, C. Xu et al. Maleated Natural Rubber Prepared through Mechanochemistry and Its Coupling Effects on Natural Rubber–Cotton Fiber Composites. *Journal of Polymer Research,* 17 (2010): 213–219.
62. W. Wongsorat, N. Suppakarn, and K. Jarukumjorn. Effects of Compatibilizer Type and Fiber Loading on Mechanical Properties and Cure Characteristics of Sisal Fiber–Natural Rubber Composites. *Journal of Composite Materials* (2013).
63. H. Ismail, A. Rusli, and A.A. Rashid. Effect of Filler Loading and Epoxidation on Paper Sludge-Filled Natural Rubber Composites. *Polymer–Plastics Technology and Engineering,* 45 (2006): 519–525.
64. R. Tantatherdtam, T. Tran, S. Chotineeranat et al. Preparation and Characterization of Cassava Fiber-Based Polypropylene and Polybutylene Succinate Composites. Paper presented at 47th Kasetsart University Annual Conference, Thailand, March 2009.
65. N. Tinh, Z. Eugene, and M.B. Edward. Thermal Analysis of Lignocellulosic Materials. I: Unmodified Materials. *Journal of Macromolecular Science,* C20 (1981): 1–65.
66. J. Gassan and A.K. Bledzki. Thermal Degradation of Flax and Jute Fibers. *Journal of Applied Polymer Science,* 82 (2001): 1417–1422.
67. V.A. Alvarez and A. Vázquez. Thermal Degradation of Cellulose Derivatives–Starch Blends and Sisal Fibre Biocomposites. *Polymer Degradation and Stability,* 84 (2004): 13–21.
68. X.C. Ge, X.H. Li, and Y.Z. Meng. Tensile Properties, Morphology, and Thermal Behavior of PVC Composites Containing Pine Flour and Bamboo Flour. *Journal of Applied Polymer Science,* 93 (2004): 1804–1811.
69. G.J. Griffin. Biodegradable Fillers in Thermoplastics. In *Fillers and Reinforcements for Plastics.* Advances in Chemistry Series. Washington, DC: American Chemical Society, 1974, pp. 159–170.
70. J.I. Weon. Effects of Thermal Ageing on Mechanical and Thermal Behaviors of Linear Low Density Polyethylene Pipe. *Polymer Degradation and Stability,* 95 (2010): 14–20.
71. R. Chollakup, W. Kongtud, and F. Delor-Jestin. Photo- and Thermo-Degradation of Polyethylene–Palm Fiber Composites. In *Proceedings of 51st Kasetsart University Annual Conference.* Bangkok, 2013, pp. 68–76.
72. N.M. Stark and L.M. Matuana. Surface Chemistry and Mechanical Property Changes of Wood Flour–High Density Polyethylene Composites after Accelerated Weathering. *Journal of Applied Polymer Science,* 94 (2004): 2263–2273.
73. D. Dudić, V. Djoković, and D. Kostoski. High Temperature Secondary Crystallisation of Aged Isotactic Polypropylene. *Polymer Testing,* 23 (2004): 621–627.
74. S.L. Levan. *Concise Encyclopedia of Wood and Wood-Based Materials.* Elmsford: Pergamon Press, 1989.
75. R. Seldén, B. Nyström, and R. Långström. UV Aging of Poly(Propylene)–Wood Fiber Composites. *Polymer Composites,* 25 (2004): 543–553.
76. V.G. Geethamma, L.A. Pothen, B. Rhao et al. Tensile Stress Relaxation of Short Coir Fiber-Reinforced Natural Rubber Composites. *Journal of Applied Polymer Science,* 94 (2004): 96–104.

77. V.G. Geethamma and S. Thomas. Diffusion of Water and Artificial Seawater through Coir Fiber-Reinforced Natural Rubber Composites. *Polymer Composites,* 26 (2005): 136–143.

78. V.G. Geethamma, G. Kalaprasad, G. Groeninckx et al. Dynamic Mechanical Behavior of Short Coir Fiber-Reinforced Natural Rubber Composites. *Composites Part A,* 36 (2005): 1499–1506.

79. H. Ismail and R.M. Jaffri. Physicomechanical Properties of Oil Palm Wood Flour-Filled Natural Rubber Composites. *Polymer Testing,* 18 (1999): 381–388.

80. H. Ismail, R.M. Jaffri, and H.D. Rozman. Oil Palm Wood Flour-Filled Natural Rubber Composites: Fatigue and Hysteresis Behaviour. *Polymer International,* 49 (2000): 618–622.

81. H. Ismail, N. Rosnah, and H.D. Rozman. Curing Characteristics and Mechanical Properties of Short Oil Palm Fibre-Reinforced Rubber Composites. *Polymer Composites,* 38 (1997): 4059–4064.

82. M. Jacob, S. Thomas, and K.T. Varughese. Novel Woven Sisal Fabric-Reinforced Natural Rubber Composites: Tensile and Swelling Characteristics. *Journal of Composite Materials,* 40 (2006): 1471–1485.

83. F. Chigondo, P. Shoko, B.C. Nyamunda et al. Maize Stalk as Reinforcement in Natural Rubber Composites. *International Journal of Scientific & Technology Research,* 2 (2013): 263–271.

84. S. Raghavendra, P. Lingaraju, B. Shetty et al. Mechanical Properties of Short Banana Fiber-Reinforced Natural Rubber Composites. *International Journal of Innovative Research in Science, Engineering, and Technology,* 2 (2013).

85. J. Bras, M.L. Hassan, and C. Hassan-Bruzesse. Mechanical, Barrier, and Biodegradability Properties of Bagasse Cellulose Whisker-Reinforced Natural Rubber Nanocomposies. *Industrial Crops and Products,* 32 (2010): 627–663.

86. D. Pasquinia, E.M. Teixeira, A.A.S. Curvelob et al. Extraction of Cellulose Whiskers from Cassava Bagasse and Their Applications as Reinforcing Agent in Natural Rubber. *Industrial Crops and Products,* 32 (2010): 486–490.

87. D. De and B. Adhikari. Effect of Grass Fiber Filler on Curing Characteristics and Mechanical Properties of Natural Rubber. *Polymer Advanced Technology,* 15 (2004): 708–715.

88. M.K. Joshy, L. Mathew, and R. Joseph. Effect of Alkali Treatment on the Properties of Unidirectional Isora Fibre-Reinforced Epoxy Composites. *Plastics, Rubber and Composites,* 36 (2007): 259–266.

89. H.M.D. Costa, L.L.Y. Visconte, R.C.R. Nunes et al. Effect of Coupling Agent and Chemical Treatment on Rice Husk Ash-Filled Natural Rubber Composites. *Journal of Applied Polymer Science,* 76 (2000): 1019–1027.

14

Characterization and Strength Modeling of Laminated Bio-Based Composites

Peggi L. Clouston, Sanjay R. Arwade, and Alireza Amini

CONTENTS

14.1 Introduction

This chapter focuses on a specific type of structural bio-based composite made by laminating layers of natural fiber veneer, strands, or other small elements to form timber- and lumber-like products (Figure 14.1). Categorized generically as structural composite lumber (SCL), the various product types have different fiber species, adhesives, and element sizes, shapes, and arrangements.

The fibers of all products are predominantly unidirectional, which is a strategy for optimizing longitudinal structural properties for framing type applications. SCL is commonly used as beams, joists, columns, studs, scaffolding, and, formwork in residential construction (new homes, multi-story residences, and repairs and remodeling) and non-residential structures such as schools, stores, restaurants, and warehouses.

Sustainability and emerging trends suggest that future generations of bio-based laminates will be aimed at the industrial sector for end uses such as automotive parts, sports equipment, and even propeller and wind turbine components. There is strong support from influential players for sustainable material use; for example, the U.S. Department of Energy (DOE) recently issued a *Technological Road Map for Plant/Crop-Based Renewable Resources 2020.*[1] This vision document contains an aggressive plan for the development of more industrial products from agricultural and forestry feedstocks (e.g., crops, grasses, and wood), "to achieve at least 10% of basic chemical building blocks arising from plant-derived renewables by 2020, with development concepts in place by then to achieve a further increase to 50% by 2050."

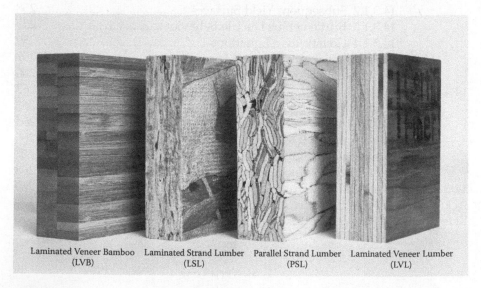

Laminated Veneer Bamboo Laminated Strand Lumber Parallel Strand Lumber Laminated Veneer Lumber
(LVB) (LSL) (PSL) (LVL)

FIGURE 14.1
Structural composite lumber products (ASTM 2012).

Wood laminates in particular display excellent potential to fulfill this need, having been successfully used for years in high-performance industrial applications. Besides their hundred-year-plus history in the boat hull industry,[2] wood laminates have a prominent place in aviation history. In the 1940s, during World War II, wing spars and other structural aircraft members were made from laminated veneers of Sitka spruce as well as balsa and cedar plywood.[3]

Because of their superior strength-to-weight properties and remarkable fatigue performance, wood laminates are used commonly in production of small to intermediate-sized wind turbine blades.[4] Many types of high-performance sports equipment, such as skis and hockey sticks were also made of wood laminates. However, over the years, as in the marine and aircraft industries, laminates in sports equipment have been replaced by non-renewable materials such as aluminum, fiberglass, and carbon fiber composites.

As a result of recent advances in adhesive technology and improved precision in computer numerically controlled (CNC) machinery a movement back to industrially engineered wood products and other natural fiber composites may already be underway. For example, several recent high-end bicycle producers such as Renovo Bicycles, Erba Cycles, and Sylvan Cycles are promoting wood and bamboo as their top-of-the-line frames. Also, in 2012, the world's first glued–laminated wood wind turbine tower was built in Hanover, Germany. The tower is 100 m tall and prevented the use of about 300 tons of sheet steel.[5]

Key to this green material movement is the development of reliable mechanics-based computational prediction tools. Computer models that can calculate a material's mechanical response can be powerful product design aids and spur innovations in many areas of bio-based composite design. For example, new product designs can focus on environmental innovation (e.g., profiling or lowering density for more efficient usage of natural fiber) or economic innovation (e.g., strategically using lower quality or less expensive species). Moreover, tremendous savings in equipment, materials, and labor can be realized through a strategic, planned design process that will decrease time to market for new products.

In 2000, Clouston proposed a methodology (partially summarized in Clouston and Lam 2001 and 2002 and republished in Clouston 2010)[6-8] to predict mechanical behavior of wood laminates up to and beyond failure under multiaxial stress states. The method has evolved through many other studies[9-12] and is the foundation upon which the methods described in this chapter are based.

This chapter presents a framework for predicting the mechanical behaviors of laminated bio-based composites. The intent of the work is to lay a foundation of modeling techniques to help foster a new era of natural fiber laminate products. The methodology is necessarily probabilistic and materially nonlinear with inherent size effect to address the unique characteristics of natural fiber laminates. To date, the modeling techniques have been validated through comparisons with experimental results evaluating tension, compression, and bending scenarios for various angle-ply laminate lay-ups and two species-types of parallel strand lumber.

14.2 Method and Materials Overview

14.2.1 Method: Key Considerations

Laminated bio-based composites exhibit anisotropic, materially nonlinear, and heterogeneous behavior. They display different stress-strain responses in tension and compression and are susceptible to size effect in brittle modes of failure. The complex spatial heterogeneity gives rise to many characteristics distinct from those of advanced composite materials. Because of these complexities, the predictive model necessarily incorporates several modeling strategies. It follows a continuum mechanics approach based on measured meso-structure parameters and mechanical behavior of the laminate. Its three key features are anisotropy, nonlinearity, and stochasticity.

14.2.1.1 Use of Anisotropic Multiaxial Strength Criterion

Strength criteria are employed to estimate the combinations of stress components that cause the onset of material failure (or yielding for ductile behavior). Although most engineering applications involve multiaxial stress states, most design procedures and many research studies predict yield or failure based on conditions of uniaxial stress. This simplified approach is sufficient when one stress governs and the other stresses are small enough to be neglected, but the method can be inaccurate when combined stress conditions exist. In these circumstances, multiaxial criteria are more appropriate.

Further, to be representative of the organized and two-dimensional layered structure of a laminate, a criterion should be orthotropic (or transversely isotropic). Numerous orthotropic strength criteria are available as found in many introductory texts on composite mechanics (for example, Gibson 1994).[13] Each criterion has strengths and limitations and no single criterion is suitable for all materials. However, one made popular by Tsai and Wu (1971),[14] received widespread attention due to its simplicity and generality. This important theory for bio-based laminates accounts for differences in tension and compression strengths.

14.2.1.2 Modeling of Material Nonlinearity with Plasticity Theory

Many studies of wood mechanical behavior as well as research standards applying to wood (e.g., ASTM D143 and D198)[15,16] use linear elastic analysis.

This works well for conventional wood products and their applications, for example, sawn lumber in bending because failure in these cases is typically brittle, precipitated by strength-reducing flaws like knots or slope-of-grain.[17,18]

However, due to the minimized and controlled dispersion of flaws in bio-based laminates, significant yielding occurs in bending due to early and ductile compression failure. It is argued that the plastic-like damage that occurs beyond the proportional limit may be significant and should be considered in the constitutive model. Although the failure mechanisms for bio-based laminates are not the same as for true work-hardening materials, it is presumed that the prediction methods of plasticity theory can be adopted. Thus, the proposed methodology is formulated within the framework of incremental plasticity theory to predict nonlinear behavior beyond initial yield.

14.2.1.3 Simulation of Stochastic and Probabilistic Behaviors with Inherent Size Effects

Homogeneity is a common assumption for many studies because it facilitates simplified deterministic analyses; however, bio-based materials are not homogeneous. Grown in nature and subjected to variable seasons and growth conditions, they are inherently heterogeneous and heterogeneity has deterministic and random components.

Composites reduce that heterogeneity somewhat by dispersing defects throughout layers but point-to-point material properties can still vary significantly. Moreover, the presence of critical strength-reducing defects spread randomly throughout a material leads to a phenomenon known as size effect, whereby the mean strength of a material decreases as the stressed volume increases.

To account for this effect in a predictive model, a stochastic approach is logical: constitutive properties can be expressed in statistical terms as random variables and generated appropriately throughout a member. Monte Carlo simulation is used to facilitate the procedure and a probabilistic solution is obtained.

14.2.2 Materials

In the sections that follow, modeling techniques are discussed in the context of application to two distinct wood laminate meso-structures. We define *meso-structure* as that level of physical structure between micro- and macro- that represents a composite member comprising: (1) a wood–adhesive phase only (without voids) or (2) a wood–adhesive phase and a phase with voids. The modeling techniques are validated through comparison with experimental results for various loading scenarios on two materials: (1) [±θ°] angle-ply laminates and (2) parallel strand lumber (PSL).

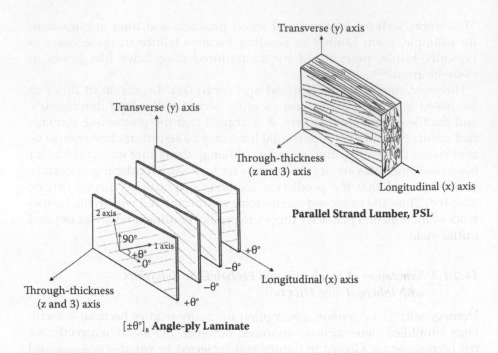

FIGURE 14.2
Angle-ply laminates (exploded view) and parallel strand lumber.

Referencing Figure 14.2, the meso-structure of a [±θ°] angle-ply laminate has a constant fiber angle per layer and no voids. These laminates serve as good control cases for validating model assumptions. They are also similar in their meso-structures to laminated veneer lumber, laminated veneer bamboo, and possible future products. Application of the techniques is a straightforward extension of the methods described. The meso-structure of PSL, however, is comparatively complex: a three-dimensional staggered stacking sequence of 3 mm thick × 19 mm wide strands of varying fiber direction and density. Moreover, upon compaction of the strands, distinct voids of various sizes and shapes occur between them.

Imaging and characterization of grain angle direction and void structure are critical components of modeling PSLs and other laminates with voids. In 2007, Clouston[9] carried out predictive analyses on Douglas fir PSL. The grain angles of each strand were measured statistically and treated as random variables in the model. Voids were accounted for in a smeared approach in which a percentage void content was assigned to each element based on statistical measurement. Reasonable comparisons were found between predicted and experimental results (3.3, 5.5, and 6.4% differences for compression, tension, and bending, respectively). Section 14.4 describes in detail other advanced methods of modeling the void phase of PSL.

14.3 Laminates without Voids

14.3.1 Constitutive Model

Incremental plasticity theory with an evolving three-dimensional strength surface is employed in describing the stress–strain behaviors (load paths) of the layers within a laminate. A load path consists of four fundamental behavioral regimes: elastic, elastoplastic, post-failure brittle, and post-failure ductile. In the first regime, the load path is linear-elastic until the initial yield surface is reached. At this point, one of two events can happen depending on the stress state. If the stress state is brittle, the model follows the post-failure brittle regime and loses both stiffness and strength. If it is ductile, the model undergoes work hardening (elastoplastic regime) using subsequent yield surfaces.

At the upper bound of work hardening, again, one of two events can happen. If the stress state is deemed to correspond to brittle behavior, the model follows a post-failure brittle response. But if the stress state is ductile, the model follows a post-failure ductile response; it loses all stiffness but retains strength. The onset of ductile or brittle behavior (transition between regimes) is identified by initial and subsequent yield surfaces.

14.3.1.1 Initial Yield Surface

The general expression for a yield surface of an orthotropic plastic material is:

$$f_{yield} = \bar{\sigma}^2\left(\sigma_i, \alpha_i, M_{ij}\right) - k^2 = 0 \tag{14.1}$$

where $\bar{\sigma}$ is the effective (or equivalent) stress and k is called the threshold stress. The effective stress is a function of the following in index notation: the applied stresses (σ_i), the terms α_i which delimit the offset of the origin of the yield surface, and the terms M_{ij} which represent strength tensors that define the shape of the yield surface. If the plane stress form of the Tsai-Wu criterion is used, the initial yield surface is ellipsoidal and expressed as follows using index notation:

$$M_{ij}\left(\sigma_i - \alpha_i\right)\left(\sigma_j - \alpha_j\right) - k^2 = 0 \tag{14.2}$$

The applied stresses (σ_i or σ_j, where i and $j = 1, 2, 6$) denote stresses with respect to the principal material directions: 1 (normal stress longitudinal to the fiber direction), and 2 (normal stress transverse to the fiber direction), and 6 (shear stress in the 1, 2 plane). The strength tensors (M_{ij}) and threshold stress (k) are written in terms of normal and shear strengths in the principal directions determined from uniaxial loading: longitudinal tension and compression (X_t and X_c) transverse tension and compression (Y_t and Y_c), in-plane shear (S) as well as an interaction parameter (F_{12}) as follows:

$$M_{ij} = X_t X_c \begin{bmatrix} \dfrac{1}{X_t X_c} & F_{12} & 0 \\[2mm] F_{12} & \dfrac{1}{Y_t Y_c} & 0 \\[2mm] 0 & 0 & \dfrac{1}{S^2} \end{bmatrix}; \quad k^2 = X_t X_c + M_{ij} \alpha_i \alpha_j \qquad (14.3)$$

The terms α_i are derived from simultaneous solution of the equations:

$$\frac{1}{X_t} - \frac{1}{X_c} = -2\left(\frac{\alpha_1}{X_t X_c} + F_{12}\alpha_2 \right)$$

$$\frac{1}{Y_t} - \frac{1}{Y_c} = -2\left(F_{12}\alpha_1 + \frac{\alpha_2}{Y_t Y_c} \right) \qquad (14.4)$$

The interaction parameter (F_{12}) characterizes the interaction of the normal stresses (σ_1 and σ_2) and must therefore be evaluated under a biaxial load-ing condition. Many different methods for evaluation have been suggested, but no standard approach has been agreed upon. F_{12} is known to be diffi-cult to assess because of its high sensitivity to experimental variation.[14,19,20] Stemming from a finding in Clouston et al.[20] suggesting that data from 15-degrees-off-axis tests on LVL had more tolerance for experimental inaccu-racies for calculating F_{12} than 30, 45, or 60 degrees, a nonlinear least squares minimization routine involving a $[\pm 15°]_s$ angle-ply laminate was used.

14.3.1.2 Subsequent Yield Surfaces

In the elastoplastic regime, the strain-hardening response is modeled by specifying a new yield surface at each stage of plastic deformation, known as a subsequent yield surface. In general, it is managed by a hardening param-eter that affects any or all of the variables: α_i, M_{ij}, and k.

The hardening model is said to be isotropic if only parameter k varies with hardening, while α_i and M_{ij} remain constant through plastic flow. In this case, the subsequent yield surfaces expand uniformly around the original yield surface. A model is referred to as kinematic if only α_i vary with the hardening parameter. In this case, the subsequent yield surfaces do not change shape but instead translate in stress space as a rigid body.

In the current methodology, a special type of general anisotropic hard-ening model is used whereby k varies proportionally with the hardening parameter (taken as equivalent to effective plastic strain), α_i remains con-stant, while only two strength variable components of M_{ij} (X_c and Y_c) vary non-proportionally with plastic deformation. This model leads to simultaneous expansion and distortion of the yield surface upon plastic strain. A similar approach was taken by Vaziri et al.[21] and is explained in Clouston.[8]

14.3.1.3 Brittle versus Ductile Behavior

The constitutive behavior of wood is typically brittle under uniaxial tensile loading and ductile under uniaxial compressive loading. However, when wood experiences certain multiaxial states of tensile, compressive, and shear stresses at a failed integration point (e.g., high transverse compression stress with moderate shear stress) it is not clear whether the stress state is most appropriately modeled as brittle or ductile.

In this methodology, the distinction is made through a set of phenomenological conditions based on fundamental wood behavior and chosen to provide good comparisons with empirical data. For example, one condition is that if the transverse tensile stress exceeds the transverse tensile strength independent of other stresses at the point of failure, brittle failure ensues and the post-failure brittle regime is followed. If, however, the combination of stresses is deemed predominantly ductile, the point initially follows the elastoplastic regime, and then follows the post-failure ductile regime.

When brittle failure is detected at an integration point, the stresses are programmed to reduce gradually to facilitate convergence of the finite element procedure. When ductile failure is detected, work hardening ensues. The threshold stress (k) and compressive strengths (X_c and Y_c) change with plastic strain and the yield surface is updated. In the post-failure ductile regime, the yield quantities (X_c and Y_c) reach their ultimate values and the stress level remains constant. At this point, stresses can only traverse the ultimate yield surface until both equilibrium and the constitutive relations are satisfied.

14.3.1.4 Constitutive Equations

The relationship between stress and strain for a laminate is defined through the incremental form of Hooke's law as $d\sigma'_i = Q'_{ij} d\varepsilon'_j$ to account for material nonlinearity. For a three-dimensional stress state, $i, j = 1, 2, \ldots, 6$ and the stiffness matrix (Q'_{ij}) of a laminate contains nine independent components: Young's moduli, shear moduli, and Poisson's ratio. The stiffness matrix is either elastic, elastoplastic (reflecting degradation of the elastic stiffness matrix due to plastic flow), or post-failure (ductile or brittle), depending on stress level. Prime symbols indicate that values have been transformed from local ply coordinates (1, 2, 3) to global laminate coordinates (x, y, z) (see Figure 14.2) via standard transformation laws for Cartesian tensors.

14.3.2 Probabilistic Finite Element Formulation

Stiffness and strength in the principal axes of the strands (X_t, Y_t, etc.) per Equation (14.3) are random variables in the model and represented by probability distributions. While the tensile properties were generated as log-normal distributions, the compressive properties were generated as correlated bivariate standard normal values.

All strength and stiffness properties were assumed to be the same within layers but were generated independently across layers with the exception of transverse tension strength. This strength was regenerated for each integration point reflective of the high sensitivity of transverse strength to heterogeneity of the material. In this way, arrays of strengths and stiffnesses were generated anew for each finite element run (replication) via Monte Carlo simulation, resulting in a probabilistic solution.

Statistical parameters (mean and standard deviation) were obtained a priori for input to the strength criterion. The parameters for uniaxial tension and compression were established experimentally through tests on small unidirectional laminates. Compression properties parallel and perpendicular to grain were represented nonlinearly by four defining variables to constitute a tri-linear curve: (1) elastic modulus prior to yielding, (2) yield stress, (3) elastoplastic tangent modulus, and (4) ultimate stress.

Parameters were regenerated each as correlated to the previous parameter (yield stress to elastic modulus, tangent modulus to yield stress, etc.) according to a bivariate standard normal distribution. The tensile strengths were adjusted for size effect in accordance with the Weibull weakest link theory. In the longitudinal direction, strengths were adjusted from the experimental gauge length to the model gauge length, but in the transverse direction, they were adjusted from representing the tested volume to representing the tributary area surrounding one Gaussian point. The procedures are outlined in detail in Clouston and Lam.[7]

Shear strength S, shear modulus G, and the interaction parameter F_{12} were obtained semi-empirically through a nonlinear least square minimization of error between simulated and experimental compression strengths of a $[\pm15]_s$ angle-ply laminate. Specifically, the error was minimized with respect to the mean and standard deviation of the three unknown variables as described in Clouston and Lam.[6]

Mesh size and boundary conditions for the analyses varied based on load configuration. Solid block elements were employed and element thickness was consistent with layer thickness to facilitate different strand properties per layer. For each simulation, the problem was solved with a modified Newton-Raphson method by prescribing incrementally larger displacement while assuming initial stiffness for each increment.

14.3.3 Experimental Comparison

Comparisons of experimental and predicted results were based on small specimens with cross sections of 10.5 mm × 19 mm. The lengths of the tension and compression specimens were between 40 mm and 120 mm; the bending specimen length was 190 mm. On average for each treatment,

TABLE 14.1

Comparison of Experimental and Simulated Results for Angle-Ply Laminates in Compression

		Initial Stiffness (MPa)			Ultimate Stress (MPa)		
Lay-Up	Statistic	Experiment	Simulation	Difference (%)	Experiment	Simulation	Difference (%)
[±15]$_s$	Mean	6505.3	6672.9	2.6	51.6	51.3	0.6
	COV (%)	26.3	10.7	59.3	14.4	13.8	4.3
	Minimum	3243.5	3787.7	16.8	39.5	39.1	1.0
	Maximum	10908.0	7973.9	26.9	66.2	68.5	3.4
	Count	31	500	–	31	489	–
[±30]$_s$	Mean	3052.0	2742.0	10.2	20.3	21.1	3.9
	COV (%)	27.5	8.3	69.8	11.1	13.8	24.3
	Minimum	1295.1	1708.6	31.9	16.7	13.8	17.4
	Maximum	4567.9	3536.7	22.6	23.5	30.8	31.1
	Count	39	500	–	39	487	–

TABLE 14.2

Comparison of Experimental and Simulated Results for Angle-Ply Laminates in Tension

		Ultimate Stress (MPa)		
Lay-Up	Statistic	Experiment	Simulation	Difference (%)
[±15]$_s$	Mean	39.4	36.4	8.2
	COV (%)	16.7	17.5	4.6
	Minimum	24.8	21.0	18.1
	Maximum	51.7	54.8	5.7
	Count	39	366	–
[±30]$_s$	Mean	21.7	21.4	1.4
	COV (%)	9.0	13.2	46.7
	Minimum	15.5	10.7	31.0
	Maximum	25.6	29.6	15.6
	Count	41	336	–

32 experimental tests were conducted and 450 simulations were performed. Statistical summaries of the uniaxial results for initial stiffness and ultimate stress are provided in Table 14.1 (compression) and Table 14.2 (tension). The full range of load displacement behavior in bending is compared visually in Figure 14.3 (for [±15°]$_s$ laminates) and Figure 14.4 (for [±30°]$_s$ laminates).

The compression results for the [±15°]$_s$ laminate were the most accurate of all comparisons because this test was used to semi-empirically estimate the parameters S, G, and F_{12}. The percent difference, particularly for ultimate stress, is understandably very low—in the range of 0.6 to 4.3. For the other uniaxial comparisons, the difference in predicted versus simulated ultimate

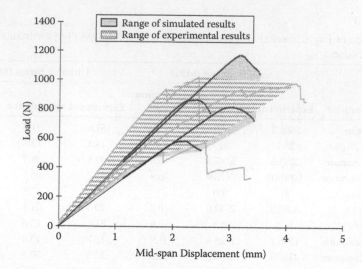

FIGURE 14.3
Comparison of experimental and simulated load displacement range for [±15°] angle-ply laminate in bending.

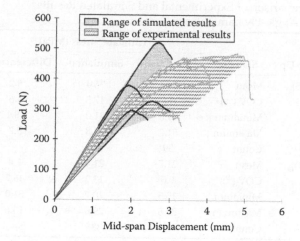

FIGURE 14.4
Comparison of experimental and simulated load displacement range for [±30°] angle-ply laminate in bending.

stress is quite reasonable, especially considering the mean values: not more than 8.2% difference for the [±15°]ₛ laminate in tension.

The extreme values for ultimate stress were less well predicted with a maximum difference of 31.1% for the maximum value of the [±30°]ₛ laminate in compression. Mean initial stiffness values for compression loading were quite reasonable at 2.6% and 10.2% differences for the [±15°]ₛ and [±30°]ₛ laminates, respectively.

Figures 14.3 and 14.4 illustrate the full range of load displacement curves in three-point bending for the two angle-ply laminates. The experimental and simulated ranges clearly overlap, although the percent difference in coefficient of variation (COV) for the [±30°]$_s$ laminates is somewhat high (103%). The mean values for initial stiffness are very close, however, (maximum 14.8% different) as are those for ultimate load (maximum 9.8% different). Importantly, the prediction model is adept at simulating ductile behavior prior to abrupt brittle failure, as observed in experiments for all load configurations.

14.4 Laminates with Voids

This section addresses explicit modeling of the void phase of PSL as follows:

1. A method for generating a three-dimensional measurements of the void phase in a sample of PSL is described

2. An approach to statistical characterization of the void phase is introduced that quantifies the sizes, shapes, and locations of voids within a composite

3. Finite element modeling is used that includes explicit representations of the voids intended to explore the relationship of void structure and macroscopic material performance

4. The finite element simulations are validated against sets of experimental measurements

14.4.1 Void Structure Measurement

In this section, a PSL is considered to consist of a wood–adhesive phase and a void phase. The geometry of the individual wood strands is neglected although models for strand geometry and models for void phases could be superimposed on one another. The void structure in a PSL object occupying a domain $\Omega \subset \Re^3$ is given by the indicator function $I_{[x \in \Omega_{void}]}(x)$, where $x \in \Omega$ and $\Omega_{void} \subset \Omega$ is the region of the PSL object occupied by the void phase. The indicator function takes a value of 1 at position x in the void phase. Thus, $I_{[x \in \Omega_{void}]}(x)$ is a three-dimensional binary map of the PSL void structure.

The challenge in measuring and visualizing the void phase Ω_{void} is that it is embedded and thus hidden in the opaque solid phase $\Omega_{solid} = \bar{\Omega}_{void} = \Omega - \Omega_{void}$. The overbar indicates the complement. Although tomographic techniques such as computerized tomography (CT) scanning can be used to measure the void phase,[22,23] traditional and robust serial sectioning methods are also viable options.

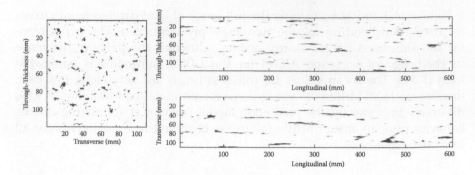

FIGURE 14.5
Void structure of PSL billet determined by serial sectioning. Black pixels indicate void phase.

Serial sectioning methods are well suited to PSL because the material can be easily worked and machined and the length scales associated with the void phase are not so short as to require the extremely high resolution imaging required, for example, if one were trying to characterize the void structure in a fine-grained rock sample.

Example — Serial sectioning procedures were applied to a 102 mm square × 610 long 2.0E Eastern species PSL sample manufactured by Weyerhaeuser. The sample was sawn with a 6 tpi blade into 2.5-mm-thick transverse-through thickness (T-TT) sections with an average kerf width of 1.25 mm. Digitization of the T-TT plane samples was performed by a flat-bed scanner at a pixel size of 0.25 mm in the T and TT directions. Digitized sections were reconstructed in MATLAB® to a three-dimensional array containing binary (0,1) values indicating the presence of a void phase (1) or a solid phase (0). Figure 14.5 shows sections through the reconstructed three-dimensional void structures that clearly present a highly anisotropic void structure.

14.4.2 Meso-Structural Characterization

Probabilistic models for void structures can be used to instantiate finite element models that contain explicit representations of the structures. Such models can also be used to interrogate the effects of meso-structural param-eters on macroscopic responses (effect of void fraction on material stiffness, void shape on fracture strength). Several methods exist for characterizing the structures and geometries of randomly heterogeneous materials.[24] Here we characterize the void structure by volume fraction, void size distribution, and principal moments of inertia of individual void volumes.

The void volume fraction is simply the ratio $\phi = \dfrac{V_{\Omega void}}{V_\Omega}$. One extension of this is to consider the void fraction a stochastic process $\phi(x_L)$, where x_L is the position along the PSL body in the L direction. In doing so, the features of the void structure, such as the correlation length of the void fraction may be discerned readily. While Ω_{void} is the total domain occupied by the void phase, the individual voids occupy sets of subdomains $\{\Omega_{void,i}\}$.

The distribution of void sizes is represented statistically by the mean μ_{void} and standard deviation σ_{void} parameters. Experience shows, however, that second moment characterization is inadequate due to skewness of the distribution and a heavy upper tail. By assigning a fictitious unit mass density to each of the void volumes $\{\Omega_{void,i}\}$, void shape can be partially characterized by the set of principal moments of inertia (I_{11}, I_{22}, I_{33}) found by computing principal values of the matrix formed from the mass moments of inertia:

$$
\begin{bmatrix}
I_{LL} & I_{LT} & I_{LTT} \\
I_{LT} & I_{TT} & I_{TTT} \\
I_{LTT} & I_{TTT} & I_{TTTT}
\end{bmatrix}
= \int_{V_{void,i}}
\begin{bmatrix}
y^2 + z^2 & -xy & -xz \\
-xy & x^2 + z^2 & -yz \\
-xz & -yz & x^2 + y^2
\end{bmatrix} dV
\quad (14.5)
$$

where (x,y,z) is a coordinate system aligned with the L, T, TT material directions with its origin at the centroid of $\Omega_{void,i}$. The shape and orientation of the void are provided through ratios of the principal moments of inertia and the directions of the matrix defined by Equation (14.5).

Example — The measured void structure described in Section 14.4.2 example and the structure of a second, nominally identical billet were characterized statistically via the methods described above. Figure 14.6 shows how the void fraction varies along the length of a specimen and the marginal distribution of the void fraction in each section along with Gaussian and beta best fit distributions. Table 14.3 shows the mean and variance of the moments of inertia of the voids. The tabulated results that do not show the principal moments of inertia since the L and first principal axis are closely

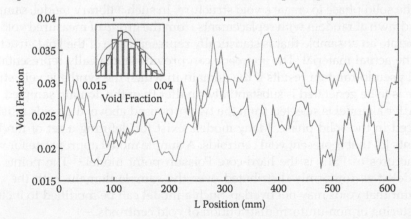

FIGURE 14.6

Variations of void fractions in two tested billets of PSL along the longitudinal direction and histogram of section void fractions with best fits to Gaussian and beta distributions.

TABLE 14.3

Void Moment of Inertia Statistics

Property	Mean (mm⁵)	Standard Deviation (mm⁵)
I_{LL}	500	50,000
I_{TT}	40,000	2,000,000
I_{TTTT}	40,300	2,010,000

aligned in most cases, clearly show the highly elongated nature of the voids in the L direction.

14.4.3 Meso-Structural Model

One approach to modeling void structures is to idealize the geometry of each void as an ellipsoid and model the void structure as an aggregate of ellipsoids embedded in a wood phase. An ellipsoid is defined by its three axis lengths or any two of the axis lengths and the volume. Since the goal is to match as closely as possible the void structure of the actual material, and the voids are characterized by moments of inertia and volume, the problem of fitting an ellipsoid to an actual void is over-determined. By eliminating the two minor axes from the system of equations (effectively assuming that they are related by a constant) and then solving for the remaining major axis, the issue is resolved.

The equivalent ellipsoid model can then be calibrated to the point of generating ensembles of ellipsoidal voids that have means, medians, and variances that agree with the experimental values within a few percentages.

An alternative to the equivalent ellipsoid model is to treat the set of measured voids as a library of void shapes that can be sampled and then placed in the solid phase to create a void structure. In such a library model, samples are drawn at random with replacements from the group of measured voids to generate an ensemble that is statistically representative of the void structure in the actual material. The approach can provide statistically representative and pseudo-random results if the domain in which the synthetic void structure is to be generated is substantially smaller than the domain sampled.

After a model is selected from the two proposed above, the voids must be placed in the solid phase. Many models exist for defining a set of random points $\{c_i\}$ that represent void centroids. A simple model appropriate for void structures of PSLs is the hard-core Poisson point model.[25] The points are modeled as uniformly distributed over the sample domain with the constraint that voids may not overlap. Such a model can be modified to include clustering or non-uniform distribution of void centroids.

Example — In the samples described above, a total of approximately 15,000 voids were characterized and no evidence of clustering or non-uniform distribution was detected. The total material volume of the samples was

1.6×10^8 mm³, providing a void library that is more than sufficient for prisms roughly 50 mm × 50 mm × 100 mm used in the computational studies.

14.4.4 Compression and Tension Modeling

These methods are useful for simulating the nonlinear compressive behaviors of PSLs with voids and tensile behaviors up to the point of specimen failure. In both cases, a regular mesh of 27-node solid finite elements is used to model the wood phase and elements at locations inside void volumes are deleted from the mesh to represent the void phase. The solid phase is modeled as orthotropic without distinguishing the properties of individual strands. Hill's orthotropic yield criterion is used:

$$f_{yield} = F\left(\sigma_{bb} - \sigma_{cc}\right)^2 + G\left(\sigma_{cc} - \sigma_{aa}\right)^2 + H\left(\sigma_{aa} - \sigma_{cc}\right)^2$$
$$+ 2L\sigma_{ab}^2 + 2M\sigma_{bc}^2 + 2N\sigma_{ac}^2 - 1 = 0 \qquad (14.6)$$

The F, G, H, L, M, and N parameters correspond to material strengths. For compression analysis, the wood phase is assumed to be perfectly plastic after yield, and in tension the wood phase is assumed to be perfectly brittle after yield. In tension, fracture initiates immediately when the state of stress satisfies the Hill criterion.

Example — In compression, models were developed for prismatic rectangular samples of dimensions 25 mm × 25 mm × 76 mm. Boundary conditions corresponding to uniform, uniaxial, and compressive stresses were applied to models with the long dimensions in each of the three principal material directions. Figure 14.7 shows finite element-derived stress–strain curves for models of PSL that compare the response of an actual meso-structure to that

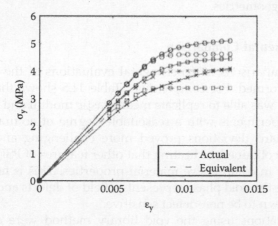

FIGURE 14.7
Compression simulations of measured and synthetic PSL meso-structures.

TABLE 14.4

Comparison of Tensile Simulations with Actual Meso-Structures and
Virtual Meso-Structures Generated from Void Library Model

Loading Direction	Mean (MPa)				Standard Deviation (MPa)			
	Strength		Ultimate Strain		Strength		Ultimate Strain	
	Measured	Library	Measured	Library	Measured	Library	Measured	Library
Longitudinal	44.2	41.3	0.32	0.27	11.6	6.0	0.10	0.04
Transverse	0.82	0.86	0.25	0.26	0.18	0.14	0.05	0.04
Through-thickness	0.94	0.95	0.70	0.74	0.31	0.25	0.22	0.18

of a virtual meso-structure generated by the equivalent ellipsoids method. The agreement is quite good, with prediction error in the mean ranging from 2 to 15%, the best performance obtained in the L direction, and the worst performance obtained in the T direction.

The equivalent ellipsoid model performed poorly when exercised in tension using an element deletion algorithm based on a brittle element failure criterion. In tensile simulations, model size was selected to match the physical specimens tested by Arwade et al.,[12] namely, 51 mm × 38 × mm × 15 mm (2 in. × 1.5 in. × 0.60 in.) for T and TT loading and 102 mm × 24 mm × 15 mm (4 in. × 1 in. × 0.60 in.) for L loading.

The void library model was used in each case to generate synthetic void structures and compare them to models that use the measured void structure. For each loading direction, 15 models were run for the measured and virtual void structures to obtain suitable statistical characterizations of the response. Table 14.4 shows that the void library model performs reasonably well in generating meso-structures that behave similarly to models that use measured void geometries.

14.4.5 Experimental Comparison

Two sets of comparisons to experimental evaluations of the mechanics of PSLs were performed. For compression, Table 14.5 shows that the virtual ellipsoid model was able to replicate mean elastic moduli and yield stresses observed in experiments with a reasonable degree of accuracy. Matching observed standard deviations proved more challenging and satisfactory results were not obtained, indicating that other features of PSL may be driving uncertainty in compressive material properties. This is not necessarily surprising since the void phase represents a field of defects and compressive behavior is known to be non-defect sensitive.

Tensile simulations using the void library method were conducted on models of varying lengths to determine whether a size effect in tensile strength could be discerned. Table 14.6 shows that the numerical simulations

TABLE 14.5

Comparison of Experimental Compressive Mechanical Properties to Simulated Properties Developed Using Equivalent Ellipsoid Model

Loading Direction	Mean Yield Stress (MPa)		Standard Deviation (MPa)		Coefficient of Variation (%)	
	Experiment	Simulation	Experiment	Simulation	Experiment	Simulation
Tangential	3.32	3.72	0.22	1.20	6.7	44.7
Through-thickness	2.92	2.51	1.37	0.58	47.0	24.2
Longitudinal	52.0	50.4	8.0	1.0	15.4	2.1

TABLE 14.6

Comparison of Size Effect Power Law Exponents Observed in Experiments and Simulations

Loading Direction	Absolute Value of Slope	
	Experiment	Simulation
Tangential	0.36	0.32
Through-thickness	0.30	0.32
Longitudinal	0.09	0.21

predict a size effect that is very nearly equal to that observed in tests except in the *L* direction, which is less defect sensitive due to the elongation of voids in that direction. This is a particularly encouraging result given that tensile properties are expected to be defect sensitive and that tensile failure is brittle.

14.5 Conclusions

Bio-based laminates present excellent potential for becoming the next generation of industrial products. Reliable computational tools are essential in this pursuit. This chapter describes modeling strategies for predicting the mechanical responses of bio-based laminates—those with and those without macrovoids in their meso-structures.

The overarching methodology is based on a continuum mechanics approach with an orthotropic elastoplastic constitutive model. Stiffness and strength properties of laminates are modeled as stochastic random variables and probabilistic solutions for laminate strength are obtained. In cases where macrovoids are present in a laminate, the void structure is first measured and then modeled as being embedded into the wood phase.

Two methods for characterizing and implementing voids are described: (1) as an aggregate of equivalent ellipsoids (characterized by volume fraction, void size distribution, and principal moments of inertia) or (2) as a library of measured void shapes that can be sampled and placed into the wood phase.

The numerical strategies were applied and compared with experimental results for various loading configurations with two distinct materials: (1) $[\pm\theta°]_s$ angle-ply laminates and (2) PSLs. Favorable agreement between simulated and experimental data demonstrates the effectiveness of the techniques described.

Acknowledgments

This work was supported in part by the National Science Foundation through grant CMMI-0826265. We are also grateful to Dan Pepin, the wood shop manager at the University of Massachusetts Building and Construction Technology Department, and former undergraduate students Meghan Krupka and Saranthip Rattanaserikiat for their contributions.

References

1. Inverizon International Inc. *The Technology Roadmap for Plant/Crop–Based Renewable Resources 2020*. Executive Steering Group, 1999.
2. APA–Engineered Wood Association. Milestones in the History of Plywood. 2013. http://www.apawood.org/level_b.cfm?content=srv_med_new_bkgd_plycen.
3. C.G. Drury. *Handbook of System Reliability in Airframe and Engine Inspection*. U.S. Department of Transportation, Federal Aviation Administration, 2005.
4. S. Lieblein, M. Gaugeon, G. Thomas et al. Design and Evaluation of Low-Cost Laminated Wood Composite Blades for Intermediate Size Wind Turbines: Blade Design, Fabrication Concept, and Cost Analysis. Rocky River, Ohio: Technical Report Services, 1982.
5. Advantages of the Timbertower. 2013. http://www.timbertower.de/en/product/advantages/
6. P.L. Clouston and F. Lam. Computational Modeling of Strand-Based Wood Composites. *Journal of Engineering Mechanics*, 127 (2001): 844–851.
7. P.L. Clouston and F. Lam. A Stochastic Plasticity Approach to Strength Modeling of Strand-Based Wood Composites. *Composites Science and Technology*, 62 (2002): 1381–1395.
8. P. Clouston. *Strength Modeling of Structural Composite Lumber*. Lulu.com. 2010.
9. P. Clouston. Characterization and Strength Modeling of Parallel-Strand Lumber. *Holzforschung*, 61 (2007): 394–399.

10. S. Arwade, P. Clouston, and R. Winans. Measurement and Stochastic Computational Modeling of the Elastic Properties of Parallel Strand Lumber. *Journal of Engineering Mechanics*, 135 (2009): 897–905.
11. S. Arwade, R. Winans, and P. Clouston. Variability of the Compressive Strength of Parallel Strand Lumber. *Journal of Engineering Mechanics*, 136 (2010): 405–412.
12. S.R. Arwade, P.L. Clouston, and M.T. Krupka. Length Effects in Tensile Strength in the Orthogonal Directions of Structural Composite Lumber. *Journal of Testing and Evaluation*, 39 (2011): 576–582.
13. R.F. Gibson. *Principles of Composite Material Mechanics*. New York: McGraw Hill, 1994.
14. S.W. Tsai and E.M. Wu. A General Theory of Strength for Anisotropic Materials. *Journal of Composite Materials*, 5 (1971): 58–80.
15. ASTM D143: Standard Test Methods for Small Clear Specimens of Timber. West Conshohocken, PA: Americal Society for Testing and Materials, 2009.
16. ASTM D198: Standard Methods of Static Tests of Timbers in Structural Sizes. West Conshohocken, PA: American Society for Testing and Materials, 2009.
17. B. Madsen and A.H. Buchanan. Size Effects in Timber Explained by a Modified Weakest Link Theory. *Canadian Journal of Civil Engineering*, 13 (1986): 218–232.
18. J. Barrett, F. Lam, and W. Lau. Size Effects in Visually Graded Softwood Structural Lumber. *Journal of Materials in Civil Engineering*, 7 (1995): 19–30.
19. J.C. Suhling, R.E. Rowlands, M.W. Johnson et al., Tensorial Strength Analysis of Paperboard. *Experimental Mechanics*, 25 (1985): 75–84.
20. P. Clouston, F. Lam, and J. Barrett. Interaction Term of Tsai-Wu Theory for Laminated Veneer. *Journal of Materials in Civil Engineering*, 10 (1998): 112–116.
21. R. Vaziri, M.D. Olson, and D.L. Anderson. A Plasticity-Based Constitutive Model for Fibre-Reinforced Composite Laminates. *Journal of Composite Materials*, 25 (1991): 512–535.
22. M. Sugimori and F. Lam. Macro-Void Distribution Analysis in Strand-Based Wood Composites Using X–Ray Computer Tomography Technique. *Journal of Wood Science*, 45 (1999): 254–257.
23. Q. Wu, B. Zhang, L. Wang, and G. Han. Application of 3D X-Ray Tomography with Finite Element Analysis for Engineering Properties of Strand-Based Composites. In Proceedings of the 8th Pacific Rim Bio-based Composites Symposium, Advances and Challenges in Biocomposites (2006): 20.
24. S. Torquato. *Random Heterogeneous Materials: Microstructure and Macroscopic Properties*, Vol. 16. Heidelberg: Springer, 2002.
25. D.P. Bentz, E.J. Garboczi, and K.A. Snyder. A Hard-Core–Soft-Shell Microstructural Model for Studying Percolation and Transport in Three-Dimensional Composites. National Institute of Science and Technology, 1999.

15

Micromechanical Modeling of Bio-Based Composites

Michael May and Deborah Mohrmann

CONTENTS

15.1 Introduction

Large composite structures are usually modeled with shell elements following a continuum mechanics approach. Continuum mechanics describe the mechanical response of materials using approximations that assume homogeneous materials and thus neglect phenomena occurring at the microscale. However, composite materials are intrinsically heterogeneous at different length scales due to the composition of various constituents such as fibers and resins.

Micromechanics consider these heterogeneities by explicitly modeling the characteristics of the microscale such as fibers, resin, voids, and inclusions. Therefore, numerical simulation of composites at the microscale is a useful tool for improving the general understanding of composites because processes occurring at microscale may influence mechanical responses at larger scales. The fundamental understanding gathered from micromechanical modeling can also help improve the architecture of composites and assess novel combinations of resins and fibers and effects of manufacturing defects.

This chapter describes the underlying principles of micromechanical simulation of bio-based composites. It is structured as follows. First, experimental methods that allow the identification of relevant micromechanical properties required as inputs for micromechanical simulations are described, then advice is given for identifying and setting up representative volume elements (RVEs) for micromechanical simulation of composites.

Homogenization strategies are highlighted, allowing the derivation of macroscopic properties from micromechanical simulations. The methodologies presented are demonstrated with practical examples.

15.2 Experimental Identification of Micromechanical Properties

This section focuses on experimental techniques allowing the generation of input data required for high quality micromechanical simulations of bio-composites. The micromechanical properties required for such simulations of composite materials are the deformation and failure behavior of the individual constituents (fiber and matrix).

The properties of the fiber–matrix interface are crucial for predicting the behaviors of bio-composites. The differences in the natures of constituents and their interactions require consideration of different types of failures: fiber failure (tensile, buckling, splitting), matrix failure (tension, compression, shear), and interface failure (debonding). The last two failure modes are designated interfiber failures.[1] Due to the wide spectrum of properties that must be tested to generate a complete picture of material properties, experiments investigating the micromechanical level of a composite must be planned very carefully. The strengths of the constituents, fracture toughness, failure mechanisms, and fatigue lives of fibers, matrices, and interfaces should be tested.

Using all experimental data available from micromechanical experiments, good verification simulations can be performed it is possible to predict average behavior at the lamina level as a function of constituent properties and local conditions.[1] We now describe in detail the most important and popular methods for determining the quasi-static mechanical properties of the constituents of composites (fibers and matrices) and their interfaces.

15.2.1 Mechanical Properties of Fibers

The most important fiber properties to examine are tensile strength, Young's modulus, elongation at break, and density. Because detailed knowledge of properties of fibers is vital for the textile industry and engineers working with fiber-reinforced materials, several testing methods for single fibers have been developed. Generally, only one type of experiment is required for determining tensile strength and elongation at break. The Young's modulus can be calculated using the test results generated.

European Committee for Standardization's DIN EN ISO 5079[2] published in February 1996 defines the conditions for determining breaking force and elongation at break of individual fibers. The data can be used to calculate the tensile strength and Young's modulus if fiber diameter is known. The diameter may be determined using a field emissions gun scanning electron microscope (FEG-SEM).

There are certain requirements for devices used for tensile tests on single fibers. First, the system must operate on the principle of constant testing velocities in a range of 5 to 20 mm/min. Second, it must be capable of measuring and recording tensile forces and corresponding changes in length. Another important component is clamping. Individual fibers must be clamped and stretched to rupture at a constant rate of extension.[2] A fiber should not slide within a clamp during testing. The clamps must have special surfaces to prevent sliding. Conversely, fiber damage induced through excessive clamping pressure during testing must be avoided[3] so the pressure of the clamps holding fibers must be adjusted precisely.

Examples of devices used commonly for single-fiber testing are the Fafegraph Hr[3] and the Favimat Robot[3,4] (both from Texttechno). The Favimat is designed especially for testing single fibers and the system can measure fineness, strength, and elongation of single fibers in a single test procedure.[4]

Single-fiber testing can also be performed with conventional tensile testing machines. Eichhorn and Young[5] described testing of single ultimate fibers (8 mm) of hemp. To calculate tensile strength, the assumption is that the fiber cross section is roughly circular and remains unchanged during the test.

Another avenue for obtaining information about fiber properties is to deform them under the microscope of a Raman spectrometer.[5] Raman spectroscopy can be used for material identification, characterization of the state of a material, and also for measuring local mechanical stresses.[6] To characterize a fiber with Raman spectroscopy, the fiber is placed under the microscope of an appropriate spectrometer device. The fiber is then loaded with a tensile force in specific steps and Raman spectra are recorded for every step. The recorded spectra can be evaluated to determine stress and strain levels at each step.[5]

Standard methods for measuring the linear densities of textile fibers can be found in ASTM D1577-07 published in 2012.[7] Two possibilities are described for determining single-fiber density: weighing and using a vibroscope. Both methods are appropriate and can be used if further tests of fibers are required after density measurements are performed. Properties of natural fibers that are typically used in composites are described in the tables in Section 15.2.4.

15.2.2 Mechanical Properties of Resins

The most critical mechanical properties of polymers used as resins for composites are determined by testing under tension, compression, shear, and bending loading conditions. Further tests may be done to identify impact properties. For a basic characterization of a polymer, tests at quasi-static velocities are performed using servo-hydraulic machines. The most popular test methods for characterizing the properties of a polymer are described below.

A typical specimen for tensile testing of resin is defined in standard DIN EN ISO 527-2.[8] The so-called dog-bone specimen cited in the standard is used for the characterization of resins with standard servo-hydraulic testing machines. Raman spectroscopy, described in Subsection 15.2.1, can be used to characterize tensile properties of pure resin specimens as well.

An example for tensile testing of a biopolymer is given by Graupner et al.[9] Their paper gives a comparison of samples of pure poly(lactic acid) (PLA) and natural fiber-reinforced PLA. The tensile force and the elongation at break were measured, allowing the tensile strength and Young's modulus to be determined.

The Charpy impact test was developed originally to compare the impact performances of notched metallic bars loaded by a swinging pendulum. However, this test is also used to assess composite materials; see Graupner et al.[9] and Sohn and Hu.[10] However, some uncertainty surrounds kinetic energy dissipation because the energy of a broken sample is still contained. For this reason, this impact test can be done only for comparative purposes and does not deliver inputs for micromechanical simulations.

Another method of testing biopolymers is described by Yang et al.[11] The tensile properties of samples of PLA were examined with an Instron 4302 tensile tester. Yang's group describes other tests relating to thermal and other properties of polymers.[11]

Specimen geometries used to analyze the compressive properties of pure resins are given in DIN EN ISO 6036.[12] Notched and un-notched rectangular specimens are tested in a standard servo-hydraulic device. The bending properties of resins are typically determined using rectangular strip specimens subjected to three- or four-point bending conditions. Details can be found in standard DIN EN ISO 178.[13]

Arcan's classical specimen[14] is often used to evaluate the shear properties of composites and polymers. Over the years, the specimen geometry has been modified several times. A recent example of a possible geometry is described by Pucillo et al.[15] They discuss a new optimal geometry of the Arcan specimen evaluated by three-dimensional parametric finite element analysis; the results are then validated experimentally.

The use of local strain measurement techniques is strongly recommended for determining true high-quality stress–strain data required for predictive micromechanical simulations of resins. One option is applying strain gauges on the surfaces of specimens. However, the maximum strain that may be measured with gauges is on the order of 4%. Depending on the amorphous or crystalline state of the resin, the strain to failure of PLA can reach 6.8%.

Digital image correlation (DIC) techniques are the methods of choice for measuring true strains in resins. These non-contact measurement techniques use one or two cameras to trace the displacements of random speckles applied to the surface of a specimen. Comparison of the images recorded by the cameras reveals speckle displacement and therefore the true strain at any location on a specimen. Consequently, material data for simulation can be gathered directly at or near the location of failure and is thus more accurate than data from the cross head that blurs these localized effects

One camera is used for monitoring in-plane motion. If three-dimensional measurements are required because of out-of-plane movements, a set of two synchronized cameras can be used to obtain displacement and strain data in all spatial directions. Commercially available DIC systems are the GOM Aramis, and Limess Vic. Mechanical properties of typical and often used biopolymers and conventional polymers, are listed in the tables in Subsection 15.2.4.

15.2.3 Mechanical Properties of Interfaces

In the region of contact between fiber and matrix, three different phases can be distinguished: fiber, matrix, and the interphase between them. The interphase is defined as a region of contact between the fiber and the matrix called the interface and a region of finite thickness extending on both sides of the interface in both the fiber and the matrix.[16]

The fiber–matrix interface in composite materials plays a very important role in determining composite mechanical properties.[17] As a consequence, many tests are available for evaluating the properties of the interface and the fiber–matrix adhesion, for example, the fiber pull-out, embedded fiber fragmentation, embedded fiber compression, and microindentation methods. They measure fiber–matrix adhesion and fiber–matrix failure modes. They also allow measurement of the energy involved in fractures of fiber–matrix interfaces.[16]

The single-fiber pull-out method is one of the oldest and most popular ways to measure interfacial shear strength. Broutman was among the first to introduce this method in 1969.[18] He was followed by Kelly in 1970 who described in detail the different phases that occur as a fiber is pulled out of a resin (initial debonding, crack propagation, completion, and pull-out).[19] In the early days of this testing method, a single fiber was pulled from a massive block of resin. Fracture of the fiber often occurred within the block, making the test extremely difficult.

As a result, later methods involve pulling a fiber out of a microdroplet or a thin disk of resin.[16] A very good overview of the evolution of the single-fiber pull-out test is given in Difrancia et al.[20] They provide detailed descriptions of the historical development of the experiments and performance parameters. An example of a single-fiber pull-out test performed on a bio-composite appears in Mukhopadhyaya et al.[21] The performance of a composite containing polypropylene as the matrix and sisal fibers as reinforcements in a single-fiber pull-out test is reported. The fibers used for the test were treated in various ways and the effects of treatments on the adhesions of fibers and matrices are discussed.[21]

Several additional methods can measure interface properties. An example is the cruciform specimen test[17] developed to investigate interfacial failure criteria under combined stress states. When failure occurs along an interface, the stress state at this point is normally triaxial—a combined state of normal and shear stresses exists. The methods mentioned above are not suitable for testing such combined stresses. For that reason, the cruciform specimen test was devised. Using a special cruciform sample geometry, an elastic finite element analysis and two assumed interfacial failure criteria (normal and shear strengths) can be determined, providing that every equivalent interfacial stress is identical for all geometries of the specimens.[17] In 2012, Koyanagi et al. published a paper dealing comparing glass–epoxy interface strength measured by a single-fiber pull-out test and the cruciform specimen test. Reasonably consistent results were obtained for both tests.[22]

The fiber–matrix interface can also be investigated by Raman spectroscopy, as described above. A microdroplet of epoxy resin is produced around the fiber along the gauge length and the fiber is put into the spectroscope. To determine the stress profile of the system through the microdroplet, the fiber is scanned and Raman spectra are generated. After each scan along the fiber, the strain of the material is increased incrementally and the process is repeated.[5] It is possible to obtain a complete strain spectrum for an interface.

It is also possible to visualize the stress distributions generated by fibers embedded in polymeric matrices if the polymer can transmit light. Additionally, the resin must have birefringent properties when it is mechanically loaded. For stress analysis, a sample of pure resin or a sample in which fibers are embedded is irradiated with circular polarized or simply polarized light and loaded in uniaxial tension. Depending on the loading state, changes in photoelastic effects appear, mainly isochromatic areas and isoclines. According to the type of light used, the color of the light (white) or the intensity of the light (monochromatic) changes. Based on correlating the stress–strain curve and the isoclines of pure matrix sample, stress distribution around the fibers in the sample can be determined.[23]

15.2.4 Typical Micromechanical Properties

This subsection provides an overview of mechanical properties required for micromechanical simulations for some of the most frequently used fibers and biopolymers. Table 15.1 lists typical mechanical properties of fibers. Table 15.2 lists mechanical properties of resins.

TABLE 15.1

Typical Mechanical Properties of Fibers

Fiber	Density (g/cm³)	Diameter (μm)	Tensile Strength (MPa)	Young's Modulus (GPa)	Elongation at Break (%)
Abaca	1.5	10–30	400–813	31.1–33.6	2.9–10
Bamboo	0.6–1.1		140–1000	11–89	
Flax	1.5	5–600	345–1500	27–39	2.7–3.2
Hemp	1.4–1.48	10–500	550–900	35–70	1.6–4
Jute	1.3–1.49	10–250	393–800	13–26.5	1.16–1.8
Kenaf	1.5–1.6	2.6–4	350–930	40–53	1.6
Sisal	1.45–1.5	8–200	468–700	9.44–22	2.0–7
Ramie	1.5–1.6	11–80	400–938	24.5–128	1.2–3.8
Pineapple	0.8–1.6	20–80	413–1627	1.44–82.5	1.6–14.5
Coir	1.15–1.46	100–460	131–200	4–22	15–40
Cotton	1.5–1.6	12–38	287–800	5.5–12.6	7–8
Lyocell		11.4 ± 3.4	1019.8 ± 256.3	8.94 ± 4.42	12.5 ± 3.9
Viscose			691	20.2	
E–Glass	2.5–2.55	10–25	2000–3500	70–73	2.5–3.7
Carbon	1.4–1.78	5–10	3400–4800	230–425	1.4–1.8

TABLE 15.2

Typical Mechanical Properties of Resins

Polymer	Density (g/cm³)	Young's Modulus (GPa)	Tensile Strength (MPa)	Elongation at Break (%)
PLA	1.21–1.25	0.35–3.5	21–63.5	2.5–6.0
PHB	1.18–1.26	3.5–4.0	24–40	5.0–9.0
PHBV	1.23–1.25	0.5–2.14	20–27.3	7.0–25.0
PP	0.9–1.16	1.1–1.6	29.2–40	12–40

15.3 Micromechanical Simulation Using Representative Volume Elements

Micromechanical simulation is an essential tool for describing and understanding material behavior of composites. Resolving the microstructure of a composite material requires the use of extremely fine meshes. As a consequence of the huge computational costs associated with this approach, numerical modeling of large composite structures while resolving real microstructures is now and will be in the near future virtually impossible. Consequently, micromechanical simulations are limited to small sections representative of a complete composite. The following sections describe the requirements of these so-called representative volume elements (RVEs) and their identification.

15.3.1 RVE Requirements and Identification

Representative volume elements allow us to describe a material point in a heterogeneous continuum by explicitly modeling the inhomogeneities at small scales. Two main size requirements are involved. The RVE must be small enough from a macroscopic view to be treated as a material point. Conversely, it must be large enough to reveal the constituents, voids, and inclusions and demonstrate distributions and volume fractions at the microscale. The Hashin inequality defines the RVE length scale:[24]

$$L_{const} \ll L_{RVE} \ll L_{macro} \tag{15.1}$$

L_{const} is the length scale of the constituents at the microscale, L_{RVE} is the length scale of the RVE, and L_{macro} is the length scale of the macroscale. If the Hashin inequality is satisfied, the Hill principle[25] is unconditionally valid. The Hill principle states that the work done at the macroscale is equal to the work done at the microscale integrated over the volume V of the RVE.

$$\bar{\sigma} : \bar{\varepsilon} = \frac{1}{V} \int \sigma : \varepsilon \, dV \tag{15.2}$$

Therefore, macroscopic quantities (denoted with overbars) can be related to microscopic quantities. The average strain theorem states that macroscopic strains $\bar{\varepsilon}$ can be calculated from microscopic strains ε by integration over the volume V of the RVE.

$$\bar{\varepsilon} = \frac{1}{V} \int \varepsilon \, dV \tag{15.3}$$

In similar fashion, the average stress theorem postulates that macroscopic stresses $\bar{\sigma}$ are related to microscopic stresses σ.

$$\bar{\sigma} = \frac{1}{V} \int \sigma \, dV \tag{15.4}$$

As noted earlier, the RVE must be large enough to show the morphology of the composite microstructure including the orientation and distribution of fibers, fiber volume fraction, voids, and inclusions. High-quality quantity information can be obtained from micrographs or non-destructive x-ray computed tomography (CT). An example of CT analysis of a bio-based composite with random fiber distribution appears in Figure 15.1.

FIGURE 15.1
X-ray CT image of microstructure of hemp–flax–reycled PE–epoxy composite.

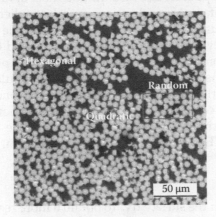

FIGURE 15.2
Micrograph of carbon–epoxy composite showing possible types of RVEs.

CT provides full 3D information on the internal structures of composites and is thus a very powerful tool for generating representative volume elements. The 3D CT data can be loaded directly into a finite element (FE) program, thus allowing simulation of "real" composites. Figure 15.2 is a micrograph of a unidirectional (UD) carbon–epoxy composite. It is particularly interesting because it features several types of RVEs used frequently in the literature to describe UD composite materials: quadratic fiber distribution,[26,27] hexagonal fiber distribution,[28] and random fiber distribution.[29,30]

Although all of these RVEs are supposed to be representative of the composite microstructure, the results obtained during numerical analyses may vary. All types of RVEs for UD composites predict the same elastic constants in fiber direction if the fiber volume fraction is the same. However, it has been shown that the transverse stiffness calculated using different types of RVEs may vary by about 50%. Comparing the results to available experimental data for stiffness and failure, it seems as if random RVEs provide the best correlation if the RVEs are big enough. A statistical analysis of micrographs of UD composites has shown that the distances and orientations between fibers are representative of a whole composite when the RVE consists of 25 randomly placed fibers.[31]

Although bio-based composites may be unidirectional,[32–34] the fibers may be distributed randomly such that the resulting composite is of isotropic[35] or quasi-isotropic[36] nature in which case it is suggested to generate the RVE from CT data. Alternatively, RVEs can be generated using random sequential adsorption (RSA) or Monte Carlo (MC) procedures. A review of these techniques is given in Harper et al.[37]

15.3.2 Boundary Conditions

Depending on the type of RVE (UD continuous fiber or random discontinuous fiber composite), different considerations must be applied in applying boundary conditions (Figure 15.3). However, the Hill principle (Equation (15.2)) can be applied for any type of boundary condition applied to the representative volume element. For UD composites, two different types of boundary conditions must be considered for micromechanical simulation: homogeneous displacement conditions and periodic displacement conditions.

For a simple case of homogeneous boundary conditions, the displacement field can be approximated by linearly correlating the displacement at the boundary of the RVE to the applied macroscopic strain:

$$u = x \cdot \bar{\varepsilon} \tag{15.5}$$

where x is a point on the surface dV. To obtain a more realistic representation of a microstructure, the use of periodic boundary conditions is strongly recommended. Periodic boundary conditions allow capturing of local variations in displacement at the boundaries of the RVE, which is especially important

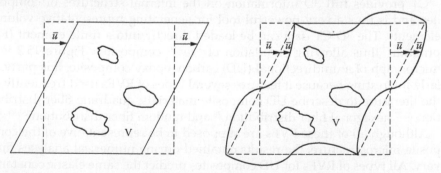

FIGURE 15.3
Application of boundary conditions in periodic media. Left: homogeneous boundary conditions. Right: periodic boundary conditions.

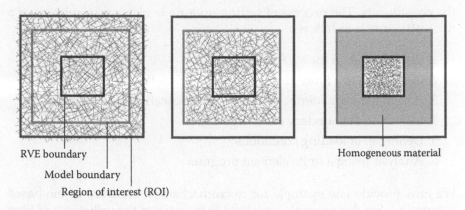

RVE boundary

Model boundary

Region of interest (ROI)

Homogeneous material

FIGURE 15.4
Application of boundary conditions in non-periodic random media.

in shear load cases. Typically, these local variations are consequences of inhomogeneous distribution of the constituents within the RVE. Periodicity requires that the fluctuations along opposing edges are identical:

$$\tilde{u} = \tilde{u}' \tag{15.6}$$

The total displacement is given as the superposition of the homogeneous displacement field and the fluctuating displacement field:

$$u = x \cdot \bar{\varepsilon} + \tilde{u} \tag{15.7}$$

Consequently, the following displacement condition must be enforced at two opposing points x and x':

$$u - x \cdot \bar{\varepsilon} = u' - x' \cdot \bar{\varepsilon} \tag{15.8}$$

The random nature of discontinuous fiber composites does not allow the application of periodic boundary conditions. The microstructure cannot be described periodically as it is possible for UD composites. Harper et al.[38] suggest surrounding the RVE in which the microstructure is resolved with a boundary of homogeneous material. Loads or displacements are then applied to the external boundary of the material and automatically transferred into the RVE. An example is given in Figure 15.4.

15.3.3 Numerical Modeling

Finite element simulation is a viable tool for predicting the effective properties (strength and stiffness) of composites based on the properties of the constituents. The process of setting up a finite element simulation of a microstructure is as follows:

1. Identification of the RVE
2. Generation of a finite element mesh based on the RVE identified
3. Assignment of material properties to the constituents and interface
4. Definition of boundary conditions
5. Definition of loading conditions
6. Analysis using a finite element program

We now provide one example for micromechanical analyses of bio-based composites. For this example, we chose to investigate the influences of fiber type and fiber volume fraction on the mechanical properties of UD bio-based composites subjected to tensile loading in the fiber direction.

In this case, we are interested in the longitudinal properties of the composite only. As shown earlier, the spatial distribution of parallel fibers does not influence simulation results. We therefore opt for the simplest RVE, a quarter fiber and surrounding matrix. Three fiber volume fractions (25, 36, and 49%) were realized. The mechanical properties for four types of fibers (Cordenka EHM, Cordenka 1840, Enka viscose, and Lyocell regenerated cellulose) were taken from the literature.[39]

Each fiber was modeled using a linear-elastic plastic material model with strain hardening. The properties of PLA were extracted from stress–strain data from Suryanegara et al.[40] PLA was modeled using a linear-elastic perfect-plastic material model. Figure 15.5 and Table 15.3 summarize the mechanical properties of the resin and the fibers used in the simulations.

Perfect bonding between the fiber and the resin was assumed. Therefore, the interface was not modeled explicitly. An approach for modeling the interface connecting resin and fiber would be the use of cohesive interface elements. These elements have been used successfully for modeling damage initiation and failure at distinct interfaces in composite materials; see Wisnom[41] for a recent review. They are now available as options in many commercial finite element software packages.

For this particular loading case, it was not necessary to apply periodic boundary conditions because capturing the oscillations on the edges of the RVE is important only for transverse and shear load cases. Finite element codes such as ABAQUS and LS-DYNA offer options to define equations that allow correlations of displacements at certain nodes via Equation (15.8). In each computation step, the total set of equations for all nodes at the boundaries is solved implicitly.

The nodes on the back face of the model were constrained in the z-direction. A constant velocity in the z-direction was applied to the nodes on the front face of the model. Figure 15.6a compares the predicted engineering stress–strain

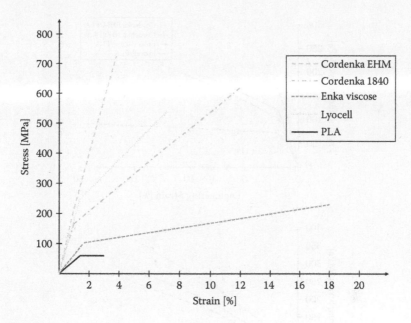

FIGURE 15.5
Stress–strain data used for micromechanical simulations.

TABLE 15.3

Mechanical Properties Used for Micromechanical Simulations

Material	Young's Modulus (GPa)	Tensile Strength (MPa)	Poisson Ratio	Elongation at Break (%)
Cordenka EHM	32.2	710	0.2	3.9
Cordenka 1840	16.9	660	0.2	12.7
Enka viscose	9.4	220	0.2	17.2
Lyocell	15.2	540	0.2	7.0
PLA	3.0	59	0.36	3.1

curves* for variable fiber types and a constant fiber volume fraction of 25%. The influence of the different fiber types on the mesoscopic response is clearly visible. The right portion of the figure compares the predicted force displacement curves for Cordenka EHM-reinforced PLA and a variable fiber volume fraction. As expected, the predicted engineering stress–strain curves show increasing performance with increasing fiber volume fraction.

---

* Engineering stress and strain data are based on the original cross-section before testing and do not account for changes in cross-sectional area due to Poisson's contraction or damage accumulation. True stress and strain data consider changes of cross-sectional area during the test.

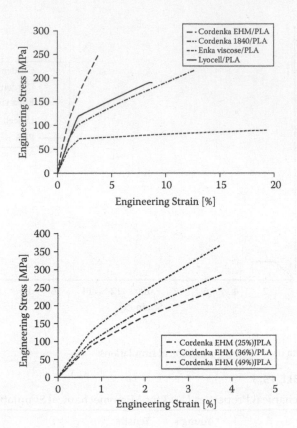

FIGURE 15.6
Engineering stress–strain curves for bio-composites obtained from micromechanical simulation.

15.3.4 Analytical Modeling

As an alternative to the direct application of the finite element method, the effect of the microstructure can also be described analytically. This section gives a brief overview of analytical methods for describing composite materials.

Classical rules for calculating elastic constants are the Voigt and Reuss-models that allow calculation of effective composite properties based on the properties of the constituents and their volume fractions. Self-consistent methods such as the composite cylinder assemblage (CCA), self-consistent scheme (SCS), and generalized self-consistent scheme (GSCS) define axial-symmetric models of fibers and surrounding matrices, allowing the calculations of transversely isotropic material properties suitable for UD composites.

Based on the self-consistent methods, Halpin and Tsai derived analytical formulations of effective elastic constants for continuous fiber and discontinuous (short) fiber-reinforced composites.[42,43] The Eshelby[44] and Mori-Tanaka methods[45] are very general analytical solutions. An alternative that has drawn attention in recent years is the method of cells (MOC) developed by

Aboudi and its derivatives known as the generalized method of cells (GMC) and the high-fidelity generalized method of Cells (HFGMC).[46] These methods have been applied successfully to modeling carbon composites subjected to quasi-static and impact loading; see May et al.[47] An application to modeling UD-bio-composites is possible. More detailed summaries of analytical micromechanics can, for example, be found in Daniel and Ishai[1] or Nossek.[48]

15.4 Conclusions

Micromechanical simulations require good knowledge of the micromechanical properties of the constituents, the fiber–matrix interface, and the architecture. Information about micromechanical properties can be obtained from mechanical testing. The architecture can be analyzed using computed tomography or micrographs.

With this information available, analytical or numerical modeling approaches can be followed for predicting the properties of composites based on micromechanical information. Micromechanical modeling of bio-based composites can be a powerful tool for a priori assessments of various combinations of fibers and resins and composite architectures, thus reducing the required amount of mechanical testing and associated costs.

References

1. I.M. Daniel and O. Ishai. *Engineering Mechanics of Composite Materials*. New York, Oxford University Press, 2006.
2. European Committee For Standarization. DIN EN ISO 5079. Determination of Maximum Force and Elongation at Break of Textile Fibers. Berlin, German Institute for Standardization, 1995
3. R.D. Reumann. Faserzugdeformationsverhalten. In *Prüfverfahren in Der Textil und Bekleidungstechnik*, R.D. Reumann, Ed. Heidelberg, Springer, 200, pp. 180–184.
4. C.D. Delholm, X. Cui, and D.P. Thibodeaux. Single-Fiber Testing via Favimat. Paper presented at National Cotton Council Beltwide Cotton Conference, January 2010.
5. S.J. Eichhorn and R.J. Young. Deformation Micromechanics of Natural Cellulose Fibre Networks and Composites. *Composites Science and Technology*, 63(2003): 1225–1230.
6. W.H. Müller. Mikromechanik: Wann Weniger Mehr Ist. *GAMM*, 2 (2012): 6–11.
7. ASTM D1577-07. Standard Test Methods for Linear Density of Textile Fibers. West Conshohocken, PA, American Society for Testing and Materials, 2012.

8. European Committee For Standarization. DIN EN ISO 527-2. Determination of Tensile Properties of Plastics. Test Conditions for Moulding and Extrusion Plastics. Part 2: Test Conditions for Molding and Extrusion Plastics. Berlin, German Institute for Standardization, 1993.

9. N. Graupner, A.S. Herrmann, and J. Müssig. Natural and Man-Made Cellulose Fibre-Reinforced Poly(Lactic Acid) (PLA) Composites: An Overview of Mechanical Characteristics and Application Areas. *Composites Part A,* 40 (2009): 810–821.

10. M.S. Sohn and X.Z. Hu. Impact and High Strain Rate Determination Characteristics of Carbon Fibre Epoxy Composites. Theoretical and Applied Fracture Mechanics, 25, no. 1 (1996); 17–29.

11. S.L. Yang, Z.H. Wu, W. Yang et al. Thermal and Mechanical Properties of Chemical Crosslinked Polylactide (PLA). *Polymer Testing,* 27 (2008): 957–963.

12. European Committee For Standarization. DIN EN 6036. Fibre-Reinforced Plastics Test Method: Determination of Notched, Unnotched, and Filled Hole Compression Strength. Berlin: German Institute for Standardization, 1996.

13. European Committee for Standarization. DIN EN ISO 178. Plastics: Determination of Flexural Properties. Berlin: German Institute for Standardization, 2003.

14. M. Arcan, Z. Hashin, and A. Voloshin. Method to Produce Uniform Plane-Stress States with Applications to Fiber-Reinforced Materials. *Experimental Mechanics,* 18 (1978): 141–146.

15. G.P. Pucillo, M. Grasso, F. Penta et al. On the Mechanical Characterization of Materials by Arcan-Type Specimens. *Engineering Fracture Mechanics,* 78 (2011): 1729–1741.

16. L.T. Drzal, P.J. Herrera-Franco, and H. Ho. Fiber–Matrix Interface Tests. In *Comprehensive Composite Materials,* A. Kelly and C. Zweben, Eds. Oxford: Elsevier, 2000, pp. 71–111.

17. S. Ogihara and J. Koyanagi. Investigation of Combined Stress State Failure Criterion for Glass Fiber–Epoxy Interface by the Cruciform Specimen Method. *Composites Science and Technology,* 70 (2010): 143–150.

18. L.J. Broutman. Interfaces in Composites, STP 452. Philadelphia, American Society for Testing and Materials, 1969.

19. A. Kelly. Interface Effects and the Work of Fracture of a Fibrous Composite. *Proceedings of Royal Society of London A,* 319 (1970): 95–116.

20. C. Difrancia, T.C. Ward, and R.O. Claus. Single-Fibre Pull-Out Test. 1: Review and Interpretation. *Composites Part A,* 27 (1996): 597–612.

21. S. Mukhopadhyaya, R. Pal, V. Narula et al. Study of Interface Behavior in Sisal Fibre Composites: Single-Fibre Pull-Out Test. *Indian Journal of Fibre & Textile Research,* 38 (2013): 87–91.

22. J. Koyanagi, H. Nakatani, and S. Ogihara. Comparison of Glass–Epoxy Interface Strengths Examined by Cruciform Specimen and Single-Fiber Pull-Out Tests under Combined Stress State. *Composites Part A,* 43 (2012): 1819–1827.

23. H. Voß. *Visualisierung Von Spannungsverteilungen in Faserverstärkten Kunststoffproben.* Freiburg, Hochschule Offenburg, 2012.

24. Z. Hashin. Analysis of Composite Materials. *Journal of Applied Mechanics,* 50 (1983): 481–505.

25. R. Hill. Self-Consistent Mechanics of Composite Materials. *Journal of the Mechanics and Physics of Solids,* 13 (1965): 213–222.

26. C.T. Sun and R.S. Vaidya. Prediction of Composite Properties from a Representative Volume Element. *Composites Science and Technology,* 56 (1996): 171–179.

27. R. Rolfes, G. Ernst, M. Vogler et al. Material and Failure Models for Textile Composites. In *Mechanical Response of Composites*. Computational Methods in Applied Sciences Series. Heidelberg, Springer, 2008, pp. 27–56.
28. A.A. Gusev, P.J. Hine, and I.M. Ward. Fiber Packing and Elastic Properties of a Transversely Random Unidirectional Glass–Epoxy Composite. *Composites Science and Technology*, 60 (2000): 535–541.
29. C. González and J. Llorca. Mechanical Behavior of Unidirectional Fiber-Reinforced Polymers under Transverse Compression: Microscopic Mechanisms and Modeling. *Composites Science and Technology*, 67 (2007): 2795–2806.
30. S.M. Kilchert, M. May, and S. Hiermaier. Modellierung Des Einflusses Von Unregelmäßigkeiten in Der Mikrostruktur Auf Das Versagen Von Ud Composites. Paper presented at NAFEMS Deutschsprachige Konferenz, Bamberg, Germany, 2012.
31. S.N. Kilchert, M. Nossek, M. May, et al. Influence of the Microstructure on the Macroscopic Responses of Long-Fiber Composites. In *Proceedings of Joint Sheffield–Cambridge Conference on Deformation and Fracture of Composites*. Cambridge, Queens' College, 2011.
32. M. Shibata, K. Ozawa, N. Teramoto et al. Biocomposites Made from Short Abaca Fiber and Biodegradable Polyesters. *Macromolecular Materials and Engineering*, 288 (2003): 35–43.
33. R. Burgueño, M. Quagliata, G.M. Mehta et al. Sustainable Cellular Biocomposites from Natural Fibers and Unsaturated Polyester Resin for Housing Panel Applications. *Journal of Polymers and the Environment*, 13 (2005): 139–149.
34. L.J. Da Silva, T.H. Panzera, A.L. Christoforo et al. Numerical and Experimental Analyses of Biocomposites Reinforced with Natural Fibres. *International Journal of Materials Engineering*, 2 (2012): 43–49.
35. M.S. Huda, L.T. Drzal, A.K. Mohanty et al. Effect of Fiber Surface Treatments on the Properties of Laminated Biocomposites from Poly(Lactic Acid) (PLA) and Kenaf Fibers. *Composites Science and Technology*, 68 (2008): 424–432.
36. S.M. Lee, D. Cho, W.H. Park et al. Novel Silk–Poly(Butylene Succinate) Biocomposites: Effect of Short Fibre Content on Their Mechanical and Thermal Properties. *Composites Science and Technology*, 65 (2005): 647–657.
37. L.T. Harper, C. Qian, T.A. Turner et al. Representative Volume Elements for Discontinuous Carbon Fibre Composites. 1: Boundary Conditions. *Composites Science and Technology*, 72 (2012): 225–234.
38. L.T. Harper, C. Qian, T.A. Turner et al. Representative Volume Elements for Discontinuous Carbon Fibre Composites. 2: Determining Critical Size. *Composites Science and Technology*, 72 (2012): 204–210.
39. M. Reinhardt, J. Kaufmann, M. Kausch et al. PLA–Viscose Composites with Continuous Fibre Reinforcement for Structural Applications. *Procedia Materials Science*, 2 (2013): 137–143.
40. L. Suryanegara, A.N. Nakagaito, and H. Yano. Effect of Crystallization of PLA on the Thermal and Mechanical Properties of Microfibrillated Cellulose-Reinforced PLA Composites. *Composites Science and Technology*, 69 (2009): 1187–1192.
41. M.R. Wisnom. Modelling Discrete Failures in Composites with Interface Elements. *Composites Part A*, 41 (2010): 795–805.
42. J.C. Halpin. Effects of Environmental Factors on Composite Materials. DTIC Document, 1967.

43. J.C. Halpin. Stiffness and Expansion Estimates for Oriented Short Fiber Composites. *Journal of Composite Materials*, 3 (1969): 732–734.
44. J.D. Eshelby. Elastic Field Outside an Ellipsoidal Inclusion. *Proceedings of Royal Society of London A*, (1959): 561–569.
45. T. Mori and K. Tanaka. Average Stress in Matrix and Average Elastic Energy of Materials with Misfitting Inclusions. *Acta Metallurgica*, 21 (1973): 571–574.
46. J. Aboudi. Generalized Method of Cells and High-Fidelity Method of Cell Micromechanical Models: A Review. *Mechanics of Advanced Materials and Structures*, 11 (2004): 329–366.
47. M. May, N. Nossek, N. Petrinic et al. Adaptive Multi-Scale Modeling of High Velocity Impact on Composite Panels. Composites Part A: Applied Science and Manufacturing, 58, no. 1 (2014): 56–64.
48. M. Nossek. Multiscale Modeling of Impact Loads on Fiber Composite Laminates: Method Development, Parameter Identification and Application. München, Universität der Bundeswehr, 2010.

16

Life Cycle Assessments of Bio-Based Composites: A Review

Thimothy Thamae and Caroline Baillie

CONTENTS

16.1 Introduction

For nearly two decades, the composite community has turned its attention to the development of bio-based composites, often called natural fiber composites (NFCs). From inception, bio-based composite activities have been running on their environmental credentials when compared to conventional composites.[1] Consequently, a number of life cycle assessments (LCAs) of NFCs have emerged. LCA is a tool for evaluating the environmental impacts associated with products throughout their life cycles (from extraction of raw materials to the ends of their lives—from the cradle to the grave).[2,3] Generally, these studies attempt to provide the bases for the environmental claims often associated with NFCs. A range of very broad to very specific reasons are cited as the driving factors behind LCAs. We begin this chapter by taking a brief look at a few of the reasons.

NFCs are seen as exerting huge influence due to their scale, variety of product applications, and growth potential.[4] Due to the reliance of these composites on natural fibers and recent developments as bio-based resins, they are viewed as "eco-friendly" and often marketed as such. Eco-friendly labeling is becoming increasingly attractive to consumers. Therefore, it is logical that such an influential group of products running on environmental credentials be scrutinized with appropriate tools to substantiate the perceived environmental claims. Life cycle analyses (LCAs) serve as the best available tools for that purpose.[5,6]

In a typical LCA used to assess a product containing NFCs, one question often asked is whether the product is genuinely environmentally friendly *because* it is made from natural materials.[1,7] For instance, while natural fibers are seen as "natural," their cultivation may involve heavy use of "unnatural" substances such as pesticides, artificial fertilizers, and fossil-fuel dependent machinery—the abilities of all these components to pollute are well documented. As another example, natural fibers are seen as lightweight in comparison to dense, heavy conventional counterparts such as fiberglass. Thus, NFCs are seen as contributing to good fuel efficiency in automotive applications. However, it has been shown that lightweight fibers based on carbon and magnesium carry huge environmental burdens due to the way in which they are produced. Are natural fibers different?

Other reasons for conducting LCAs of NFCs often cited by academics are needs need to fill gaps in the literature. Pervaiz and Sain[8] suggested that while plenty of technical studies show NFCs as viable replacements of conventional composites, only a few studies provide an environmental basis for the perceived need. Also, in such a large field as NFCs, more publications would be needed to reflect specific focal areas. For instance, while providing an argument for a need to carry out an LCA of hardboard, González-García et al.[4] argued that no LCAs addressed green hardboards.

Even within the same NFC area, some authors cite specific fibers or matrices as having not received attention in previous LCAs.[9] Following a similar argument, Müssig et al.[10] observed the increasing use of bio-based instead of petroleum-based resins and concluded that LCAs for fully bio-based composites were needed.

In many cases, very broad arguments have also been presented to justify why many scholars and practitioners carry out LCAs of NFCs. These arguments range from contributing to regional environmental efforts, meeting international and environmental obligations (e.g., the Kyoto Protocol)[6] to influencing governmental policies that encourage the use of greener products.[1,11]

In this chapter, we take a closer look at recent literature on the LCAs of NFCs. We detail the attempts made in a number of studies to adhere to the basics of LCA while trying to present pictures of the environmental profiles of NFCs. The LCA steps of goal definition, functional unit and system boundaries, data, and impact assessment are compared and common trends are revealed. Subsequently, we show what a number of LCAs in the area of NFCs reveal about the environmental credentials of these products.

16.2 Methodological Approaches of Studies

16.2.1 Goal and Scope

According to UNEP (1996)[12] guidelines, the goal of an LCA must be stated precisely to leave no room for ambiguity. A clearly stated goal helps define and delineate the boundaries of a study. For most studies, LCA goals may include, for example, comparisons of two products, product and process development, decisions on buying, structuring, and building up information, eco-labeling, environmental product declarations, and decisions on regulations.[12]

Academic LCAs, like those examined in this study, often focus on a few of those areas, especially environmental product declarations and product comparisons. In examining the present LCA literature on NFCs, it is clear that the stated goals can be characterized by three main attributes (Table 16.1):

1. Those with broad overriding goals such as assessing the ecological impact of a product.[9]
2. Those that start with broad statements but include specific sub-goals. For example, after stating the main goal as assessment of the impact of replacing fiberglass composites with wood fiber composites, Thamae and Baillie[1] cited a sub-goal as identifying major environmental impact categories.
3. Those that are more specific, for example, assessing greenhouse gas (GHG) emissions of NFCs.[11]

It has been observed the first two choices of goal definition utilize a number of environmental impact indicators. In the third choice, a specific environmental impact indicator is considered sufficient. For example, only one indicator (GHG emissions or energy use) is involved. Thus, the extent of goal definition tends to show how an LCA becomes holistic. However, in most cases, being holistic does not necessarily mean being accurate; in fact, the opposite may be true in many cases.

Generally, the levels of goal definitions above are confined within two main areas. Most LCAs seek to make declarations concerning the environmental credentials of NFCs. They also seek to compare NFCs against conventional composites. Fiberglass composites are most commonly compared with NFCs, reflecting their widespread use in the composite industry (Table 16.1).

TABLE 16.1

Examples of Goals Stated in Recent Literature on Life Cycle Analyses of NFCs

Level	Initial Goal	Compared with	Reference
Broad	Assess ecological impact of using Curauá composites for automobile applications	Fiberglass composites	9
	Analyze manufacture of green hardboard from LCA perspective	Conventional hardboard	4
	Assess environmental impact of using jute fiber composites and their necessary technical treatments	Fiberglass composites	7
	Define energy and environmental profile of insulating NFC board	Other insulating boards	6
	Analyze wood fiber composite performances	PP	13
	Assess impacts of using recycled instead of virgin PP in NFCs	Recycled and virgin PP	19
Broad and with sub-goals	Assess impacts of replacing fiberglass composite with wood fiber composite; identify major environmental impact categories; identify processes and substances that contribute more to major impact categories	Fiberglass composites	1
	Compare environmental impacts of sugarcane bagasse–PP composites with those of talc–PP composites; identify key environmental parameters and phases over entire lives of composite materials; study various end-of-life disposal scenarios of NFCs	Talc–PP composites	14
Specific	Investigate carbon sequestering abilities of natural fibers used in NFCs	Fiberglass composites	8
	Investigate GHG emissions of petrochemical and natural fiber mattresses	Petrochemical and natural fiber mattresses	11

16.2.2 Functional Unit and System Boundaries

16.2.2.1 Functional Unit

A functional unit is often defined as a measure of performance of a product.[3] It forms the basis upon which all inputs, outputs, and processes are related. In practice, it measures the amounts of a product or service needed to perform a particular function. Clarifying the functional unit is very important in cases where environmental impacts of two or more products are compared. LCA results largely depend on the choice and definition of a functional unit. Only products that perform the same function should be compared. Nevertheless, the choice of a functional unit may be subjective since one product can fill more than one function (e.g., a single product can perform both technical and social functions: "it is strong" and "it looks and feels good"). No doubt the second function is the most subjective but no less important.

As seen in the definition, a functional unit may consist of two related factors: amount and function. In the literature covering LCAs of NFCs, authors relate these two factors in a number of different ways, sometimes with and often without justification.

The most common scenario is where a function is defined and the amount needed to satisfy the function is provided. For instance, for a particular composite, we might ask what mass, volume, or area is needed to satisfy a certain strength requirement? A number of studies followed this approach.

Xu et al.[13] defined a functional unit as "material service density," which meant the volume of material leading to a specific strength. Clearly, different volumes of the different materials compared (NFCs based on polypropylene [PP] and pure PP in this case) would be needed to satisfy the requirement. Alves et al.[7] similarly defined a functional unit as an engine cover of 0.35 m² needed to achieve the required mechanical and structural performance.

Some authors are even more specific. Adente et al.[6] defined a functional unit as a mass of insulation board that provided a thermal resistance R of 1 (m³/KW). In a functional unit, sometimes strength is expressed in terms of life span of a product. The underlying assumption is that stronger products have longer life spans and vice versa. In Thamae and Baillie,[1] the functional unit was defined as a car door panel of 992 cm³ that would service a car body's life span of 200,000 km. Similar arguments were made by Luz et al.[14]

Interestingly, any number of variables can be used in combination to define a functional unit. Glew et al.[11] identified some of these variables, including durability, life span, quality, comfort, cost, and size. As an example of how these variables may be used in a combination, Glew et al. expressed their own functional unit as 1 m³ of a pocket spring mattress that could last for 10 years and cost £1500 (price is assumed to express quality in this case). Again, note that the ability of a mattress to withstand certain repeated use or stress is implied in its life span (10 years).

After stating this kind of functional unit, a question is: What features of the NFC under study and the composites with which the NFC is compared would help it meet the stated requirement? Such features could be fiber content variation, matrix properties, and choices of sizes and shapes.

Other examples of implied functions are shown by Zah et al.[9] For example, when a functional unit as a basis for comparison of composites is based on equal stability, it is immediately clear that stability is a function. However, when a unit is based on equal weight, other functions may be implied even not clearly stated.[8] According to Zah's group,[9] equal weight comparisons are often used when the look or feel of a composite is important or when cost reduction is a vital factor. When a unit is based on equal volume, the researchers may be interested on how well the composites will help reduce weight. The work of Zah et al.[9] shows how LCA results are affected by the choice of a functional unit.

16.2.2.2 System Boundaries

Drawing of system boundaries delineates the extent of an assessment. LCA practitioners decide which inputs, outputs, and processes to include or exclude.[2] Goedkoop and Oele[15] explained why this stage is a problematic when they stated, "In an LCA of milk cartoons, trucks are used; to produce trucks, steel is needed; to produce steel, coal is needed; to produce coal, trucks are needed. One cannot trace all inputs and outputs." Drawing clear system boundaries will help clarify the boundaries of data collection.

In the surveyed literature, most studies make general statements in this area. The statements generally suggest that LCAs cover entire life cycles (cradle to grave analyses).[1,4,6,7,9] An entire life cycle is often seen as three main stages: production, use, and end-of-life stages; the relationships of the stages are normally shown as diagrams. Perhaps due to the widespread reliance of most studies on commercial databases, few studies go into the details of system boundaries to seriously answer tough questions that must be addressed before data collection.

Some studies focus on only one of the three main life cycle stages. After analyzing previous studies, Xu et al.[13] concluded that the environmental credentials of NFCs compared to their conventional counterparts were not conclusive when only the production phase was considered. Furthermore, it became clearer that NFCs were environmentally superior when use and end-of-life scenarios were included. Therefore, they argued that a focus on the production stage could prove a key factor in understanding the environmental standings of NFC, that is, a focus on this stage alone could clear the confusion by providing more conclusive results. They carried out an LCA to ascertain the environmental standing of wood fiber-reinforced PP composites by conducting an LCA for the production phase only.

It is clear from this analysis that it is possible to select boundary conditions that reinforce the analyst's desired outcome such as might be expected in marketing a particular material or product. Therefore, refinement of standards may allow better comparisons to be made and thus enhance the reliability of LCA data in predicting environmental impacts.

16.3 Data Inventory

A list of emissions and use of resources per functional unit is called an inventory. Due to a large number of factors involved, LCAs are very data intensive

TABLE 16.2

Sources of LCA Dated Cited in Surveyed Studies

Cited Sources	Reference
Ecoinvent database, system	9
Published literature, local company records	10
Onsite measurements, Ecoinvent database, published literature, ETH-ESU 96 database	4
Ecoinvent and IDEMAT databases	19
Company records, Ancel database, government reports, experimental data, published literature	7
GEMIS, ANPA, and Bousted databases, field data, published literature	6
BUWAL 250 database	
Office of National Statistics, Econinvent database	11
BUWAL 250, Ecoinvent, ETH-ESU, IDEMAT, Franklin USA 98 databases, field data	18

and require extensive resources. Clearly, most academic LCAs cannot afford such extensive data collections. The problem has been solved through consultation of a large number of commercial databases, most of which are embedded within LCA software. Understandably, these databases cannot be specific enough to address every situation despite the fact that some tend to contain extensive information about generic products. Thus, they may have to be supplemented with system specific data.

Other options include a careful selection of similar processes from the databases (e.g., by using data about glass wool production if fiberglass production data are not available) or identifying and omitting less significant processes.[16] Other studies supplement commercial databases by including data from published literature, experimental results, on-site field analyses, government reports, and other sources as shown in Table 16.2.

16.4 Impact Assessment

Not all LCA analysts conduct impact assessments because such assessments involve even more difficulties in obtaining reliable results. These assessments determine the impacts of products on predetermined categories affecting three main areas (ecosystem quality, human health, and resource depletion).[17]

The argument is that since different substances in the same quantities may exert different degrees of impacts on a specific environmental attribute, it is important to know their relative contributions. For instance, the global warming potential of nitrous oxide (N_2O) is 310 times that of carbon dioxide (CO_2), so an impact assessment should go beyond stating the quantities of these substances and explain their significance.

TABLE 16.3

Impact Assessments and Software Used in Surveyed Literature

Impact Assessment Method	Software Used	Reference
CML, Ecoindicator 99	Simapro®	18
UBA/IFEU	Umberto®	10
CML 2 Baseline 2000	Simapro®	4
Cumulative Energy Demand	Simapro®	19
Ecoindicator 99	Simapro®	7
CML 2001	Gabi®	14
Ecoindicator 99	Simapro®	13
Ecoindicator 99 (Egalitarian Version)	Simapro®	1

However, trying to connect inventory results to impacts on the environment has proved to be a big challenge. The results may not demonstrate spatial, temporal, and dose–response relationships or explore threshold dimensions that represent problems in real-life situations. Thus, these impact assessment methods have to rely on simplification.[17] One problem with impact assessment methods is that they use the criterion that a higher quantity of a substance exerts a greater impact on the environment. However, beyond certain thresholds, the quantity of a material will not change the damage caused. Nevertheless, the results still give a picture that no other available data can provide.

Most of the surveyed studies involved a number of public impact assessment methods that normally come with LCA software. Also, most of the methods used are said to be based on a damage-oriented approach. CML and Ecoindicator 99 were the most widely cited in the surveyed literature and Simapro software the most widely used (Table 16. 3). Some studies did not refer to any specific impact assessment method.

16.5 Lessons Learned

In this section, we consider a number of the surveyed studies and what they reveal about the nature of NFCs and their environmental impacts. The studies reveal some common trends and interesting differences.

16.5.1 Relationships of LCA Results and Functional Units

We have shown that an appropriate choice of a functional unit is important because it constitutes a major influence on the results of an LCA. The surveyed literature shows the importance of this component of the analysis.

Zah et al.[9] studied the environmental implications of replacing fiberglass with Curauá fiber as a reinforcement in polypropylene (PP) matrices in automotive applications. They found that when equal volumes of the composites were compared, the Curauá composites revealed better environmental properties than glass composites in all impact categories because they were 25% lighter. However, when stabilities were compared, the composites showed no significant differences. When the two composites were compared on the basis of equal mass, they still displayed no significant differences in most categories with the exception of four toxicology issues on which the Curauá composites were superior.

Other studies showed similar results. For instance, Xu et al.[13] studied the benefits of replacing pure PP with wood fiber–PP composites for a number of applications in the construction and automotive areas. They concluded the NFC performed better than pure PP on the basis of equal mass. When equal volumes and strengths were the bases, the NFCs performed even better environmentally.

In the case, where jute fibers were to replace fiberglass in a polyester resin composite for use in lighting columns, the functional units chosen were composites having the same bending stiffness. To satisfy this precondition, the jute fiber composite had to be thicker.[18] As a result, the jute composite required more polymer matrix than the fiberglass composite and thus produced more impact in the climate change category than the fiberglass composite.

16.5.2 Dominance of Use Phase in Automotive Applications

Another interesting phenomenon is the influence of the use phase in automotive applications. In many cases, this component dominates environmental impacts of whole life cycles and demonstrates why NFCs perform better in such applications. The lightness of NFCs contributes greatly to fuel efficiency during the use phase. The longer the life span of a car, the better the performance of NFCs. Zah et al.[9] observed that use phase dominated all impact categories for both Curauá and fiberglass composites. They found that the use phase contributed to 89% of the impact in the climate change impact category in the case of NFCs and contributed to 84% in the case of fiberglass composites.

Thamae and Baillie[1] reported similar observations. They demonstrated that the use phase contributed to 86% of the total environmental impact in the case of wood fiber–PP composites and 85% in the case of fiberglass composite in automotive applications.

In the case of Zah et al., it is the contribution of the impacts to use phase to a specific impact category. That is, climate change category. In the case of Thamae and Baillie, it is the contribution of impacts of use phase to the total environmental impacts of a composite. In this case, the total impacts were a combined contribution of three phases, i.e., for NFC, assembly (13.9%), use (86%), disposal (0.094%).

The dominance of the use phase stands in sharp contrast to the end-of-life phase in most cases. Although they are often included in the analyses, end-of-life scenarios often make insignificant contributions to the impacts over an entire life cycle[9,13] with some exceptions.

16.5.3 Role of Matrix

The role of the matrix in influencing the environmental impacts of NFCs is so significant that some authors have suggested that matrix rather than the fiber should receive the most attention.[5] In one study, the performance of kenaf and thermoplastic polyester fibers press-melted together into a composite was compared with performances of other conventional insulating boards.[6] In the case of kenaf composites, polyester fibers scored the highest environmental impacts (39% of the total).

In all other conventional boards, plastic matrices produced the greatest impacts on the environment. Again, in a comparison of wood fiber–PP composites against pure PP studied by Xu et al.,[13] the PP matrix dominated the impact. Pervaiz and Sain[8] and Thamae and Baillie[1] concluded that one reason NFCs performed better than conventional composites such as those containing fiberglass was the low density of natural fibers. Natural fibers occupy more space per composite volume. As a result NFCs use less polymer in matrices than conventional composites.

The polluting nature of traditional matrices led a number of authors to turn their attention to the use of bioplastics and waste plastics as replacements for petroleum-based virgin plastics. Müssig et al.[10] compared a bus body component based on hemp fibers in a PTP™ bio-resin matrix against a fiberglass–polyester composite in the same application. The hemp–PTP composites exhibited better environmental performance in all impact categories. In the case of fiberglass polyester composite, polyester resin was seen as the largest contributor to environmental impacts in all categories. The environmental benefits of replacing conventional resins with bio-based resins were also noted by González-García et al.[4] They assessed the impact of replacing a phenol–formaldehyde resin with one made from a wood-based phenolic material and phenol oxidizing enzyme.

16.5.4 NFCs and Environment

Our review shows that NFCs generally perform better than conventional composites in the area of environmental impact. The main reasons include less intense energy required during production, low density leading to fuel economy in automotive applications, and less use of polluting matrices.

However, it is important to note that NFCs seem to make most of their contributions on impact categories related to agricultural activities. For instance, Curauá fibers performed worse than fiberglass in the eutrophication impact category when composites of both fibers (under the basis of similar strength) were compared. The reason was the role of fertilizers needed for Curauá cultivation—the fertilizers are known for their impacts on eutrophication.

In the study of Müssig et al.,[10] a bio-based PTP resin composite performed worse than a polyester matrix composite in the areas of eutrophication, human toxicity, and acidification due to the agricultural nature of the components of the PTP composite. Simões et al.[18] found that a jute composite was also worse than a fiberglass counterpart in the impact category of land use.

16.6 Conclusions

Life cycle assessment is still the most widely used tool for assessing the environmental impact of a product from the production of the required raw materials to the end of its life despite a number of limitations as described in this chapter.

Ever since the emergence of NFCs as viable alternatives to conventional products such as fiberglass composites, NFCs were recommended because of their environmental benefits long before justifications were developed. However, an increasing number of recent studies compare environmental impacts of NFCs with the impacts of conventional composites. These studies indicate that NFCs generally show better environmental properties than most of their conventional counterparts. However, it is possible that the fragile and subjective LCA system opens the door to exaggerated claims involving choices and even manipulations of methodologies. Clear requirements for LCAs and transparent databases would start to mitigate this problem but the solution requires wider development of publicly available databases and broader ranges of data.

References

1. T. Thamae and C. Baillie. Life Cycle Assessment of Wood–Polymer Composites: A Case Study. In K. Oksman and M. Sain, Eds., *Wood–Polymer Composites*. Cambridge: Woodhead Publishing, 2008.
2. R. Murphy. *Life Cycle Assessment*. In C. Baillie, Ed., *Green Composites: Polymer Composites and the Environment*: Cambridge: Woodhead Publishing, 2003.
3. Society of Environmental Toxicology and Chemistry. Guidelines for Life-Cycle Assessment: A Code of Practice from Workshop at Sesimbra, Portugal, 31 March–3 April 1993.
4. S. González-García, G. Feijoo, C. Heathcote et al. Environmental Assessment of Green Hardboard Production Coupled with a Laccase-Activated System. *Journal of Cleaner Production*, 19 (2011): 445–453.
5. P.A. Fowler, J.M. Hughes, and R.M. Elias. Biocomposites: Technology, Environmental Credentials, and Market Forces. *Journal of the Science of Food and Agriculture*, 86 (2006): 1781–1789.

6. F. Ardente, M. Beccali, M. Cellura et al. Building Energy Performance: LCA Case Study of Kenaf-Fibre Insulation Board. *Energy and Buildings,* 40 (2008): 1–10.

7. C. Alves, P.M.C. Ferrão, A.J. Silva et al. Ecodesign of Automotive Components Making Use of Natural Jute Fiber Composites. *Journal of Cleaner Production,* 18 (2010): 313–327.

8. M. Pervaiz and M.M. Sain. Carbon Storage Potential in Natural Fiber Composites. *Resources Conservation and Recycling,* 39 (2003): 325–340.

9. R. Zah, R. Hischier, A.L. Leão, and I. Braun. Curauá Fibers in the Automobile Industry: A Sustainability Assessment. *Journal of Cleaner Production,* 15 (2007): 1032–1040.

10. J. Müssig, M. Schmehl, H.B. Von Buttlar et al. Exterior Components Based on Renewable Resources Produced with SMC Technology: Considering a Bus Component Example. *Industrial Crops and Products,* 24 (2006): 132–145.

11. D. Glew, L.C. Stringer, A.A. Acquaye et al. How Do End of Life Scenarios Influence the Environmental Impacts of Product Supply Chains? Comparing Biomaterial and Petrochemical Products. *Journal of Cleaner Production,* 29–30, (2012): 122–131.

12. UNEP. *Life Cycle Assessment: What It Is and How to Do It.* Paris, United Nations Environmental Programme, 1996.

13. X. Xu, K. Jayaraman, C. Morin, and N. Pecqueux. Life Cycle Assessment of Wood Fibre-Reinforced Polypropylene Composites. *Journal of Materials Processing Technology,* 198 (2008): 168–177.

14. S.M. Luz, A. Caldeira-Pires, and P.M.C. Ferrão. Environmental Benefits of Substituting Talc by Sugarcane Bagasse Fibers as Reinforcement in Polypropylene Composites: Ecodesign and LCA Strategy for Automotive Components. *Resources Conservation and Recycling,* 54 (2010): 1135–1144.

15. M. Goedkoop and M. Oele. *Introduction to LCA with Simapro.* Amersfoort, Product Ecology Consultants, 2004.

16. T. Corbière-Nicollier, B. Gfeller-Laban, L. Lundquist et al. Life Cycle Assessment of Biofibres Replacing Glass Fibres as Reinforcement in Plastics. *Resources, Conservation and Recycling,* 33 (2001): 267–287.

17. J.J. Daniel and M.A. Rosen. Exergetic Environmental Assessment of Life Cycle Emissions for Various Automobiles and Fuels. *Exergy: An International Journal,* 2 (2002): 283–294.

18. C.L. Simões, L.M.C. Pinto, and C.A. Bernardo. Modelling the Environmental Performance of Composite Products: Benchmark with Traditional Materials. *Materials & Design,* 39 (2012): 121–130.

19. A. Bourmaud, A. Le Duigou, and C. Baley. What Is the Technical and Environmental Interest in Reusing a Recycled Polypropylene–Hemp Fibre Composite? *Polymer Degradation and Stability,* 96 (2011): 1732–1739.

17

Bio-Based Composites: Future Trends and Environmental Aspects

V.P. Sharma

CONTENTS

17.1 Introduction

Plastics have transformed the quality of life and molded modern civilization. It is beyond expectation to visualize any activity in which polymers do not play key roles in medicine and biomedical instrumentation, automobiles, electronics, entertainment, architecture, defense, housing, disposable food packaging, transportation, communications, and aerospace applications. Plastics may be super tough, rigid or flexible, transparent to opaque, allow selective permeation or act as barriers to meet specific customer demands and expectations.

Bio-based composites and nanocomposites have attracted the attention of academicians and entrepreneurs interested in developing advanced materials for multiple applications, for example, hybrid polymers, inorganic fullerene [IF]-like nanoparticles providing high impact resistance, super-tribological behavior, and hollow quasi spherical shapes. Hybrid polymers can self-assemble into nanoscale aggregates in selective solvents and form nanostructures in bulk.

Development of novel fluoropolymers for membranes and their fabrication with specialized features have been reported. The composite material industry in Brazil, Russia, India, and China (BRIC countries) alone is expected to be worth over U.S. $27 billion by 2016. The annual growth rate for the past 3 years has been considerable is projected at 18 to 20% per year. Production of composite materials promotes rural economic development by providing steady incomes for farmers and rendering social benefits. It also uses fewer fossil fuels than production of petrochemical plastics, even considering the fuel needed to plant and harvest the required materials.

Globally, the growing population and consumption of plastics reveal an increasing pattern as plastics offer cost-effective alternatives despite great concerns about non-biodegradability and health and environmental issues. The manufacture of plastics from hydrocarbons derived from non-renewable petroleum resources has raised questions concerning their sustainability. Conversely, the processing of many natural materials (glass, paper, jute, wood, metals) consumes far more energy. These materials require greater consumption of fossil fuels and may generate toxic leachates.

Future opportunities focus on making plastics from biomass and other renewable resources. Biomaterials are based on the direct uses or biotechnological modifications of starches, sugars, plant oils, cellulose and natural fibers from soy, corn, sugar beet, sugar cane, and bananas, and special rubber and lignin biomolecules. This group of biomaterials includes durable and biodegradable bio-based plastics, wood polymer composites (WPCs), natural fiber-reinforced plastics, and natural rubbers.

The use of polymer matrix composites as structural parts is expected to be a key technology of the future. Several technical issues, e.g., structural health monitoring, optimization, miniaturization, modeling, characterization, and non-destructive inspection must be resolved before such materials can be implemented into products. A single type material for a given structure is rarely realistic because of the principal stresses and the different functions of product structures.

Green chemistry using renewable raw materials has a business value that is complementary (as opposed to alternative) to traditional chemistry. The environmental performances of some renewable materials offer clear advantages over what synthetic plastics can offer (Figure 17.1). This chapter is a review of new bio-building blocks and biopolymers. We distinguish three types of products: those that already are commercialized or close to

FIGURE 17.1
Composites from natural raw materials.

commercialization, those under development, and those for which fundamental findings were published recently.

Sugar-based chemicals, for example, can be used as building blocks to produce new monomers, polymers, and additives for the commercial plastics industry. The monomeric products can be incorporated into the backbones of new polymers, converted to low molar mass additives for thermoplastics and thermosets, or marketed as specialty chemicals.

Despite their varying maturity levels, all these products are used actively in bio-based plastic industries. The compositions of bio-based composites (environmentally friendly resins and reinforcements) make them very efficient for applications that do not require high resistance. In construction, they can be used for interior components and decorations. New interest is emerging in developing new textures and appearances of construction materials.

Thermoplastic packaging materials have in recent decades faced increasing demands to improve their environmental profiles. As a consequence, significantly increased amounts of renewable materials have been used in packaging materials because they often exhibit green profiles due to their biodegradability.[1-3]

Reduced material consumption is also an important factor. Packaging materials with reduced wall thicknesses and efficient mechanical and/or gas barrier properties generate reduced carbon footprints and combine environmental friendliness with cost efficiency.

Another route to packaging materials with increased environmental profiles is the establishment of effective recycling procedures. Littering should be significantly reduced if users know the value of post-consumer packaging. Conventional thermoplastic packaging materials can be turned into biodegradable materials by the addition of prodegradant additives during processing into packaging products.[4-6] Prodegradants based on renewable materials such as fatty acid salts and environmentally friendly metals like iron are especially interesting. Thermoplastics containing such additives are fully recyclable.[7]

Carbon fibers have attracted worldwide attention as strong and lightweight composite materials for applications in the aerospace and automotive industries and as renewable energy resources. Carbon fibers are excellent reinforcing materials for inclusion in fiber-reinforced composites because of their mechanical strength of 5000 MPa, modulus value of 250 GPa, and density of 1.76 g·cm^3. Carbon fiber was invented in 1871 by using cellulose as a precursor.

Union Carbide developed industry applicable carbon fibers in the 1960s. In the early 1970s, the use of carbon fibers for aerospace and military applications started. Besides superior specific strength (strength/density), carbon fibers exhibit excellent electrical conductivity, shielding effect, and heat resistance properties.[8-10]

Three raw materials (precursors) are required for the production of carbon fibers. The first is cellulose, first used in 1871. Pitch and polyacrylonitrile (PAN) are two other materials required for production. Due to the final fiber properties required, carbon fiber produced from PAN is commonly used and constitutes 90% of the total carbon fiber available in the market.[9,10] The production capacity of carbon fiber is increasing as new composite manufacturing techniques are devised; thus the need for composites and carbon fibers is also increasing.

However, the prices of carbon fibers are not declining and the cost factor limits their widespread use.[11] One factor in the high cost is the precursor material. Research for suitable precursor materials is limited in academia. The research findings of PAN producers are not available to the public.

In the current carbon fiber industry, precursor cost is one of the main obstacles hindering widespread use of carbon fibers.[12] The PAN precursor material is a terpolymer of acrylonitrile, vinyl acetate and itaconic acid. The precursor processing stages include polymerization, PAN–solvent (dimethylacetamide) mixture preparation, and fiber spinning. After the PAN precursor is generated, the fibers are transformed to carbon fibers via oxidation and carbonization followed by certain surface treatments.

17.2 Development of Biodegradable Polymers

Transgenic tobacco and potato plants can accumulate high levels of cyanophycin, a possible source for polyaspartate. This work opens the way to the future production of biodegradable plastics using a plant-based production system (Figure 17.2). Several problems must be overcome first: growth retardation as a result of cyanophycin accumulation in plant cytosols and need for a co-production system for economic reasons. Biodegradable plastics can degrade completely in landfills, composters, or sewage treatment plants by the actions of naturally occurring microorganisms. Truly biodegradable plastics leave no toxic, visible, or distinguishable residues after degradation.

This full biodegradability feature contrasts sharply with those of most petroleum-based plastics that are essentially indestructible in a biological context. Because of the ubiquitous use of traditional petroleum-based plastics, their persistence in the environment, and their fossil-fuel derivation, more desirable alternatives are being explored. In this section, issues of waste management of traditional and biodegradable polymers are discussed in the context of reducing environmental pressures and carbon footprints.

There is a stigma associated with transgenic plants, especially food crops, plant-based biodegradable polymers produced as value-added co-products, and marginal land crops such as switchgrass (*Panicum virgatum* L.). Transgenic plants have the potential to become viable alternatives to petroleum-based plastics and are environmentally benign and carbon-neutral sources of polymers. The increased uses of renewable bio-based materials present several

FIGURE 17.2
Choices of natural products for degradable plastics or bioplastics.

potential benefits over finite resources for reducing green house gas emissions and other environmental impacts over whole life cycles. Biodegradable packaging materials are most suitable for single use disposable applications where post-consumer waste can be composted locally.

17.3 Strategic Bioeconomy Planning

Bioeconomy plans include a bio-based industries sector in which some oil-derived plastics and chemicals are replaced by new or equivalent products derived at least partially from biomass. Some of these bio-based products are available now, but fulfilling their societal potential requires greater public awareness and increasing their market share to enable these products to help mitigate climate change.

Each year over 140 million tons of petroleum-based polymers are produced worldwide and used to produce plastics such as polyethylene, polystyrene, polyvinyl chloride, polyurethane, and others. Oil, coal, and natural gas serve as raw materials for manufacturing these plastics that impact every facet of our lives. Approximately 10% of all the oil and gas produced and imported by the United States is used to produce synthetic plastics, and the market is expected to grow up to 15% per year. While these products hold up well in our fast-paced, throw-away society, they can remain in the environment 2000 years and longer.

In 2005, the U.S. Environmental Protection Agency (EPA) stated that of the 4.4 million tons of synthetic plastics discarded in the U.S., only 5.7% was recovered and recycled; the remainder ended up in landfills, lakes, and oceans. This statistic highlights the need for all-natural renewable replacements for petroleum-based plastics.

17.4 Bio-Based Composites from Agricultural Waste Residues

The main objective of this research was to study the potential of waste agricultural residues such as sunflower stalks, cornstalks, and bagasse fibers as reinforcements for thermoplastics as alternatives to wood fibers. Commercial and industrial bio-based products consist in whole or in significant part of biological products, renewable domestic agricultural materials, or forestry residues. Examples are adhesives, construction materials and composites, fibers, paper, packaging, fuel additives, landscaping materials, compost and fertilizer, lubricants and functional fluids, plastics, paints and coatings, solvents and cleaners, sorbents, and plant inks. Bio-based polymer composites derived from corn stovers and feather meals and used as double coating materials for controlled release and water retention urea fertilizers were synthesized by Yuechao Yang et.[13] and a few entrepreneurs.

17.5 Plastics Recyclability: New Opportunities

Plastics are inexpensive, lightweight, durable materials that can be molded readily into a variety of products that find use in a wide range of applications. Because of their utility, the production of plastics has increased markedly over the past 60 years. However, the high levels of use and difficulties of disposal generate several environmental problems. Around 4% of non-renewable oil and gas production worldwide serves as feedstocks for plastics and a further 3 to 4% is expended to provide energy to manufacture them. Most plastics are used in disposable packaging and other short-lived products that are discarded within a year of manufacture. These observations indicate that our current use of plastics is not sustainable.

In addition, because of the durability of the polymers involved, substantial quantities of discarded end-of-life plastics continue to accumulate in landfills and natural habitats worldwide. Recycling is one of the most important solutions available to reduce these impacts and represents one of the most dynamic areas in the plastics industry today. Recycling provides opportunities to reduce oil consumption, carbon dioxide emissions, and huge quantities of waste requiring disposal.

Here, we briefly set recycling into context against other waste reduction strategies, namely, decreasing material use through down gauging and product reuse, the use of alternative biodegradable materials, and energy recovery as fuel. While plastics have been recycled since the 1970s, the quantities recycled vary by geography, type, and application. Recycling of packaging materials has seen rapid expansion over recent decades in a number of countries. Advances in technologies and systems for the collecting, sorting, and reprocessing recyclable plastics are creating new opportunities for recycling. The combined efforts of the public, industry, and governments may make possible the diversion of most plastic wastes from landfills to recycling over the upcoming decades.

Plastic solid waste (PSW) presents challenges and opportunities to societies regardless of their sustainability awareness and technological advances (Figure 17.3). The four routes of PSW treatment are primary (re-extrusion), secondary (mechanical), tertiary (chemical), and quaternary (energy recovery) schemes and technologies. Primary recycling involving the re-introduction of clean scraps of single polymers to extrusion processes to generate products of similar material is commonly applied on processing lines but rarely applied by recyclers because of quality issues.

Municipal solid waste management (MSWM) is a major environmental problem of Indian cities. Improper management of municipal solid waste (MSW) creates hazards for inhabitants. Various studies reveal that about 90% of MSW is disposed of unscientifically in open dumps and landfills, endangering public health and the environment. Delhi is the most densely populated and urbanized city of India. The annual growth rate in population during from 1991 to 2001) was 3.85%—almost double the national average. Delhi is also a commercial hub, providing employment opportunities and accelerating the pace of urbanization, resulting in a corresponding increase in MSW generation.

The inhabitants of Delhi generally generate about 7,000 tonnes of MSW per day of MSW and the amount is projected to range from 17,000 to 25,000 tonnes daily by 2021. The population explosion and sustained drive for economic progress and development caused remarkable increases and changes in the quantities and characteristics of MSW generation over the past 20 years.

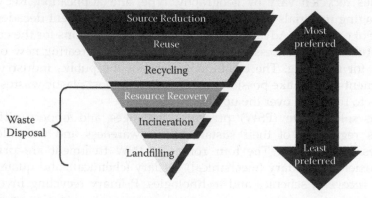

FIGURE 17.3
Plastic waste disposal: how safe?

In Portugal, 96% of MSW was collected mixed (4% was separately collected); 68% was disposed of in landfills, 21% was incinerated at waste-to-energy plants, 8% was treated at organic waste recovery plants, and 3% was delivered to sorting. The average generation rate of MSW was 1.32 kg per capita per day. Concerted efforts are needed to develop holistic measures and change mind-sets to combat plastic waste management. The extent of waste management differs from region to region around the globe.

17.6 Bio-Based Plastic Bottles and Economics

Bio-based polyethylene furanoate [PEF] bottles are now in production. A monomer of polyethylene terephthalate [PET], namely TPA, can be replaced by a bio-based version. A number of companies are working to develop cost-effective processes to produce bio-based TPA as a direct drop-in replacement for petroleum-based TPA. Moreover, PEF is said to have superior barrier and other performance properties.

The global bioplastics market will see a fivefold increase in production volume by 2016, from 1.2 million tons to an anticipated 6 million tons. Few companies in developed countries manufacture 100% plant-based water bottles that are toxin-free and carbon-neutral. Production of bottles containing plastic additives such as phthalate plasticizers, colorants, stabilizers such as bisphenol A (BPA), and organometallic compounds continues. A few manufacturers (e.g., Ingeo) developed biopolymers currently used in more than 20 types of lifestyle products.

Mesophase carbon fibers were invented in the 1970s and created great expectations in the carbon materials community as low-cost fibers with superior properties.[13,15]

Nowadays, geopolitical realities, diminishing supplies and greatly enhanced knowledge of sustainability are causing a shift to renewable feedstocks. For assessing the long term bioeffects of nanomaterials on living organism and the ecosystem the In vitro models are being developed to understand the adverse effects of engineered nanomaterials. Moreover, French Agency for Food, Environmental and Occupational Health and Safety (ANSES) during May 2014 has prepared a review to clarify scientific understanding and demonstrate the toxic effects of some nanomaterials on living organisms and the environment. The EN standards 13432, ASTM D 6400 and CFR Code of Federal Regulations (CFR section 177.1640), ISO, OECD, IS updated specifications for specific aspects should be consulted for compliance of International/National guidelines to assure that a product is compostable in the marketplace and meet the customers' demands or expectations.

Market analysts are predicting tremendous growth in the years ahead for biobased industry. The renewable chemical market is estimated to reach 83.4 billion by 2018. Biobased chemicals share may grow from 2% to 22% by 2025. In view of this adequate care is needed for strengthening and monitoring for better future of this sector.

Carbon fibers are also reusable, recyclable, and compostable in 80 days. One gallon of oil is saved by every 72 plant-based bottles produced; 65% less energy and fuel are required for production. The water required for manufacture is vapor-distilled for purity and when possible sourced within 500 miles of its destination to lower its environmental impact.

According to a market forecasts, partially bio-based PET will continue to lead the field; it currently accounts for ~40% of global bioplastics production capacity. Asia is predicted to be home to 46.3% of global bioplastics production capacity by 2016; South America will account for just over 45%, driven mainly by feedstock availability. Other factors impacting growth include robust market demand growth, relative scarcity of oil and gas, and supportive government policies.

However, several factors may inhibit the potential of biorenewable materials in the Asia Pacific region. Prices remain high, since application and technology development are ongoing processes. The low scale-up of manufacturing capacity also increases per-unit costs. In addition, the inferior performance attributes of bioplastics such as moisture absorption, low heat deflection temperature, and reduced resistance to chemical attacks limit their application range. Furthermore, the poor execution of eco-labeling policies and insufficient composting facilities in Asia Pacific countries will continue to restrict the potential applications of bioplastics.

Few bioplastics manufacturers develop novel thermoplastic formulations and resins using feedstock from renewable sources. Bioresins may derive up to 70% of their feedstocks from aquatic biomass obtained from nitrogen and phosphorus-rich wastewater blended with various concentrations of polyethylene (PE), PP, ethylene vinyl acetate (EVA), polylactic acid (PLA), thermoplastic starch (TPS), and polyhyroxyalkanoates (PHAs) for use in many end-use applications.

Companies are targeting agricultural and industrial operations such as livestock farms and wastewater treatment facilities as sources of low-cost nutrients for high productivity aquatic biomass cultivation. Few industries are developing customized bioplastics formulations for industrial, commercial, and retail applications. Biopolymers may be stable in landfills without releasing statistically significant quantities of methane.

For PLA and other bioplastics to be considered fully biodegradable, they must be able to biodegrade in marine environments. Initial composting studies have shown slight diminutions in the metabolic rates of compost organisms when PLA was added. This may be interpreted as slight toxicity of PLA. PHA is superior because it meets the ASTM standard for biodegradability in marine environments, whereas PLA may not degrade for a thousand years. According to experts, all land-based products and activities add nutrients to marine environments that are already on the way to eutrophication in urban and agricultural coastal areas. Biodegradable packaging materials are most suitable for single-use disposable applications where post-consumer waste can be composted locally.

17.7 Packaging Wastes and Sustainability

Packaging waste forms a significant part of municipal solid waste and led to increasing environmental concerns, resulting in strengthening of regulations aimed at reducing the amounts of such wastes generated. Among other materials, several oil-based polymers are used currently in packaging applications. All are virtually non-biodegradable and some are difficult to recycle or reuse because they are complex composites having varying levels of contamination.

Recently, significant progress has been made in the development of bio-degradable plastics, largely from renewable natural resources, to produce biodegradable materials with functionalities similar to those of oil-based polymers. These expanded uses of renewable (rather than finite) resources produces several potential benefits for greenhouse gas balances and other environmental impacts over whole life cycles.

It is intended that the use of biodegradable materials will contribute to sustainability and reduce the environmental impacts of disposal of oil-based polymers. There is a need to understand microstructure–mechanical property relations and characterization methods for natural fibers, processing-related issues, breeding methods for improving the fibers, interfaces in composites containing natural fibers, treatments to enhance durability, mechanical performance, and fiber–matrix adhesion, development of new bio-polymers, and performance levels of biopolymers and bio-nano reinforcements. One limitation facing bio-based plastics such as PLA is that they cannot be mixed with conventional plastics such as PET because the materials are not compatible for recycling purposes.

17.8 Biopolymers and Landfills: Future Strategies

Biopolymers operate on the concept of cradle-to-grave approach to zero waste. Their manufacture is based on environmental sustainability concepts. However, these systems are still emerging and developing. The environmental impact of greenhouse gas releases on landfills is not significant. Few biopolymers earned certifications for sustainable agricultural practices in growing feedstocks or for production from agricultural waste or recyclable materials. Nature Works is looking at sustainability from a 360-degree perspective: from sustainable agriculture to facilitating sustainable end-of-life scenarios. Consideration should be given to municipal zoning requirements, national health and safety requirements, and guidelines for transportation and use.

There is a need to improve the marketing strategies, quality, and brand visibility using state-of-the-art communication channels. Economic development may be feasible by creating value chains for delivery of services or finished products by forming alliances, clustering, and zealous networking. In the future, bioeconomies and bioplastics will be based on fundamentally less expensive and better distributed technologies. Our choices for structuring systems around biological technologies will determine the pace and effectiveness of innovation.

Bio-based plastics and composites have applications in food and household packaging and health care-related products. They are organic, environmentally friendly, new generation moieties. Continuous efforts are needed to select ideal raw materials from natural systems, improvise with technological innovations, share knowledge through networking and education, increase global usage of biocomposites, and remove barriers to further advancements. The future should present constant upgrading of polymeric composites to improve cost efficiency, viability, and environmental impacts.

References

1. D.D. Edie. Carbon Fiber Processing and Structure–Property Relations. In *Design and Control of Structures of Advanced Carbon Materials for Enhanced Performance*, B. Rand, S.P. Appleyard, and M.F. Yardim, Eds. Nato Science Series. Heidelberg, Springer, 2001, pp. 163–181.
2. D.D. Edie. The Effect of Processing on the Structure and Properties of Carbon Fibers. *Carbon*, 36 (1998): 345–362.
3. F. Gironi and V. Piemonte. Life Cycle Assessment of Polylactic Acid and Polyethylene Terephthalate Bottles for Drinking Water. *Environmental Progress & Sustainable Energy*, 30 (2011): 459–468.
4. H.H. Khoo and R.H. Tan. Environmental Impacts of Conventional Plastic and Bio-Based Carrier Bags. *International Journal of Life Cycle Assessment*, 15 (2010): 338–345.
5. C.K. Hong and R.P. Wool. Development of a Bio-Based Composite Material from Soybean Oil and Keratin Fibers. *Journal of Applied Polymer Science*, 95 (2005): 1524–1538.
6. J. M. Saiter, L. Dobircau, and N. Leblanc. Are 100% Green Composites and Green Thermoplastics the New Materials for the Future? *International Journal of Polymer Science*, ID 280181 (2012): 7.
7. J. Rincones, A.F. Zeidler, M.C.B. Grassi et al. The Golden Bridge for Nature: The New Biology Applied to Bioplastics. *Polymer Reviews*, 49 (2009): 85–106.
8. A.K. Mohanty, S. Vivekanandhan, S. Sahoo et al. Advanced Green Composites and Nanocomposites: New Industrial Biomaterials from Biorefinery. Paper presented at Eighth Joint Canada–Japan Workshop on Composites. École de Technologie Supérieure, Montréal, Québec, Canada; Industrial Materials Institute, Boucherville, Québec, Canada, July 2010.
9. Å. Larsen and E. A. Kleppe. Recycling Plastics with Nor–X Degradable in Full-Scale Industrial Recycling Process. Paper presented at IdentiPlast Conference, Brussels, April 2007.
10. A.P. Ambekar, P. Kukade, V. Mahajan et al. Bioplastics: A Solution. *Popular Plastics & Packaging*, 55 (2010): 30–32.
11. S. Bradbury and B. Bracegirdle. *Introduction to Optical Microscopy*. New York: BIOS Scientific Publishers, 1998.
12. J.C. Chen and I.R. Harrison. Modification of Polyacrylonitrile (PAN) Carbon Fiber Precursor via Post-Spinning Plasticization and Stretching in Dimethyl Formamide (DMF). *Carbon*, 40 (2002): 25–45.
13. Y. Yang, Z. Tong, Y. Geng et al. Biobased Polymer Composites Derived from Corn Stover and Feather Meals as Double Coating Materials for Controlled Release and Water Retention Urea Fertilizers. *Journal of Agricultural and Food Chemistry*, 61 (2013): 8166–8174.

Index